Elements of Finite Mathematics

4,00

Elements of Finite Mathematics

A. J. Pettofrezzo

Florida Technological University

Wadsworth Publishing Company, Inc.
Belmont, California

Designer: Gary Head

Editor: Eva Strock

Technical Illustrator: John Foster

ISBN 0-534-00304-4

L. C. Cat. Card No. 73-81782

Printed in the United States of America

1 2 3 4 5 6 7 8 9 10—78 77 76 75 74

Preface

This textbook was written to introduce the undergraduate college student to some of the basic concepts of modern mathematics. The author believes that this task can be accomplished best by restricting the content to those topics dealing with *finite mathematics*. Finite mathematics is that mathematics which does not involve infinite sets, the limit concept, continuity, and related studies.

The topics of logic, sets, relations, Boolean algebra, counting and probability, and matrices are fundamental to the study of fields other than mathematics. The undergraduate major in economics, business administration, political science, psychology, sociology, biology, statistics, and education, as well as other disciplines, can benefit from an exposure to such topics. It is this author's opinion that involved applications of the material presented in this text are best developed and explored more naturally in the courses within a student's major field. From this point of view, the text prepares the student for understanding the applications when they are presented; hence, the applications considered are of an elementary nature.

Because the subject of finite mathematics comprises a broad area of mathematics, it was essential that a selection of topics be made. Those topics considered appropriate for an elementary course are treated. Most of the topics are included because they are common to all finite mathematics courses, but others are included because of the author's personal interests. For example, more discussion is given to binary Boolean algebra and switching circuits than is usually the case in most finite mathematics books.

The only prerequisite necessary for the student is a familiarity with the real number system, which is usually obtained with 1 or $1\frac{1}{2}$ years of high school algebra. This book contains more than enough material for a full-semester course which meets three times per week. Chapters 1, 2, and 4 are essential; Chapters 3, 5, and 6 include highly useful and interesting material that can be selected according to the needs of the students and the time available. It is not suggested that a full period of classroom time be devoted to each section; in some cases, several of the beginning sections of a chapter may be combined.

Chapter 1, Statements and Logic, studies the truth values of compound statements, presents symbolic representations for statements and the operations defined on them, and identifies several valid two-premise argument forms as well as several common fallacious argument forms.

Chapter 2, Sets, contains the language and notation of the algebra of sets. This chapter presents the relations and operations on sets and considers the properties of mathematical relations in general as well as various ways of representing relations and functions.

Chapter 3, Boolean Algebra and Applications, introduces binary Boolean algebra and examines its properties. An isomorphism between the binary Boolean algebra and the algebra of switching circuits is established to demonstrate the use of this Boolean algebra to simplify existing networks and to design networks satisfying certain closure conditions. Finally, the relationship that exists among the algebra of statements, the algebra of sets, and binary Boolean algebra is demonstrated.

Chapter 4, Elements of Probability, studies the sophisticated counting procedures used in answering questions related to the probability of an event. Empirical and mathematical probabilities are discussed. The concept of a sample space is emphasized in relation to the algebra of probability, including the addition law, the multiplication law, and conditional probability. Probability functions are introduced.

Chapter 5, Algebra of Matrices, considers those basic concepts of matrix algebra which are important in the study of statistics, psychology, economics, biology, and mathematics. It also considers matrices as elements of an algebra and probability matrices.

Chapter 6, Linear Programming and Game Theory, presents solutions to the problem of finding the maximum and minimum values of a linear function over a convex polygon and considers applications. Also included is an introduction to the study of matrix games, their values and optimal strategies.

More than the usual number of illustrative examples are included in this text as an aid to the student in mastering the concepts presented. Also, a sufficient number of exercises are included, which either supplement the theory presented or provide the student with an opportunity to reinforce his understanding. A Keyed Summary is presented at the end of every chapter, which lists the most important concepts that have been presented. Each chapter ends with a section called Building Your Mathematical Vocabulary, which lists the terms and expressions that have been introduced or defined. Finally, Answers (to selected odd-numbered exercises) are provided at the end of the book.

The material in this textbook was classsroom-tested at Florida Technological University by several groups of students. A special note of thanks is due them for laboring with the preliminary versions. The author wishes to express his appreciation to Scott, Foresman and Company for graciously granting permission to use material contained in "Elementary Mathematics" ("Elementary Mathematics: Number Systems and Algebra," Anthony J. Pettofrezzo and Donald W. Hight, Scott, Foresman and Company, Glenview, Ill., 1970). I also wish to thank my wife Betty, who gave willingly of her time and ability to type the manuscript.

A. J. P.

Contents

5 *Algebra of Matrices 233*

6 *Linear Programming and Game Theory 299*

Answers 361

Index 393

1

Statements and Logic

1.1

Truth Values of Statements

Statements are usually classified grammatically as imperative, interrogative, exclamatory, or declarative. In mathematics we consider only declarative statements that actually or hypothetically must be either true or false.

***The* truth value *of a true statement is* true;**
***the* truth value *of a false statement is* false.**

A *simple statement* expresses a single idea; that is, it contains one subject and one predicate. Our knowledge of arithmetic and elementary mathematics should enable us to determine the truth values of many simple statements. For example, the following simple statements are true:

Three plus five is eight.
Two is less than six.
A triangle has three sides.

The following statements are false:

Two times three is five.
Four is greater than seven.
A square has five sides.

Example 1

Use your experience and common knowledge to determine the truth values of statements (*a*) to (*d*):

(*a*) France is in Europe.
(*b*) Lincoln was President of the United States.
(*c*) Chicago is east of New York.
(*d*) The moon is a star.

(*a*) True (*b*) true (*c*) false (*d*) false

Example 2

Use your knowledge of arithmetic and elementary mathematics to determine the truth values of statements (*a*) to (*d*):

(*a*) Seven minus three is four.
(*b*) Every rectangle is a square.
(*c*) Zero times any number is zero.
(*d*) No number greater than ten exists.

(*a*) True (*b*) false (*c*) true (*d*) false

In discussing statements it is convenient to let lowercase letters such as p, q, and r represent them. Therefore, we may write

p: A square is a rectangle

to mean "p represents the statement 'A square is a rectangle.'" In this instance, p is a true statement.

Exercises

In Exercises 1 through 8 use your experience and common knowledge to determine the truth value of each statement.

1. There are twelve months in a year.
2. Germany is in Africa.
3. Twelve hours is equivalent to one day.
4. Spinach is a vegetable.
5. Benjamin Franklin was President of the United States.

6. Atlanta is in Georgia.
7. Three feet is equivalent to one yard.
8. South Dakota is in Canada.

In Exercises 9 through 20 use your knowledge of arithmetic and elementary mathematics to determine the truth value of each statement.

9. A rectangle is a parallelogram.
10. The product of two and eight is ten.
11. A triangle is a quadrilateral.
12. The sum of four and seven is eleven.
13. Five is three less than two.
14. There is a number x that when added to three is equal to seven.
15. A line segment is a set of points.
16. An equilateral triangle has three congruent sides.
17. $2 + 5 = 10 - 3$.
18. $4 \times 3 < 12$.
19. $6 \times 0 = 0$.
20. $5 + 2 = 2 + 5$.

In Exercises 21 through 23 let p, q, and r represent the simple statements

$$p\colon\ 2 + 4 = 6$$
$$q\colon\ 5 \times 1 = 5$$
$$r\colon\ 3 > 8$$

21. Statements ________ and ________ are both true; statement ________ is false.
22. Write a related statement whose truth value is the opposite of the truth value of statement p.
23. Write a related statement whose truth value is the opposite of the truth value of statement r.

In Exercises 24 through 26 let p, q, and r represent the simple statements

p: A rectangle has four sides.
q: An isosceles triangle has two congruent sides.
r: A pentagon has less sides than a quadrilateral.

24. Statements ________ and ________ are both true; statement ________ is false.

25. Write a related statement whose truth value is the opposite of the truth value of statement *p*.
26. Write a related statement whose truth value is the opposite of the truth value of statement *r*.
27. Discuss the truth value of the statement "This statement is false."

1.2

Conjunctions

Often two simple mathematical statements are combined by such connecting words as "and," "or," and "if . . . , then . . . ." The resulting combination of simple statements is called a *compound statement.* One of our objectives in this chapter is to be able to determine the truth values of compound statements relative to the truth values of the simple statements involved. In this section we shall study one type of compound statement.

When the word "and" is used to connect two simple statements, the compound statement formed is called the *conjunction* of the simple statements. For example, the following compound statements are conjunctions:

T Statement 1: Two times three is six *and* two plus three is five.

F Statement 2: A square has four sides *and* a triangle is a quadrilateral.

F Statement 3: Five is less than three *and* six is greater than five.

F Statement 4: A triangle is a rectangle *and* one equals zero.

The truth values of compound statements depend completely upon the truth values of the simple statements contained within them.

> ***The* conjunction *of two simple statements is true if both simple statements are true. If at least one simple statement is false, then the conjunction is false.***

There are four possible combinations of truth values for two simple statements *p* and *q*:

p is true and *q* is true; T
p is true and *q* is false; F
p is false and *q* is true; F
p is false and *q* is false. F

Then, by definition, the truth values of the conjunction "p and q" may be summarized by the following rules:

If p is true and q is true, then "p and q" is true.
If p is true and q is false, then "p and q" is false.
If p is false and q is true, then "p and q" is false.
If p is false and q is false, then "p and q" is false.

Notice that Statement 1 on page 4 is the only conjunction whose truth value is true. Statements 2 to 4 are false; why?

Example 1

Use your experience and common knowledge to determine the truth values of conjunctions (*a*) to (*d*):

(*a*) There are 7 days in each week and Tuesday follows Monday.
(*b*) The radius of the Earth is approximately 4,000 miles and the Earth is approximately 2,400 miles from the Sun.
(*c*) Argentina is in Europe and Florida is in the southeastern section of the United States.
(*d*) There are 31 days in September and 30 days in January.

(*a*) True; both simple statements are true. (*b*) False; the second simple statement "The Earth is approximately 2,400 miles from the Sun" is false. (*c*) False; the first simple statement "Argentina is in Europe" is false. (*d*) False; both simple statements are false because there are 30 days in September and 31 days in January.

Example 2

Use your knowledge of arithmetic and elementary mathematics to determine the truth values of conjunctions (*a*) to (*d*):

(*a*) Two is an even number and twenty is a multiple of five.
(*b*) The diagonals of a square are perpendicular to each other and every quadrilateral is a parallelogram.
(*c*) Six is a factor of twenty-one and there is a number x that when added to five is equal to eight.
(*d*) A hexagon has seven sides and a pentagon has ten sides.

(*a*) True; both simple statements are true. (*b*) False; the second simple statement "Every quadrilateral is a parallelogram" is false. (*c*) False; the first simple statement "Six is a factor of twenty-one" is false. (*d*) False; both simple statements are false because a hexagon has six sides and a pentagon has five sides.

Let p and q represent the following simple statements:

p: Three is an odd number.
q: Three is a prime number.

Then "p and q" represents the conjunction "Three is an odd number and three is a prime number." Such a compound statement is usually written in a more simple and concise form:

p and q: Three is an odd prime number.

Example 3

Use your knowledge of arithmetic and elementary mathematics to determine the truth values of conjunctions (*a*) to (*d*):

(*a*) Seven is an even number greater than six.
(*b*) Ten is greater than seven but less than twenty.
(*c*) An equilateral triangle has three congruent sides and three congruent angles.
(*d*) Five and eight are prime numbers.

(*a*) The statement is equivalent to "Seven is an even number and seven is greater than six." This conjunction is false. (*b*) The statement is equivalent to "Ten is greater than seven and ten is less than twenty." This conjunction is true. (*c*) The statement is equivalent to "An equilateral triangle has three congruent sides and an equilateral triangle has three congruent angles." This conjunction is true. (*d*) The statement is equivalent to "Five is a prime number and eight is a prime number." This conjunction is false.

Exercises

In Exercises 1 through 6 use your experience and common knowledge to determine the truth value of each conjunction.

1. Napoleon was defeated at Waterloo and the British were victorious.
2. Oranges and grapefruits are products of the state of Florida.

3. Cincinnati and Chicago are in Ohio.
4. All apples are red and sweet.
5. Botany is the study of plant life, but zoology is the study of animal life.
6. Canada is north of the United States, but Mexico is south of our border.

In Exercises 7 through 12 use your knowledge of arithmetic and elementary mathematics to determine the truth value of each conjunction.

7. The product of two odd numbers is always odd and the product of two even numbers is always even.
8. The sum of two odd numbers is always odd and the sum of two even numbers is always even.
9. Congruent triangles have the same shape and size.
10. Both a triangle and a quadrilateral have less than five sides.
11. Seven is less than ten, but it is not a prime number.
12. Intersecting lines do not meet, but parallel lines do.

In Exercises 13 through 20 write each compound statement in a more simple and concise form.

13. Homer is a gentleman and Homer is a scholar.
14. I like meat and I like cheese.
15. Steve is a good athlete and Paul is a good athlete.
16. Donna is a tall girl and she is beautiful.
17. He owns a yellow car and the car is a Dodge.
18. It is cold and it is not windy.
19. Four is an even number and four is greater than three.
20. The quadrilateral is a parallelogram and the quadrilateral is not a rectangle.
21. What is the truth value of "p and q" if q is false?
22. What is the truth value of "p and q" if q is true?

1.3

Disjunctions

The connecting word "or" is used in our ordinary language in two different ways. For example, in the statement

At 2 o'clock Bert will go to the baseball game *or* he will go swimming the word "or" is used in an *exclusive sense*. Bert cannot do both things at the same time. If he goes to the ball game, that excludes the possibility of his going swimming; if he goes swimming, that excludes the possibility of his going to the ball game.

The second use of the connecting word "or" is in the *inclusive sense*. Consider the statement

For dessert Nancy will have cake *or* she will have ice cream.

The word "or" is used here in an inclusive sense because it is possible for Nancy to have both cake and ice cream at the same time. We shall consider the use of the word "or" only in the inclusive sense throughout the book unless otherwise stated.

When the word "or" is used to connect two simple statements, the resulting compound statement is called the *disjunction* of the simple statements. For example, these compound statements are disjunctions:

Statement 1: Six is a multiple of three *or* six is a multiple of two.

Statement 2: Eight is less than nine *or* nine is less than eight.

Statement 3: Ten is a prime number *or* two is the only even prime number.

Statement 4: A parallelogram has five sides *or* every triangle has more sides than a parallelogram.

Because a disjunction is a compound statement, its truth value depends completely upon the truth values of the simple statements involved.

The* disjunction *of two simple statements is true if at least one of the two simple statements is true. If both simple statements are false, then the disjunction is false.

As is the case for a conjunction, there are four possible combinations of truth values for statements p and q:

p is true and q is true;
p is true and q is false;
p is false and q is true;
p is false and q is false.

Then, by definition, the truth values of the disjunction "p or q" may be summarized by the following rules:

T If p is true and q is true, then "p or q" is true.
T If p is true and q is false, then "p or q" is true.
T If p is false and q is true, then "p or q" is true.
F If p is false and q is false, then "p or q" is false.

Notice that Statements 1 to 3 on page 8 are disjunctions whose truth values are true. Statement 4 is the only disjunction whose truth value is false; why?

Example 1

Let p and q represent the simple statements

p: Seven is less than five.
q: Seven is a prime number.

Write the compound statements (*a*) p and q; (*b*) p or q. Determine the truth value of each compound statement.

(*a*) p and q: Seven is a prime number less than five. The conjunction is false since at least one of the simple statements, p, is false (seven is not less than five). (*b*) p or q: Seven is less than five or it is a prime number. The disjunction is true since at least one simple statement, q, is true.

Example 2

Use your experience and common knowledge to determine the truth values of disjunctions (*a*) to (*d*):

(*a*) Canada or Mexico is in the Western Hemisphere.
(*b*) No vegetable is an animal or beef is a vegetable.
(*c*) California is the largest state in the United States or Rhode Island has a smaller population than California.
(*d*) Detroit is in Iowa or New Jersey.

(*a*) True; both simple statements are true. (*b*) True; at least one simple statement, the first statement, is true. (*c*) True; at least one simple statement, the second statement, is true. (*d*) False; both simple statements are false because Detroit is in Michigan.

Example 3

Use your knowledge of arithmetic and elementary mathematics to determine the truth values of disjunctions (*a*) to (*d*):

(*a*) Eighteen is divisible by three or two.

(*b*) The difference between 17 and 11 is 6 or the sum of 17 and 11 is 6.

(*c*) The sum of three and two is less than four or the product of three and two is six.

(*d*) The number 1 is a prime number or $3 + 3 = 3 \times 3$.

(*a*) True; both simple statements are true. (*b*) True; at least one simple statement, the first statement, is true. (*c*) True; at least one simple statement, the second statement, is true. (*d*) False; both simple statements are false since 1 is not a prime number and $3 + 3 \neq 3 \times 3$.

Exercises

In Exercises 1 through 6 use your experience and common knowledge to determine the truth value of each disjunction.

1. Grant or Lee was a Confederate general.
2. The most populated city in the United States is Miami or Salt Lake City.
3. The Pacific Ocean is either the largest or the second largest ocean in the world.
4. In the United States most oranges are grown in California or Florida.
5. Baltimore is the capital of Pennsylvania or the city does not exist.
6. Either bananas or tomatoes are edible.

In Exercises 7 through 12 use your knowledge of arithmetic and elementary mathematics to determine the truth value of each disjunction.

7. Eighteen contains the factor 4 or four is not an even number.
8. The number 11 is a prime number or it is greater than 20.
9. Every quadrilateral is either a parallelogram or a rectangle.
10. The product of any two whole numbers is odd or both numbers are even.
11. An angle is either a right angle or an acute angle.
12. A hexagon or a pentagon has five sides.
13. What is the truth value of "p or q" if q is true?
14. What is the truth value of "p or q" if q is false?

In Exercises 15 through 18 let

p: Lee went to the party.
q: John went to the party.
r: Murray went to the party.

Express the following statements in words:

15. (p and q) and r LEE, JOHN AND MURRAY WENT TO A PARTY.
16. (p and q) or r LEE, JOHN OR MURRY WENT TO A PARTY.
17. p and (q or r) MURRAY OR JOHN AND LEE WENT TO A PARTY.
18. (p or q) or r LEE OR JOHN OR MURRAY WENT TO A PARTY.
19. The *inclusive disjunction* "*p or q*" means "*p or q, or both*"; the *exclusive disjunction* "*p or q*" means "*p or q, but not both.*" Summarize the truth values of the exclusive disjunction.

1.4

Negations

Consider the following statements:

p: Three plus five is equal to eight.
q: Three plus five is not equal to eight.

Statement p is true, but statement q is false. When the word "not" is added to a simple statement in such a way that its truth value is changed, the resulting statement is called the *negation* of the original statement.

The* negation *of a simple statement is true if the simple statement is false; the negation is false if the simple statement is true.

We may represent the negation of a statement p by use of the word "not" in front of it. Thus, "not p" is the negation of p. In the example above, q may be represented by "not p"; p may be represented by "not q."

The negation of a simple statement is considered a compound statement with the connecting word "not." However, note that although a negation is considered to be a compound statement, only one simple statement is involved.

Example 1

State the negations of statements (*a*) to (*d*), and determine their truth values:

(*a*) A quadrilateral has four sides.

(*b*) The product of three and four is ten.

(*c*) A rose is a rose.

(*d*) Canada is not in North America.

(*a*) A quadrilateral does not have four sides (false). (*b*) The product of three and four is not ten (true). (*c*) A rose is not a rose (false). (*d*) Canada is in North America (true).

The negation of each statement *p* in Example 1 could have been stated alternately in the form

It is not the case that *p*.

For example, "It is not the case that a quadrilateral has four sides" is the negation of "A quadrilateral has four sides."

Also note in part (*d*) of Example 1 that the negation of the negation of a simple statement is the simple statement itself. For example,

p: Canada is in North America.
not *p*: Canada is not in North America.
not (not *p*): Canada is in North America.

The negation of the negation of a simple statement is called the *double negation* of the statement.

Considerable care should be taken in forming the negation of certain statements. For example, if

p: Three plus two is not seven,

then the negation of *p* is correctly stated by

not *p*: Three plus two is seven.

Note that *q*, where

q: Three plus two is five,

is not the negation of *p* since *p* and *q* have the same truth values. Similarly, neither the statement "Donna has brown eyes" nor the statement "Donna has blue eyes" is the negation of the other; why?

The use of the words "every," "all," "some," and "at least one" make the task of finding the negation of a statement a little more difficult. In mathematics and logic, "every" and "all" have the same meaning, as do "some" and "at least one." Consider the statement

All quadrilaterals are rectangles.

The negation of this statement may be stated correctly in any of the following forms:

Not all quadrilaterals are rectangles.
It is not the case that all quadrilaterals are rectangles.
Some quadrilaterals are not rectangles.
At least one quadrilateral is not a rectangle.

The last two forms indicate the use of the words "some" and "at least one" in forming the negation of a statement in which the word "all" appears. Note that the statement

All quadrilaterals are not rectangles

is not the correct negation of the given statement; why?

Example 2

State the negations of statements (*a*) to (*d*):

(*a*) Every number is greater than zero.

(*b*) All athletes are good students.

(*c*) Some athletes are good students.

(*d*) At least one number is prime.

(*a*) At least one number is less than or equal to zero. (*b*) Some athletes are not good students. (*c*) No athlete is a good student. (*d*) No number is prime.

There are many alternate forms of the negations of the statements given in Example 2.

Exercises

In Exercises 1 through 6 state the negation of each statement, and determine the truth value of each negation.

1. Lake Michigan is one of the Great Lakes of the United States.
2. There exists a number that when added to 5 gives a sum of 17.
3. Every triangle has a right angle.

4. The sum of two and seven is greater than nine.
5. China has a larger population than Israel.
6. Philadelphia is the capital of Pennsylvania.

In Exercises 7 through 16 state the negation of each statement.

7. All birds have feathers.
8. Some numbers are positive.
9. Nothing is correct.
10. No balloon is perfectly round.
11. Not every movie is worth seeing.
12. All professors are smart.
13. Each tree is green.
14. At least one athlete is not intelligent.
15. Every happy man is rich.
16. The storm damage is not extensive.
17. Determine the truth value of a conjunction in which one simple statement is the negation of the other simple statement.
18. Determine the truth value of a disjunction in which one simple statement is the negation of the other simple statement.
19. Summarize the truth values of the conjunction "(not p) and (not q)" for the possible truth values of the simple statements p and q. Compare your summary with the truth values of the disjunction "p or q."
20. Summarize the truth values of the disjunction "(not p) or (not q)" for the possible truth values of the simple statements p and q. Compare your summary with the truth values of the conjunction "p and q."

1.5

Truth Tables

The rules for determining the truth values of compound statements may be summarized in truth-table form—a *truth table* is a device that states all possible truth values of a compound statement in terms of the possible truth values of the simple statements involved.

Let p and q represent two simple statements. Then Table 1.1 represents the truth table for the conjunction "p and q," where T means "true" and F means "false."

This table requires four rows because there are four possible combinations of truth values for the simple statements p and q. From the table note that "p and q" is true if both simple statements are true; otherwise, "p and q" is false.

Table 1.1 Truth Table for a Conjunction

p	q	p and q
T	T	T
T	F	F
F	T	F
F	F	F

Table 1.2 represents the truth table for the disjunction "p or q," where p and q are simple statements. From the table note that "p or q" is true if at least one simple statement is true; otherwise, "p or q" is false.

Table 1.2 Truth Table for a Disjunction

p	q	p or q
T	T	T
T	F	T
F	T	T
F	F	F

Table 1.3 represents the truth table for the negation "not p," where p is a simple statement. This table requires only two rows since there are only two possible truth values for the simple statement p.

Table 1.3 Truth Table for a Negation

p	not p
T	F
F	T

A *logically true statement* is one whose only truth value is true; a *logically false statement* is one whose only truth value is false.

Example 1

Find the truth values of the disjunction "p or (not p)."

p	not p	p or (not p)
T	F	T
F	T	T

The simple statement p has two possible truth values: T or F. The compound statement "p or (not p)" is true for every possible truth value of p. This disjunction is an example of a logically true statement.

The logically true statement "p or (not p)" is called the *law of the excluded middle*. The law of the excluded middle expresses the fact that either the simple statement p is true or its negation "not p" is true in all cases.

Example 2

Find the truth values of the conjunction "p and (not p)."

p	not p	p and (not p)
T	F	F
F	T	F

The compound statement "p and (not p)" is false for every possible truth value of p. This conjunction is an example of a logically false statement.

Since "*p* and (not *p*)" is a logically false statement, its negation "not [*p* and (not *p*)]" is a logically true statement. The logically true statement "not [*p* and (not *p*)]" is called the *law of contradiction*,[1] which expresses the fact that a simple statement *p* and its negation "not *p*" cannot be true simultaneously.

Exercises

In Exercises 1 through 10 construct a truth table for each statement.

1. not (not *p*)
2. *p* and (not *q*)
3. not (*p* and *q*)
4. (not *p*) and *q*
5. (not *p*) or (not *q*)
6. (*p* and *q*) and *r*
7. *p* or (*q* or *r*)
8. *p* and (*q* or *r*)
9. (*p* and *q*) or *r*
10. (*p* and *q*) or (*p* and *r*)
11. Construct the truth table for the exclusive disjunction (see Exercise 19 of Sec. 1.3).

The problems in Exercises 12 and 13 can be solved by an analysis of the logical possibilities.

12. A small business office is organized with a manager, assistant manager, cashier, teller, clerk, and stenographer. The personnel employed are Mr. Brown, Mr. Smith, Miss Jones, Mrs. Johnson, Miss Williams, and Mr. Dean.
    (*a*) The assistant manager is the manager's grandson.
    (*b*) Mr. Brown is a bachelor.
    (*c*) Mr. Smith is 21 years old.
    (*d*) The cashier is the stenographer's son-in-law.
    (*e*) The teller is Miss Jones' stepsister.
    (*f*) Mr. Dean is the manager's neighbor.
    Who is the cashier?

---

[1] The brackets are necessary so that the truth value be correctly determined: the truth value of "*p* and (not *p*)" is first determined, followed by the determination of the truth value of its negation "not [*p* and (not *p*)]." In Sec. 1.6 we discuss the use of the parentheses and brackets in more detail.

13. Steve, Paul, Art, Don, and Bruce are freshman, sophomore, junior, senior, and graduate students in chemistry, engineering, medicine, law, and mathematics but not necessarily in that order.
   (*a*) The engineering student graduated from the same high school as Paul and Art.
   (*b*) Bruce graduated from the law school last year with honors.
   (*c*) Steve will become an intern next year.
   (*d*) Paul and the sophomore mathematics student room together.
   (*e*) Don is in a higher class than Paul.
   Who is the freshman?
   Who majors in engineering?

1.6

Logically Equivalent Statements

Two statements that have the same truth values are called *logically equivalent statements.* If two statements p and q are logically equivalent, we write $p \Leftrightarrow q$, which is read "p is equivalent to q." If p and q are not logically equivalent, we write $p \not\Leftrightarrow q$.

Two statements p and q are* logically equivalent *if the truth values of p and q are simultaneously both true or both false.

Consider the statements $x + 2 = 5$ and $x = 3$, where x represents some number. Each statement here is an example of an *open statement.* The truth value of each statement can be determined when the variable x is replaced by some number. Both statements are true provided that x is replaced by 3; that is, the statements $3 + 2 = 5$ and $3 = 3$ are both true. Both statements are false if x is replaced by any number other than 3; for example, the statements $8 + 2 = 5$ and $8 = 3$ are both false. Since the statements $x + 2 = 5$ and $x = 3$ have the same truth values for all possible replacements of x, these statements are logically equivalent:

$$x + 2 = 5 \Leftrightarrow x = 3$$

Truth tables are useful in determining whether or not two compound statements are logically equivalent.

Example 1

Use a truth table to show that "p and q" is logically equivalent to "q and p."

p	q	p and q	q and p
T	T	**T**	**T**
T	F	**F**	**F**
F	T	**F**	**F**
F	F	**F**	**F**

The four possible combinations of truth values of the simple statements p and q are shown in columns one and two. The third column represents the corresponding truth values of "p and q"; the fourth column represents the corresponding truth values of "q and p." Since the truth values in the third and fourth columns are identical, the statements "p and q" and "q and p" are logically equivalent.

The logical equivalence of the statements expressed in Example 1 is called the *commutative property of the conjunction of statements.*

Example 2

Use a truth table to show that "p or p" is logically equivalent to "p and p."

---

p	p or p	p and p
T	**T**	**T**
F	**F**	**F**

Compound statements may have more than one connecting word; in all such cases one of the connecting words is *dominant.* Parentheses are usually necessary to indicate which connecting word is dominant. The parentheses are the punctuation symbols of mathematics. For example, consider the compound statement "not p or q," where

p: A rectangle has five sides.
q: A triangle has three sides.

The compound statement contains the two connecting words "not" and "or." How shall we interpret the compound statement? The statement "not *p* or *q*" could be interpreted in one of two ways:

not (*p* or *q*) and (not *p*) or *q*

In the first case the dominant connecting word is "not"; the statement is the negation of a disjunction. In the second case the dominant connecting word is "or"; the statement is the disjunction of two statements, one of which is a negation. Since *p* is a false statement and *q* is a true statement, the truth value of "not (*p* or *q*)" is false. However, the truth value of "(not *p*) or *q*" is true. Therefore parentheses are necessary to obtain the desired interpretation.

In English the compound statement "not (*p* or *q*)" is read "It is not the case that either a rectangle has five sides or a triangle has three sides." The compound statement "(not *p*) or *q*" is read "A rectangle does not have five sides or a triangle has three sides."

Example 3

Use a truth table to show that "(*p* or *q*) or *r*" is logically equivalent to "*p* or (*q* or *r*)."

p	*q*	*r*	*p* or *q*	(*p* or *q*) or *r*	*q* or *r*	*p* or (*q* or *r*)
T	T	T	T	**T**	T	**T**
T	T	F	T	**T**	T	**T**
T	F	T	T	**T**	T	**T**
T	F	F	T	**T**	F	**T**
F	T	T	T	**T**	T	**T**
F	T	F	T	**T**	T	**T**
F	F	T	F	**T**	T	**T**
F	F	F	F	**F**	F	**F**

The eight possible combinations of truth values of the simple statements *p*, *q*, and *r* are shown in columns one, two, and three. The fourth column represents the corresponding truth values of "*p* or *q*"; the fifth column represents the corresponding truth values of "(*p* or *q*) or *r*." The sixth column represents the corresponding truth values of "*q* or *r*"; the seventh column represents the corresponding truth values of "p or (q or r)." Since the truth values in the fifth and seventh columns are identical, the statements "(p or q) or r" and "p or (q or r)" are logically equivalent.

The logical equivalence of the statements expressed in Example 3 is called the *associative property of the disjunction of statements.*

Exercises

In Exercises 1 through 6 use truth tables to demonstrate each logical equivalence.

1. p or $q \Leftrightarrow q$ or p (*commutative property of the disjunction of statements*)
2. p and $p \Leftrightarrow p$ (*idempotent property of conjunction*)
3. p or $p \Leftrightarrow p$ (*idempotent property of disjunction*)
4. (p and q) and $r \Leftrightarrow p$ and (q and r) (*associative property of the conjunction of statements*)
5. p and (q or r) $\Leftrightarrow$ (p and q) or (p and r) (*distributive property of conjunction over disjunction*)
6. p or (q and r) $\Leftrightarrow$ (p or q) and (p or r) (*distributive property of disjunction over conjunction*)
7. Use a truth table to show that

$$(p \text{ and } q) \text{ or } r \not\Leftrightarrow p \text{ and } (q \text{ or } r)$$

8. Use a truth table to show that

$$(\text{not } p) \text{ and } q \not\Leftrightarrow \text{not } (p \text{ and } q)$$

9. Use a truth table to show that

$$(\text{not } p) \text{ and } (\text{not } q) \not\Leftrightarrow \text{not } (p \text{ and } q)$$

In Exercises 10 through 13 let

p: My car is new.
q: My car has automatic transmission.
r: My car has air conditioning.

Write each of the following statements in symbolic form.

10. My new car has automatic transmission and air conditioning.
11. My new car has either automatic transmission or air conditioning.

12. Either my car is old or it has automatic transmission and air conditioning.
13. Either my car is new and has automatic transmission or it is old and has air conditioning.
14. Any two statements such that at least one of the four possible combinations of truth values cannot possibly occur are called *dependent statements.* Show that $x > 3$ and $x < 10$ are dependent statements. (*Hint:* consider different replacement values for x.) If all possible combinations of truth values can occur, the two statements are called *independent statements.* Show that $x(x - 2) = 0$ and $x^2 = 4$ are independent statements.
15. Any two statements such that both cannot be simultaneously true are called *inconsistent statements.* Show that $x > 5$ and $x < 1$ are inconsistent statements. If both statements can be simultaneously true, they are called *consistent statements.* Show that $x(x - 2) = 0$ and $x^2 = 4$ are consistent statements.
16. Find a pair of dependent statements that are also inconsistent (see Exercises 14 and 15).
17. Find a pair of dependent statements that are also consistent (see Exercises 14 and 15).
18. In the English sentence "Either Smith is right or Smith is not right and Jones is right," the insertion of a comma before the word "or" or before the word "and" presents two possible interpretations. Are the interpretations the same?

1.7

More About Negations

The truth values of negations of compound statements may be determined most easily by the use of truth tables. In Example 1 we consider the truth values of the negation of the conjunction "p and q"; in Example 2 we consider the truth values of the negation of the disjunction "p or q."

Example 1

Show that "not (p and q)" has the same truth values as "(not p) or (not q)."

---

p	q	p and q	not (p and q)	not p	not q	(not p) or (not q)
T	T	T	**F**	F	F	**F**
T	F	F	**T**	F	T	**T**
F	T	F	**T**	T	F	**T**
F	F	F	**T**	T	T	**T**

The four possible combinations of truth values of the simple statements p and q are shown in columns one and two. The third column represents the corresponding truth values of "p and q"; the fourth column represents the negation of "p and q." Columns five and six represent "not p" and "not q," respectively. Column seven represents the disjunction of "not p" and "not q." Since the fourth and seventh columns are identical, then "not (p and q)" has the same truth values as "(not p) or (not q)." Therefore,

$$\text{not } (p \text{ and } q) \Leftrightarrow (\text{not } p) \text{ or } (\text{not } q)$$

Example 1 shows that the negation of a conjunction is logically equivalent to the disjunction of two negations. For example, let p and q be two simple statements, where

p: Chicago is in Illinois.
q: France is in Asia.

Then the following two compound statements are logically equivalent:

not (p and q): It is not the case that Chicago is in Illinois and France is in Asia.
(not p) or (not q): Chicago is not in Illinois or France is not in Asia.

Because "France is not in Asia" is a true statement, the disjunction "Chicago is not in Illinois *or* France is not in Asia" is true. Hence, "It is not the case that Chicago is in Illinois and France is in Asia" is also true.

Example 2

Show that "not (p or q)" has the same truth values as "(not p) and (not q)."

p	q	p or q	not (p or q)	not p	not q	(not p) and (not q)
T	T	T	**F**	F	F	**F**
T	F	T	**F**	F	T	**F**
F	T	T	**F**	T	F	**F**
F	F	F	**T**	T	T	**T**

The fourth and seventh columns are identical, so "not (p or q)" has the same truth values as "(not p) and (not q)." Therefore,

$$\text{not } (p \text{ or } q) \Leftrightarrow (\text{not } p) \text{ and } (\text{not } q)$$

Example 2 shows that the negation of a disjunction is logically equivalent to the conjunction of two negations. For example, the negation of the statement "Paul played basketball *or* baseball" is the statement "Paul did not play basketball *and* he did not play baseball"; that is, "Paul played neither basketball nor baseball."

The statements of logical equivalence expressed in Examples 1 and 2 are called *DeMorgan's laws for statements.*

Example 3

Use truth tables to show that the double negation of p is logically equivalent to p.

p	not p	not (not p)
T	F	**T**
F	T	**F**

Let p be any simple statement. Since the truth values of p and "not (not p)" are the same, the statements are logically equivalent. Therefore,

$$p \Leftrightarrow \text{not (not } p)$$

The fact that p and "not (not p)" are logically equivalent statements is often referred to as the *law of double negation*.

Exercises

In Exercises 1 through 8 state the negation of each compound statement.

1. Our professor is a gentleman and a scholar.
2. Spain is a beautiful country, rich in the romance of the past.
3. He is not Mark's father and she is not Henry's mother.
4. Paul is not nominated or Steve is elected.
5. Students listen carefully or they have no interest in the course.
6. St. Louis is in Kansas, but it is not a large city.
7. Going to college is neither an interesting nor a worthwhile experience.
8. O'Malley is not nearsighted or he is not an umpire.
9. Show that

$$\text{not } [p \text{ and (not } q)] \Leftrightarrow \text{(not } p) \text{ or } q$$

10. Show that

$$\text{not } [(\text{not } p) \text{ or (not } q)] \Leftrightarrow p \text{ and } q$$

In Exercises 11 through 18 let

p: Betty types fast.
q: Betty takes dictation well.

Express each statement in words.

11. not p
12. p and q
13. p and (not q)

14. not (p and q)
15. (not p) and (not q)
16. not (p or q)
17. not [(not p) and (not q)]
18. not [(not p) or (not q)]

1.8

Implications

Let p and q represent simple statements. A statement of the form "If p, then q" is a compound statement with the connecting words "If . . . , then . . . ." This important type of compound statement in mathematics is called an *implication*. For example, these compound statements are implications:

Statement 1: *If* three is a prime number, *then* six is a composite number.
Statement 2: *If* a triangle has three sides, *then* a parallelogram is a triangle.
Statement 3: *If* $5 = 3$, *then* $5 + 1 = 6$.
Statement 4: *If* $2 < 1$, *then* $2 = 0$.

In the implication "If p, then q," the simple statement p is called the *hypothesis*; the simple statement q is called the *conclusion*.[1]

Since an implication is a compound statement, its truth value depends completely upon the truth values of the simple statements involved. As with conjunctions and disjunctions, there are four possible combinations of truth values for statements p and q:

The hypothesis p is true and the conclusion q is true.
The hypothesis p is true and the conclusion q is false.
The hypothesis p is false and the conclusion q is true.
The hypothesis p is false and the conclusion q is false.

To appreciate the reasonableness of the definition for the truth values of an implication "If p, then q" for each of these four possible combinations of truth values for statements p and q, consider the following true statement:

If each of two numbers is odd, *then* their sum is even.

Let us explore why the truth value of this implication is considered to be true. We know from our experience in arithmetic and elementary mathematics that the

---

[1] In the literature of mathematical logic the hypothesis of an implication is also called the *antecedent;* the conclusion is also called the *consequent*.

sum of two odd numbers is even, the sum of an odd number and an even number is odd, and the sum of two even numbers is even. When the hypothesis "each of two numbers is odd" is true, the truth value of the conclusion "their sum is even" must be true. It is impossible for the hypothesis of this implication to be true and the conclusion to be false. Now, suppose the hypothesis is false; that is, suppose it is not the case that each of two numbers is odd. Then both numbers could be even, or one number could be odd and the other even. In the first case, their sum is even (the conclusion must be true); in the second case, their sum is odd (the conclusion must be false). Thus it is possible for this implication to be meaningful when the hypothesis is false and the conclusion is either true or false. Therefore, in mathematics and logic we arbitrarily assign a truth value of true to an implication that has a false hypothesis regardless of the truth value of the conclusion.

An* implication *is false if the hypothesis is true and the conclusion is false; otherwise, the implication is true.

Table 1.4 represents the truth table for the implication "If p, then q," where p and q are simple statements.

Table 1.4 Truth Table for an Implication

p	q	If p, then q
T	T	T
T	F	F
F	T	T
F	F	T

Notice that Statements 1, 3, and 4 on page 26 are implications whose truth values are true. Statement 2 is the only implication whose truth value is false; why?

Example 1

Use your knowledge of arithmetic and elementary mathematics to determine the truth values of implications (a) to (d):

(a) If a geometric figure has four sides, then it does not have three sides.

(*b*) If $9 < 11$, then $9 + 2 = 11$.

(*c*) If $2 + 3 = 6$, then $2 \times 3 = 6$.

(*d*) If $7 - 2 = 5$, then $2 = 5$.

Note that in determining the truth value of an implication we need check only the truth value of the conclusion when the truth value of the hypothesis is true.

(*a*) When the hypothesis "A geometric figure has four sides" is true, the conclusion "It does not have three sides" is true. Hence, the implication is true. (*b*) The hypothesis $9 < 11$ is true and the conclusion $9 + 2 = 11$ is true. Hence, the implication is true. (*c*) Since the hypothesis $2 + 3 = 6$ is false, the implication is true. (*d*) The hypothesis $7 - 2 = 5$ is true, but the conclusion $2 = 5$ is false. Hence, the implication is false.

It is possible to express an implication in terms of a disjunction; that is, an implication is logically equivalent to a certain disjunction. To demonstrate this equivalence, consider the following situation: the Dodgers are to play the Cardinals in baseball. The following compound statements have the same meaning:

If the Dodgers do not win, *then* the Cardinals will win.
The Dodgers win *or* the Cardinals will win.

Let p represent the statement "The Dodgers do not win," and let q represent the statement "The Cardinals will win." The compound statements above may be represented by

If p, then q and (not p) or q

respectively. Thus we should expect the truth values of the implication "If p, then q" and the disjunction "(not p) or q" to be the same.

Example 2

Use truth tables to show that the implication "If p, then q" and the disjunction "(not p) or q" are logically equivalent.

---

p	q	If p, then q	not p	(not p) or q
T	T	T	F	T
T	F	F	F	F
F	T	T	T	T
F	F	T	T	T

Since the truth values in the third and fifth columns are identical, the compound statements "If p, then q" and "(not p) or q" are logically equivalent. Both statements are true except when p is true and q is false.

The truth values of the implication "If p, then q" could have been defined by means of the logical equivalence established in Example 2.

Whenever an implication "If p, then q" is true, we denote this fact by writing $p \Rightarrow q$, which is read "p implies q." For example,

Each of two numbers is odd $\Rightarrow$ their sum is even

$$3 + 5 = 8 \Rightarrow 8 > 3$$

A quadrilateral is a rectangle $\Rightarrow$ the quadrilateral is a parallelogram

$$x + 2 = 6 \Rightarrow x = 4$$

Exercises

In Exercises 1 through 6 use your experience and common knowledge to determine the truth value of each implication.

1. If General Motors makes cars, then they make Fords.
2. If one reads the *National Observer*, then he is a member of the Conservative Party.
3. If a person is over 21 years of age, then he is an adult.
4. If a person lives in San Francisco, then he lives in California.
5. If apples are vegetables, then oranges and pears are fruits.
6. If a tail is a leg, then a horse has five legs.

In Exercises 7 through 12 use your knowledge of arithmetic and elementary mathematics to determine the truth value of each implication.

7. If a geometric figure is a rectangle, then it is a square.
8. If $x + 3 = 5$, then $x = 2$.
9. If a number is an odd number, then its square is an odd number.
10. If five is equal to two, then seven is equal to seven.
11. If $x = 3$, then $x^2 = 9$.
12. If $x^2 = 9$, then $x = 3$.

In Exercises 13 through 20 write each compound statement in the "If . . . , then . . . ." form.

13. If all men were good, there would be no wars.
14. I will wear my coat if it is raining.
15. The alternate interior angles are not equal or the lines are parallel.
16. In parallelogram $ABCD$, side AB is parallel to CD.
17. A professional football player should be well paid since he possesses very special skills.
18. Either it is windy or it is not cold.
19. The sum of the angle measures of a geometric figure is not 180 degrees or the geometric figure is a triangle.
20. A failing grade means Tom will not graduate.
21. What is the truth value of "If p, then q" when p is false?
22. What is the truth value of "If p, then q" when p is true?
23. What is the truth value of "If p, then q" when q is true?
24. What is the truth value of q when "If p, then q" and p are both true?
25. What is the truth value of "If (p and q), then (p or q)" when q is false?
26. Show that "If p, then not p" is logically equivalent to "not p."
27. Show that "If p, then q" is not logically equivalent to "p and q."
28. State the negation of "If p, then q" as a conjunction of two statements.

In Exercises 29 through 34 use the results of Exercise 28 and state the negation of each implication.

29. If money is evil, then wise men are poor.
30. If the sky is clear, then it probably will not rain.

31. If the team wins, then every player performed well.
32. If no woman is elected, then the election is meaningless.
33. If every student studies well, then no failing grades will be given.
34. You will be able to earn a living if you are properly trained.
35. Construct a truth table for "If p, then (q and r)."

1.9

Necessary and Sufficient Conditions

The words "necessary" and "sufficient" are often used in discussing an implication. Consider when the implication "If p, then q" is true, that is, $p \Rightarrow q$; in such cases it is said that

p is a *sufficient condition* for q.

The word "sufficient" is used here because whenever p is true, q must also be true. Thus, the truth of statement p is sufficient (all that is needed) to conclude that q is true.

When the implication "If p, then q" is true, it is also said that

q is a *necessary condition* for p.

The word "necessary" is used here because q must necessarily be true if p is to be true. That is, p cannot be true unless q is true.

For example, consider the true implication

If two triangles are congruent, then they are similar.

To show that two triangles are similar, it would be sufficient to show that they are congruent; congruence is enough to imply similarity. To show that two triangles are congruent, it is necessary to first show that they are similar; similarity is a required condition for congruence. Hence, we may restate the given implication in the following alternate forms without any change in meaning:

A sufficient condition that two triangles be similar is that they be congruent.
A necessary condition that two triangles be congruent is that they be similar.

Hereafter these alternate forms of an implication and the implication may be used interchangeably:

The statement "p is a* sufficient condition *for q" means "If p, then q"; the statement "q is a* necessary condition *for p" means "If p, then q."

Example 1

Write each statement in "If . . . , then . . ." form.

(*a*) A sufficient condition that two lines be parallel is that they have no points in common.

(*b*) A necessary condition that a car be a Pontiac is that it is a General Motors product.

(*c*) In order to become a doctor, it is necessary to have a college education.

(*d*) In order for $x^2 = 25$, it is sufficient for $x = 5$.

(*a*) If two lines have no points in common, then they are parallel. (*b*) If a car is a Pontiac, then it is a General Motors product. (*c*) If one is to become a doctor, then one must have a college education. (*d*) If $x = 5$, then $x^2 = 25$.

In everyday language the meanings of the compound statements "q if p" and "q only if p" are often confused. The statement "q if p" states that p is a sufficient condition for q:

q if p means If p, then q

The statement "q only if p" states that p is a necessary condition for q:

q only if p means If q, then p

Finally,

q if and only if p means (If p, then q) and (if q, then p)

For example, consider the statement

x is an even number *if* x is a multiple of six.

The condition that x is a multiple of six is sufficient to show that x is an even number. Therefore, the statement means

If x is a multiple of six, then x is an even number.

In the statement

A geometric figure is a square *only if* it is a rectangle

the condition that the figure is a rectangle is necessary to show that the geometric figure is a square. Therefore, the statement means

If a geometric figure is a square, then it is a rectangle.

Example 2

Determine the necessary and sufficient conditions in implications (*a*) to (*d*):

(*a*) I gain weight only if I do not exercise.

(*b*) If prices are high, then the stock market rises.

(*c*) The weather is bad if the barometer falls.

(*d*) Algebra is difficult only if you do not study.

Sufficient condition	Necessary condition
(*a*) I gain weight.	I do not exercise.
(*b*) Prices are high.	The stock market rises.
(*c*) The barometer falls.	The weather is bad.
(*d*) Algebra is difficult.	You do not study.

Every statement involving the direct or indirect use of the quantifiers "all" or "none" can be expressed as an implication without changing its meaning. Consider the statement

All football players are quick.

This statement can be written in the form

If a person is a football player, then he is quick.

Now consider the statement

No lion is a vegetarian.

This statement can be written in the form

If an animal is a lion, then it is not a vegetarian.

Exercises

In Exercises 1 through 8 write each statement in "If . . . , then . . ." form.

1. The baseball game will be played Saturday only if it does not rain.
2. $x - 3 = 7$ only if $x = 10$.
3. To pass the course, it is necessary to attend class.

4. To graduate, it is necessary for a student to study French for 2 years.
5. All jet engines are noisy.
6. No circle has a circumference less than its diameter.
7. A necessary condition for promotion to associate professor is 5 years experience as an assistant professor.
8. A sufficient condition that $x(x - 3) = 0$ is that $x = 3$.

In Exercises 9 through 16 determine (*a*) the necessary and (*b*) the sufficient condition expressed in each implication.

9. If today is Monday, then tomorrow is Tuesday.
10. If a triangle is equilateral, it is isosceles.
11. I watch television when I am bored.
12. I play basketball on Wednesday evenings.
13. Adams will resign only if he is wrong.
14. The candidate will be elected only if he obtains the farm vote.
15. Provided all goes well, we should arrive by morning.
16. Unless plagued by injuries, the Hornets will have a good season.

In Exercises 17 through 22 state the negation of each implication.

17. If the temperature is below freezing tonight, we can go ice skating tomorrow.
18. The picnic will be postponed if it rains.
19. Our fullback can score only if the blocking is good.
20. All men are mortal.
21. All girls at our school are cute.
22. No elephant is small.
23. Write the statement "All p are q" in "If . . . , then . . ." form.
24. Write the statement "No p is not q" in "If . . . , then . . ." form.

1.10

Related Implications

In this section we shall study the relationship that exists among four forms of implication.

Let p and q represent simple statements. Other implications that are related to "If p, then q" are defined as follows:

Implication: If p, then q.
Converse of the implication: If q, then p.
Inverse of the implication: If (not p), then (not q).
Contrapositive of the implication: If (not q), then (not p).

For example:

Implication	Converse of the implication
If two plus three is five, then five is greater than two.	If five is greater than two, then two plus three is five.
If this month is September, then next month is October.	If next month is October, then this month is September.
If a quadrilateral is a square, then it is a rectangle.	If a quadrilateral is a rectangle, then it is a square.
If Robby passes the algebra course, then he gets an A.	If Robby gets an A, then he passes the algebra course.

Inverse of the implication	Contrapositive of the implication
If two plus three is not five, then five is not greater than two.	If five is not greater than two, then two plus three is not five.
If this month is not September, then next month is not October.	If next month is not October, then this month is not September.
If a quadrilateral is not a square, then it is not a rectangle.	If a quadrilateral is not a rectangle, then it is not a square.
If Robby does not pass the algebra course, then he does not get an A.	If Robby does not get an A, then he does not pass the algebra course.

From the examples, note that an implication may be true while its converse is false:

If a quadrilateral is a square, then it is a rectangle

is a true statement, but its converse

If a quadrilateral is a rectangle, then it is a square

is a false statement.

Example

State the converses of implications (*a*) to (*d*). Use your knowledge of arithmetic and elementary mathematics to determine the truth values of the converses.

(*a*) If a geometric figure has four sides, then it does not have three sides.

(*b*) If $9 < 11$, then $9 + 2 = 11$.

(*c*) If $2 + 3 = 6$, then $2 \times 3 = 6$.

(*d*) If $7 - 2 = 5$, then $2 = 5$.

(*a*) The converse "If a geometric figure does not have three sides, then it has four sides" is false. (*b*) The converse "If $9 + 2 = 11$, then $9 < 11$" is true. (*c*) The converse "If $2 \times 3 = 6$, then $2 + 3 = 6$" is false. (*d*) The converse "If $2 = 5$, then $7 - 2 = 5$" is true.

Note that the converses of the implications in parts (*a*) and (*c*) are false, but the implications are true; the converse of the implication in part (*d*) is true; the implication is false.

Consider the truth values of the four related implications shown in Table 1.5. Because the truth values in the third and sixth columns are identical, an implication and the contrapositive of the implication are logically equivalent:

$$\text{If } p, \text{ then } q \Leftrightarrow \text{if (not } q\text{), then (not } p\text{).}$$

For example, since the implication

If a quadrilateral is a square, then it is a rectangle

is true, the contrapositive

If a quadrilateral is not a rectangle, then it is not a square

is also true.

The truth values in the fourth and fifth columns of Table 1.5 are identical, so the converse of an implication and the inverse of the implication are logically equivalent:

$$\text{If } q, \text{ then } p \Leftrightarrow \text{if (not } p\text{), then (not } q\text{).}$$

Table 1.5

		Implication	Converse	Inverse	Contrapositive
p	q	If p, then q	If q, then p	If (not p), then (not q)	If (not q), then (not p)
T	T	T	T	T	T
T	F	F	T	T	F
F	T	T	F	F	T
F	F	T	T	T	T

Recall that we write $p \Rightarrow q$ to mean "If p, then q," provided that the implication is true. We may also write $q \Rightarrow p$, provided that the converse of the implication is true. The following table shows that an implication and its converse are both true when the truth values of the statements p and q are both true or both false:

p	q	If p, then q	If q, then p	(If p, then q) and (if q, then p)
T	T	T	T	T
T	F	F	T	F
F	T	T	F	F
F	F	T	T	T

However, when the truth values of the statements p and q are the same, the statements p and q are logically equivalent and we write $p \Leftrightarrow q$. Thus,

$$p \Leftrightarrow q \quad \text{means} \quad (p \Rightarrow q) \text{ and } (q \Rightarrow p).$$

Two statements are logically equivalent when each one implies the other. For example,

$$x + 2 = 5 \Leftrightarrow x = 3$$

because

$$x + 2 = 5 \Rightarrow x = 3 \qquad \text{and} \qquad x = 3 \Rightarrow x + 2 = 5$$

Exercises

In Exercises 1 through 6 state (*a*) the converse, (*b*) the inverse, and (*c*) the contrapositive of each implication. Determine the truth values of all four statements.

1. If a person lives in Dallas, then he lives in Texas.
2. If today is Monday, then yesterday was Saturday.
3. A number is a prime only if it is odd.
4. Every rectangle is a parallelogram.
5. Any number greater than five is greater than three.
6. Dodge cars are made by the Chrysler Corporation.

In Exercises 7 through 12 determine whether or not the statements are logically equivalent.

7. If Nixon is elected, Agnew will be elected. If Agnew is not elected, Nixon will not be elected.

8. I will attend the game only if you do. I will attend the game if you do.
9. All the men were brave. No man was not brave.
10. All integers are rational numbers. Some rational numbers are not integers.
11. If a parallelogram is not a rectangle, then it is not a square. If a parallelogram is a rectangle, then it is a square.
12. A necessary condition that two geometric figures be congruent is that they have the same area. A sufficient condition that two geometric figures be congruent is that they have the same area.
13. State the converse of "If p, then q" as a disjunction of two statements.
14. State the inverse of "If p, then q" as a disjunction of two statements.
15. State the contrapositive of "If p, then q" as a disjunction of two statements.
16. Show that the contrapositive of the converse of an implication is logically equivalent to the inverse of the implication.
17. Use a truth table to show that the converse of a true implication is not always true.
18. Use a truth table to show that the inverse of a true implication is not always true.
19. An anxious father said, "I'll make you my business partner only if you marry my daugher." If you marry his daughter, must he make you his partner?
20. The principal announced that "The game will not be played only if it rains."
    (*a*) If the game is not played, did it rain?
    (*b*) If the game is played, must it have rained?
    (*c*) If it rains, must the game be played?
    (*d*) If it does not rain, must the game be played?

1.11

Symbolic Representations

We have found it convenient to let lowercase letters such as p, q, and r represent statements. Logicians and mathematicians frequently use symbols to represent the connectives "and," "or," "not," "if . . . , then . . . ," and "if and only if." The most commonly used symbol for each connective is listed in the following table:

Symbol	Meaning
$\wedge$	and
$\vee$	or
$\sim$	not
$\rightarrow$	if . . . , then . . .
$\leftrightarrow$	if and only if

For example, let

p: Don is visiting Orlando.
q: Don is playing golf.

Then

$p \wedge q$: Don is visiting Orlando and playing golf.
$p \vee q$: Don is either visiting Orlando or playing golf.
$\sim p$: Don is not visiting Orlando.
$p \rightarrow q$: If Don is visiting Orlando, then he is playing golf.
$p \leftrightarrow q$: Don is visiting Orlando if and only if he is playing golf.

Example 1

Let

p: The game is televised on ABC.
q: The game is televised on CBS.
r: The game is televised on NBC.

Formulate English statements that correspond to the symbolic statements (*a*) to (*d*):

(*a*) $(\sim p) \wedge q$
(*b*) $\sim(p \vee q)$
(*c*) $p \rightarrow (\sim q)$
(*d*) $(\sim p) \rightarrow (q \vee r)$

(*a*) The game is televised on CBS, but not on ABC. (*b*) The game is not televised on ABC or CBS. (*c*) If the game is televised on ABC, then it is not on CBS. (*d*) If the game is not televised on ABC, then it is televised on CBS or NBC.

Example 2

Construct a truth table for (*a*) $(\sim p) \rightarrow q$ and (*b*) $[p \vee (\sim q)] \rightarrow r$.

(*a*)

p	q	$\sim p$	$(\sim p) \rightarrow q$
T	T	F	**T**
T	F	F	**T**
F	T	T	**T**
F	F	T	**F**

(*b*)

p	q	r	$\sim q$	$p \vee (\sim q)$	$[p \vee (\sim q)] \rightarrow r$
T	T	T	F	T	T
T	T	F	F	T	F
T	F	T	T	T	T
T	F	F	T	T	F
F	T	T	F	F	T
F	T	F	F	F	T
F	F	T	T	T	T
F	F	F	T	T	F

Exercises

In Exercises 1 through 10 let

p: It is raining.
q: Today is Saturday.
r: I will go to the movies.
s: I will go to the baseball game.

Formulate in English the statement that corresponds to each symbolic statement.

1. $\sim q$
2. $p \wedge r$
3. $q \vee (\sim r)$
4. $(\sim p) \rightarrow s$
5. $\sim (r \vee s)$
6. $q \leftrightarrow r$
7. $(\sim p) \vee (\sim r)$
8. $q \rightarrow (r \vee s)$
9. $(q \wedge s) \vee p$
10. $\sim [(\sim p) \wedge (\sim s)]$

In Exercises 11 through 22 construct a truth table for each statement.

11. $p \rightarrow (\sim q)$
12. $(\sim p) \rightarrow (\sim q)$
13. $p \vee (\sim q)$

14. $(\sim p) \wedge q$
15. $\sim(p \rightarrow q)$
16. $p \rightarrow [p \vee (\sim q)]$
17. $[(p \rightarrow q) \wedge p] \rightarrow q$
18. $[(p \rightarrow q) \wedge (\sim q)] \rightarrow (\sim p)$
19. $[(p \vee q) \wedge (\sim q)] \rightarrow p$
20. $[(p \rightarrow q) \wedge (\sim p)] \rightarrow (\sim q)$
21. $[(p \vee q) \wedge p] \rightarrow (\sim q)$
22. $[(p \rightarrow q) \wedge q] \rightarrow p$

In Exercises 23 through 26 insert parentheses wherever appropriate in order to make each statement a conjunction, if possible.

23. $p \vee q \wedge r$
24. $p \wedge q \rightarrow r$
25. $p \wedge q \wedge r$
26. $\sim p \vee q$

In Exercises 27 through 30 insert parentheses wherever appropriate in order to make each statement a disjunction, if possible.

27. $p \vee q \wedge r$
28. $\sim p \vee r$
29. $p \vee q \vee r$
30. $\sim p \wedge r$

In Exercises 31 and 32 insert parentheses wherever appropriate in order to make each statement an implication, if possible.

31. $p \vee q \rightarrow r$
32. $\sim p \vee q$

In Exercises 33 through 36 use truth tables to show each logical equivalence.

33. $\sim[(\sim p) \rightarrow q] \Leftrightarrow (\sim p) \wedge (\sim q)$
34. $\sim[(\sim p) \rightarrow q] \Leftrightarrow \sim(p \vee q)$
35. $\sim[p \rightarrow (\sim q)] \Leftrightarrow p \wedge q$
36. $p \wedge [q \vee (\sim q)] \Leftrightarrow p$

1.12

Valid Arguments

A fundamental problem of logic is the study of the validity of arguments. In everyday usage an argument is a collection of statements from which it is reasoned that some conclusion is established. More precisely, an *argument* is an implication of the form

$$\text{If } p_1 \text{ and } p_2 \text{ and } p_3 \cdots p_k, \text{ then } c.$$

The statements $p_1, p_2, p_3, \ldots,$ and p_k are called the *premises*; the statement c is called the *conclusion*.

Consider the following example of an argument expressed in verbal form:

> In 1968 the people elected Nixon or Humphrey as President of the United States. They did not elect Humphrey. Therefore, they elected Nixon.

It is convenient to list the premises of this argument in a column form and then write the conclusion beneath a horizontal line:

p_1: In 1968 the people elected Nixon or Humphrey as President of the United States.
p_2: The people did not elect Humphrey.

---

c: The people elected Nixon.

This argument is of the form

$$\text{If } p_1 \text{ and } p_2, \text{ then } c$$

that is,

$$(p_1 \wedge p_2) \rightarrow c$$

An argument is *valid* if and only if the conjunction of the premises implies the conclusion:

$$(p_1 \wedge p_2 \wedge p_3 \cdots p_k) \Rightarrow c$$

Thus, to prove an argument is valid, it must be shown that the conclusion is true whenever the conjunction of the premises is true. Note that the conjunction of the premises is true if and only if each premise is true. The argument above is an example of a valid argument.

There are basically two methods of determining whether or not an argument is valid. The first method is to use truth tables to show that the implication "If p_1 and p_2 and $p_3 \cdots p_k$, then c" is true whenever $p_1, p_2, p_3, \ldots,$ and p_k are each true. The second method is to replace the original argument by a series of equivalent arguments. In each equivalent argument one or more premises are replaced by logically equivalent statements or valid conclusions drawn from known valid arguments. The process is continued until a recognizable valid or invalid argument is obtained. Of course, we cannot use this second method until we have established some valid argument patterns by using the first method.

Example

Show that the following argument is valid:

$$\begin{array}{l} p \to q \\ \underline{p} \\ q \end{array}$$

p	q	$p \to q$	$(p \to q) \wedge p$	$[(p \to q) \wedge p] \to q$
T	T	T	T	T
T	F	F	F	T
F	T	T	F	T
F	F	T	F	T

Since the implication $[(p \to q) \wedge p] \to q$ is always true,

$$[(p \to q) \wedge p] \Rightarrow q$$

and the argument is valid.

The form of argument presented in the Example is called the *rule of detachment.* It states that if (1) the implication $p \to q$ is true and (2) the statement p is true, then the statement q is true. The rule of detachment is also referred to as *modus ponens* in the literature of mathematical logic.

We could have reduced the labor involved in demonstrating the validity of this argument by constructing the following abbreviated truth table:

p	q	$p \to q$	$(p \to q) \wedge p$	$[(p \to q) \wedge p] \to q$
T	T	T	T	T
T	F	F	...	...
F	T	...	...	...
F	F	...	...	...

The truth values of the argument need be determined only for "true" values of the premises.

There are numerous forms of valid arguments. Three other forms that are often used will be mentioned here and illustrated; their proofs are left as exercises. Other forms of valid arguments will be studied in the exercises.

The *law of the syllogism* is a valid argument of the form

$$[(p \rightarrow q) \wedge (q \rightarrow r)] \Rightarrow (p \rightarrow r)$$

For example,

$p \rightarrow q$: If Seaver can win 20 games, then the Mets will win the National League pennant.
$q \rightarrow r$: If the Mets win the National League pennant, then they will win the World Series.

$p \rightarrow r$: If Seaver can win 20 games, the Mets will win the World Series.

Disjunctive syllogism is a valid argument of the form

$$[(p \vee q) \wedge (\sim q)] \Rightarrow p$$

For example,

$p \vee q$: I am right or I am wrong.
$\sim q$: I am not wrong.

---

p: I am right.

Modus tollens is a valid argument of the form

$$[(p \rightarrow q) \wedge (\sim q)] \Rightarrow (\sim p)$$

For example,

$p \rightarrow q$: If two triangles are congruent, then they are similar.
$\sim q$: The triangles are not similar.

$\sim p$: The triangles are not congruent.

Note that the validity of an argument and the truth value of the conclusion are independent of one another. For example, by the rule of detachment, the following argument is always valid:

p_1: If today is Monday, then tomorrow is Tuesday.
p_2: Today is Monday.

---

c: Tomorrow is Tuesday.

However, the conclusion will be true provided that statement p_2 is true; otherwise, the conclusion will be false.

Exercises

In Exercises 1 through 8 use truth tables to prove that each argument is valid.

1. Law of the syllogism
2. Disjunctive syllogism
3. Modus tollens
4. $$\frac{\begin{array}{c}(\sim p) \vee q \\ p\end{array}}{q}$$
5. $$\frac{\begin{array}{c}(\sim q) \rightarrow p \\ \sim p\end{array}}{q}$$
6. $$\frac{\begin{array}{c}p \rightarrow (\sim q) \\ r \rightarrow q\end{array}}{p \rightarrow (\sim r)}$$
7. $$\frac{\begin{array}{c}p \rightarrow q \\ (\sim r) \rightarrow (\sim q)\end{array}}{p \rightarrow r}$$
8. $$\frac{\begin{array}{c}p \\ p \rightarrow q \\ q \rightarrow r\end{array}}{r}$$

In Exercises 9 through 16 two premises are given. Indicate any valid conclusion that may be drawn from the premises and state the argument used.

9. If Jerry is over 21 years of age, then he is an adult. Jerry is not an adult.
10. If she lives in Buffalo, then she is a resident of New York State. If she is a resident of New York State, she votes Republican.
11. If you like mathematics, you enjoy this course. You like mathematics.
12. If one angle of a triangle is less than 90 degrees, then the sum of the other two angles is greater than 90 degrees. The sum of the other two angles is not greater than 90 degrees.
13. The committee will be well balanced politically only if a Democrat is appointed. A Democrat is not appointed.
14. Every boy invites a girl to the dance or some girls do not attend the party. Every girl attends the party.
15. An even number x is equal to 2 or it is not a prime number. The even number x is not equal to 2.
16. Failure to pay your bill is sufficient cause for you to be denied credit. You fail to pay your bill.
17. What is the truth value of the conclusion if an argument is valid and all premises are true?
18. What can be stated about the premises if an argument is valid and the conclusion is false?

19. Prove that disjunctive syllogism is a valid argument form by using the fact that $p \vee q \Leftrightarrow (\sim p) \rightarrow q$ and $(\sim p) \rightarrow q \Leftrightarrow (\sim q) \rightarrow p$, and the rule of detachment.
20. Prove that modus tollens is a valid argument form by using the fact that $p \rightarrow q \Leftrightarrow (\sim q) \rightarrow (\sim p)$, and the rule of detachment.

1.13

Fallacious Arguments

An argument that is not valid is called a *fallacious argument* or simply a *fallacy*. To demonstrate that an argument is invalid it is sufficient to show that the implication is false for a particular set of truth values of the premises. Thus, at least one row of the truth table for the argument must be shown to be false.

Two forms of invalid arguments are often confused with the rule of detachment:

1. The fallacy of affirming the consequent;
2. The fallacy of denying the antecedent.

Fallacy of affirming the consequent:

First premise: $p \rightarrow q$
Second premise: q
Conclusion: p

The proof that this argument is invalid rests upon an examination of the truth table for $[(p \rightarrow q) \wedge q] \rightarrow p$:

p	q	$p \rightarrow q$	$(p \rightarrow q) \wedge q$	$[(p \rightarrow q) \wedge q] \rightarrow p$
T	T	T	T	T
T	F	F	F	T
F	T	T	T	F
F	F	T	F	T

The argument is invalid since row three of the implication is false.

As an example of the fallacy of affirming the consequent, consider the following:

p_1: If a geometric figure is a square, then it is a rectangle.
p_2: A figure is a rectangle.
c: The figure is a square.

Premise p_1 is a theorem of plane geometry. The geometric object $ABCD$ in Fig. 1.1 satisfies premise p_2, yet the truth of both premises does not force the conclusion that the object is a square. Indeed, it is not.

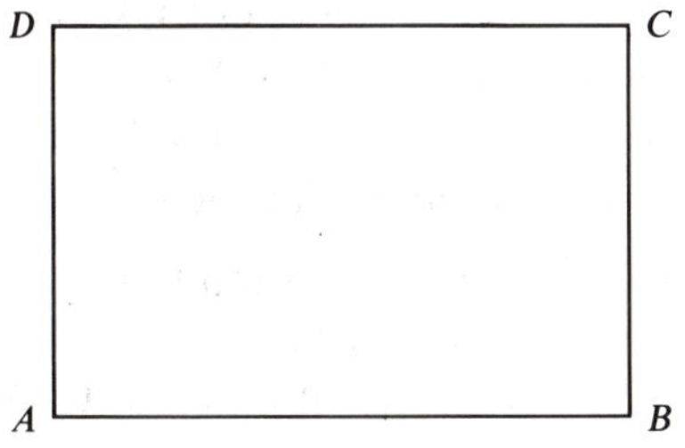

Figure 1.1

Fallacy of denying the antecedent:

$$\begin{array}{r} \text{First premise: } p \rightarrow q \\ \text{Second premise: } \sim p \\ \hline \text{Conclusion: } \sim q \end{array}$$

An examination of the values in the truth table for the argument indicates that when statement p is false and statement q is true, the implication is false. Hence, the argument is invalid.

p	q	$p \rightarrow q$	$\sim p$	$(p \rightarrow q) \wedge (\sim p)$	$\sim q$	$[(p \rightarrow q) \wedge (\sim p)] \rightarrow (\sim q)$
T	T	T	F	F	F	T
T	F	F	F	F	T	T
F	T	T	T	T	F	**F**
F	F	T	T	T	T	T

As an example of the fallacy of denying the antecedent, consider the following:

p_1: If each of two numbers is odd, then their sum is even.
p_2: At least one of two numbers is not odd.

---

c: Their sum is not even.

Premise p_1 is a fact of elementary arithmetic and hence is always true. Suppose the two numbers are 2 and 8. Then premise p_2 is also true. However, since $2 + 8 = 10$, which is an even number, the conclusion is false. An argument cannot be valid if the conjunction of all premises is true while the conclusion is false.

It should be mentioned that although an argument is invalid, the conclusion may be true or false. We are once again saying that the validity of an argument and the truth value of the conclusion are independent of one another. Consider, for example, the argument

p_1: If Nixon was elected in 1968, then all Republicans voted for him.
p_2: All Republicans voted for Nixon.
---
c: Nixon was elected in 1968.

The argument is invalid because it is an example of the fallacy of affirming the consequent. However, the conclusion is true.

One form of invalid argument often confused with disjunctive syllogism will now be considered:

Invalid disjunctive syllogism:

First premise: $p \vee q$
Second premise: p

Conclusion: $\sim q$

For example,

$p \vee q$: I will go to the theatre on Friday or Saturday.
p: I will go on Friday.
---
q: I will not go on Saturday.

The argument is obviously invalid because when both premises are true, the conclusion can be false—I could go to the theatre on Saturday, also.

Consider the following argument:

$p \vee q$: Either a coin falls heads or tails.
p: It falls heads.

$\sim q$: It did not fall tails.

This argument is an example of invalid disjunctive syllogism. However, the argument appears to be perfectly valid in terms of our experience. The difficulty lies in the fact that an additional premise is tacitly assumed when we consider our experience with the tossing of a coin:

$\sim (p \wedge q)$: It is not the case that a coin can fall heads and tails.

The valid argument that our experience supports can be stated as

$p \vee q$: Either a coin falls heads or tails.
$\sim (p \wedge q)$: It is not the case that a coin can fall heads and tails.
p: It falls heads.

$\sim q$: It did not fall tails.

In any argument all premises from which a conclusion is to be obtained must be explicitly stated. The truth table for the valid argument

$$\{(p \vee q) \wedge [\sim (p \wedge q)] \wedge p\} \Rightarrow (\sim q)$$

is shown below:

p	q	$p \vee q$	$\sim (p \wedge q)$	$(p \vee q) \wedge [\sim (p \wedge q)] \wedge p$	$\sim q$	$\{(p \vee q) \wedge [\sim (p \wedge q)] \wedge p\} \rightarrow (\sim q)$
T	T	T	F	F	F	T
T	F	T	T	T	T	T
F	T	T	T	F	F	T
F	F	F	T	F	T	T

Exercises

1. Use a truth table to show that invalid disjunctive syllogism is an invalid argument form.

In Exercises 2 through 8 use truth tables to determine whether or not each argument is valid.

2. $(\sim p) \vee q$
   q
   ———
   $\sim p$

3. $(\sim p) \rightarrow q$
   p
   ———
   $\sim q$

4. $(\sim p) \rightarrow q$
   $p \vee (\sim q)$
   ———
   q

5. $p \rightarrow q$
   $(\sim p) \wedge q$
   ———
   p

6. $r \rightarrow (\sim q)$
   $r \rightarrow q$
   r
   ———
   $\sim p$

7. $p \rightarrow (\sim q)$
   $q \vee r$
   p
   ———
   r

8. $\sim p$
   $(p \vee q) \wedge (\sim r)$
   r
   ———
   q

In Exercises 9 through 20 determine whether each argument is valid or invalid. If possible, identify the form of the argument.

9. If you study, you will pass the course.
   You do not study.
   ____
   You do not pass the course.

10. This piece of legislation will become law or the Republicans do not have a majority.
    The Republicans have a majority.
    ____
    This piece of legislation will become law.

11. If I am the professor, then I am always right.
    Sometimes I am wrong.
    ____
    I am not the professor.

12. If the barn door is locked, the horse is inside.
    The horse is inside.
    ____
    The barn door is locked.

13. You will have a test on Thursday or Friday.
    You have a test on Thursday.
    ____
    You will not have a test on Friday.

14. If it is snowing, then I will not go.
    It is snowing.
    ____
    I will not go.

15. If the guards do not block well, then the tailback does not make a touchdown.
    The tailback does not make a touchdown.
    ____
    The guards do not block well.

16. If it is 6 o'clock, then the office is closed.
    It is not 6 o'clock.
    ____
    The office is not closed.

17. Either #1 or #2 is false.
    #1 is true.
    ____
    #2 is false.

18. Either #3 or #4 is true.
    #4 is true.
    ____
    #3 is false.

19. If a quadrilateral is a square, then it is a rectangle.
    If a quadrilateral is a rectangle, then it is a parallelogram.
    ____
    If a quadrilateral is a square, then it is a parallelogram.

20. If every man contributes, the project cannot fail.
Some men do not contribute.
―――――――――――
The project can fail.

21. Prove that the fallacy of denying the antecedent is an invalid argument form by using the fact that $p \to q \Leftrightarrow (\sim q) \to (\sim p)$, and the fallacy of affirming the consequent.

22. Prove that invalid disjunctive syllogism is an invalid argument form by using the fact that $p \vee q \Leftrightarrow (\sim p) \to q$, and the fallacy of denying the antecedent.

1.14

The Rule of Substitution

In Sec. 1.12 we stated that a method other than the use of truth tables exists for determining the validity of an argument. This second method employs the *rule of substitution*, which states that in an argument one or more premises may be replaced by logically equivalent statements or any implied statement. By using the rule of substitution we attempt to transform a given argument to (1) one of the valid forms easily recognized, such as the rule of detachment (modus ponens), the law of the syllogism, disjunctive syllogism, or modus tollens; (2) one of the easily recognized fallacious argument forms such as the fallacy of affirming the consequent, the fallacy of denying the antecedent, or invalid disjunctive syllogism.

To apply the role of substitution we must be able to recall statements of logical equivalence. Some of the most frequently used statements of logical equivalence are listed here for reference purposes:

$$p \wedge q \Leftrightarrow q \wedge p$$
$$p \vee q \Leftrightarrow q \vee p$$
$$(p \wedge q) \wedge r \Leftrightarrow p \wedge (q \wedge r)$$
$$(p \vee q) \vee r \Leftrightarrow p \vee (q \vee r)$$
$$p \wedge (q \vee r) \Leftrightarrow (p \wedge q) \vee (p \wedge r)$$
$$p \vee (q \wedge r) \Leftrightarrow (p \vee q) \wedge (p \vee r)$$
$$\sim (\sim p) \Leftrightarrow p$$
$$\sim (p \wedge q) \Leftrightarrow (\sim p) \vee (\sim q)$$
$$\sim (p \vee q) \Leftrightarrow (\sim p) \wedge (\sim q)$$
$$p \to q \Leftrightarrow (\sim p) \vee q$$
$$\sim (p \to q) \Leftrightarrow p \wedge (\sim q)$$
$$p \to q \Leftrightarrow (\sim q) \to (\sim p)$$
$$p \leftrightarrow q \Leftrightarrow (p \to q) \wedge (q \to p)$$

(The list is by no means exhaustive.) Also recall that $p \vee (\sim p)$ and $\sim [p \wedge (\sim p)]$ are logically true statements.

In addition to replacing the premises of an argument by logically equivalent statements, we may also replace some of the premises by conclusions using basic valid arguments. Recall that the four basic forms of valid arguments which we have studied are:

$$\begin{array}{l} p \to q \\ p \\ \hline q \end{array} \qquad \begin{array}{l} p \to q \\ q \to r \\ \hline p \to r \end{array}$$

$$\begin{array}{l} p \vee q \\ \sim q \\ \hline p \end{array} \qquad \begin{array}{l} p \to q \\ \sim q \\ \hline \sim p \end{array}$$

We can use any statement of logical equivalence or any basic valid argument to reduce the number of original premises.

Example 1

Demonstrate that the following argument is valid:

$$\begin{array}{r} \text{First premise: } p \\ \text{Second premise: } r \to (\sim p) \\ \text{Third premise: } (\sim r) \to s \\ \hline \text{Conclusion: } s \end{array}$$

The second premise may be replaced by the statement $p \to (\sim r)$ since an implication and its contrapositive are logically equivalent. Using this equivalent second premise, the third premise, and the law of the syllogism, we have

$$\begin{array}{r} p \to (\sim r) \\ (\sim r) \to s \\ \hline p \to s \end{array}$$

Thus, the statement $p \to s$ may be used in place of the second and third premises. The original argument is now of the form

$$\begin{array}{l} p \\ p \to s \\ \hline s \end{array}$$

This is obviously a valid argument because it represents the rule of detachment.

Example 2

Determine the validity of the following argument:

If this is an important course, then it is worth taking. Either the course is difficult or the course is not worth taking. The course is not difficult. Therefore, this course is not important.

Let p: This is an important course.
q: The course is worth taking.
r: The course is difficult.

The argument can be stated in the following symbolic form:

$$\begin{array}{r l} \text{First premise:} & p \rightarrow q \\ \text{Second premise:} & r \vee (\sim q) \\ \text{Third premise:} & \sim r \\ \hline \text{Conclusion:} & \sim p \end{array}$$

By the use of disjunctive syllogism, the second and third premises may be replaced by the statement $\sim q$:

$$\begin{array}{l} r \vee (\sim q) \\ \sim r \\ \hline \sim q \end{array}$$

The original argument is now of the form

$$\begin{array}{l} p \rightarrow q \\ \sim q \\ \hline \sim p \end{array}$$

This is a valid argument since it represents the modus tollens form. Hence, the conclusion of this argument must be true whenever all the premises are true.

Example 3

Determine the validity of the following argument:

The football coach compliments me only if I play well. Either I make a touchdown or I do not play well. The coach did not compliment me. Therefore, I did not make a touchdown.

Let p: The football coach compliments me.
q: I played well.
r: I made a touchdown.

The argument can be stated in the following symbolic form:

First premise: $p \to q$
Second premise: $r \vee (\sim q)$
Third premise: $\sim p$

Conclusion: $\sim r$

The second premise is logically equivalent to $(\sim q) \vee r$, which in turn is logically equivalent to the implication $q \to r$. By the use of the law of the syllogism, the first premise and this equivalent form of the second premise may be replaced by the statement $p \to r$:

$$\begin{array}{c} p \to q \\ q \to r \\ \hline p \to r \end{array}$$

The original argument is now of the form

$$\begin{array}{c} p \to r \\ \sim p \\ \hline \sim r \end{array}$$

This is obviously an invalid argument because it is an example of the fallacy of denying the antecedent. The conclusion is not implied by the premises.

Exercises

In Exercises 1 through 8 use logically equivalent statements, implied statements, and basic argument forms to determine whether or not each argument is valid.

1. $$\begin{array}{l} p \to (\sim q) \\ r \to q \\ r \\ \hline \sim p \end{array}$$

2. $$\begin{array}{l} p \to q \\ q \to r \\ r \to s \\ \hline p \to s \end{array}$$

3. $$\begin{array}{l} p \to q \\ (\sim q) \vee r \\ p \\ \hline r \end{array}$$

4. $$\begin{array}{l} p \\ (\sim q) \vee r \\ (\sim p) \to q \\ \hline \sim r \end{array}$$

5. $p \to q$
$(\sim r) \to (\sim q)$
$\sim r$
―――――
$\sim p$

6. $(p \vee q) \vee (\sim r)$
r
$\sim p$
―――――
q

7. $p \to q$
$(\sim p) \vee (\sim r)$
$q \to (p \wedge r)$
―――――
p

8. $(\sim p) \to (\sim q)$
$\sim r$
$p \to s$
$q \vee r$
―――――
s

In Exercises 9 through 12 several premises and a conclusion are given. Determine whether or not each argument is valid.

9. If O'Hara is nominated by the Democratic Party, he will be elected. If the League for Better Politicians endorses O'Hara, he will not be elected. O'Hara is not endorsed by the League for Better Politicians. Therefore, the Democratic Party nominates him.
10. If the football team has a spirited practice session, then they will not do extra wind sprints. They do extra wind sprints or they have a "chalk talk." The football team does not have a spirited practice session. Therefore, they will not have a chalk talk.
11. I play cards only if I think I can win. Either I play cards or I have no money. If I think I can win, then I am happy. Therefore, if I have money, I am happy.
12. If two distinct lines are parallel, then they have no points in common. Either two distinct lines have at least one point in common or they are equidistant. The lines are not equidistant. Therefore, the lines are not parallel.
13. Show that the argument

$$\begin{array}{l} p \to q \\ r \to s \\ \underline{p \vee r} \\ q \vee s \end{array}$$

is a valid argument. This valid argument form is often called *constructive dilemma.*

14. Use the results of Exercise 13 to draw a valid conclusion from the following premises:

First premise: Either the Cubs or the Cards will win the pennant.
Second premise: If the Cubs win the pennant, the Mets will finish third.
Third premise: If the Cards win the pennant, the Mets will finish second.

15. Use the argument form called constructive dilemma (see Exercise 13) and show that the following argument is valid:

$$\begin{array}{l} p \to q \\ (\sim p) \to r \\ \hline q \vee r \end{array}$$

16. Show that the argument

$$\begin{array}{l} p \to q \\ r \to s \\ (\sim q) \vee (\sim s) \\ \hline (\sim p) \vee (\sim r) \end{array}$$

is a valid argument. This valid argument form is often called *destructive dilemma.*

17. Use the results of Exercise 16 to draw a valid conclusion from the following premises:

First premise: If a geometric figure has three sides, then it is a triangle.
Second premise: If a geometric figure has four sides, then it is a quadrilateral.
Third premise: The figure is either not a triangle or not a quadrilateral.

18. Use the argument form called destructive dilemma (see Exercise 16) and show that the following argument is valid:

$$\begin{array}{l} p \to q \\ r \to (\sim q) \\ \hline \sim (p \wedge r) \end{array}$$

Keyed Summary

In this chapter we studied simple and compound statements, truth values, and argument forms. The concepts and methods presented are part of the body of knowledge called the *algebra of statements* or *mathematical logic.*

1.1 Truth Values of Statements

The truth value of a true statement is true; the truth value of a false statement is false.

A simple statement expresses a single idea.

A simple statement can be represented by a single lowercase letter such as p, q, or r.

1.2 Conjunctions

Simple statements combined by means of an appropriate connecting word(s) are called compound statements.

The compound statement formed by using the connecting word "and" is called the conjunction of two simple statements.

The conjunction of two simple statements p and q is represented by "p and q."

The conjunction of two simple statements is true if both simple statements are true. If at least one simple statement is false, then the conjunction is false.

1.3 Disjunctions

The compound statement formed by using the connecting word "or" is called the disjunction of two simple statements.

The disjunction of two simple statements p and q is represented by "p or q."

The disjunction of two simple statements is true if at least one of the simple statements is true. If both simple statements are false, then the disjunction is false.

1.4 Negations

The compound statement formed by appropriately using the connecting word "not" is called the negation of a simple statement.

The negation of a simple statement p is represented by "not p."

The negation of a simple statement is true if the simple statement is false; the negation is false if the simple statement is true.

The negation of the negation of a simple statement is called the double negation of the statement.

1.5 Truth Tables

A truth table is a device that states all possible truth values of a compound statement in terms of the possible truth values of the simple statements involved.

Table 1.1 Truth Table for a Conjunction

p	q	p and q
T	T	T
T	F	F
F	T	F
F	F	F

Table 1.2 Truth Table for a Disjunction

p	q	p or q
T	T	T
T	F	T
F	T	T
F	F	F

Table 1.3 Truth Table for a Negation

p	not p
T	F
F	T

A logically true statement is one whose only truth value is true.

A logically false statement is one whose only truth value is false.

The logically true statement "p or (not p)" is called the law of the excluded middle.

The logically true statement "not [p and (not p)]" is called the law of contradiction.

1.6 Logically Equivalent Statements

Two statements that have the same truth values are called logically equivalent statements; that is, two statements p and q are logically equivalent if the truth values of p and q are both true or both false.

The logical equivalence of two statements p and q is represented by $p \Leftrightarrow q$.

Any two statements such that at least one of the four possible combinations of truth values cannot possibly occur are called dependent statements. If all possible combinations of truth values can occur, the two statements are called independent statements.

Any two statements such that both cannot be simultaneously true are called inconsistent statements. If both statements can be simultaneously true, they are called consistent statements.

1.7 More About Negations

DeMorgan's laws for statements:

$$\text{not } (p \text{ and } q) \Leftrightarrow (\text{not } p) \text{ or } (\text{not } q)$$
$$\text{not } (p \text{ or } q) \Leftrightarrow (\text{not } p) \text{ and } (\text{not } q)$$

Law of double negation:

$$p \Leftrightarrow \text{not } (\text{not } p)$$

1.8 Implications

The compound statement formed by using the connecting words "If ..., then ..." is called an implication.

The implication of two simple statements p and q is represented by "If p, then q." The simple statement p is called the hypothesis or antecedent; the simple statement q is called the conclusion or consequent.

An implication is false if the hypothesis is true and the conclusion is false; otherwise, the implication is true.

A true implication "If p, then q" is represented by $p \Rightarrow q$, which is read "p implies q."

Table 1.4 Truth Table for an Implication

p	q	If p, then q
T	T	T
T	F	F
F	T	T
F	F	T

The compound statements "If p, then q" and "(not p) or q" are logically equivalent.

1.9 Necessary and Sufficient Conditions

The statement "p is a sufficient condition for q" means "If p, then q"; the statement "q is a necessary condition for p" means "If p, then q."

The statement "q if p" means "If p, then q"; the statement "q only if p" means "If q, then p."

The statement "q if and only if p" means "(If p, then q) and (if q, then p)."

1.10 Related Implications

Implication: If p, then q.
Converse of the implication: If q, then p.
Inverse of the implication: If (not p), then (not q).
Contrapositive of the implication: If (not q), then (not p).

An implication and the contrapositive of the implication are logically equivalent:

If p, then $q \Leftrightarrow$ if (not q), then (not p).

The converse of an implication and the inverse of the implication are logically equivalent:

If q, then $p \Leftrightarrow$ if (not p), then (not q).

Two statements are logically equivalent when each one implies the other:

$p \Leftrightarrow q$ means $(p \Rightarrow q)$ and $(q \Rightarrow p)$.

1.11 *Symbolic Representations*

Symbols may be used to represent the connectives:

Symbol	Meaning
$\wedge$	and
$\vee$	or
$\sim$	not
$\rightarrow$	if . . . , then . . .
$\leftrightarrow$	if and only if

1.12 *Valid Arguments*

An argument is an implication of the form

If p_1 and p_2 and $p_3 \cdots p_k$, then c.

The statements p_1, p_2, p_3, . . . , and p_k are called the premises; the statement c is called the conclusion.

An argument is valid if and only if the conjunction of the premises implies the conclusion:

$$(p_1 \wedge p_2 \wedge p_3 \cdots p_k) \Rightarrow c.$$

Rule of detachment (*modus ponens*):

$$[(p \rightarrow q) \wedge p] \Rightarrow q$$

Law of the syllogism:

$$[(p \rightarrow q) \wedge (q \rightarrow r)] \Rightarrow (p \rightarrow r)$$

Disjunctive syllogism:

$$[(p \vee q) \wedge (\sim q)] \Rightarrow p$$

Modus tollens:

$$[(p \to q) \wedge (\sim q)] \Rightarrow (\sim p)$$

1.13 Fallacious Arguments

An argument that is not valid is called a fallacious argument, or simply a fallacy.

Fallacy of affirming the consequent:

$$\begin{array}{rl} \text{First premise:} & p \to q \\ \text{Second premise:} & q \\ \hline \text{Conclusion:} & p \end{array}$$

Fallacy of denying the antecedent:

$$\begin{array}{rl} \text{First premise:} & p \to q \\ \text{Second premise:} & \sim p \\ \hline \text{Conclusion:} & \sim q \end{array}$$

Invalid disjunctive syllogism:

$$\begin{array}{rl} \text{First premise:} & p \vee q \\ \text{Second premise:} & p \\ \hline \text{Conclusion:} & \sim q \end{array}$$

1.14 The Rule of Substitution

The rule of substitution states that in an argument one or more premises may be replaced by logically equivalent statements or any implied statement.

Constructive dilemma is a valid argument form:

$$\begin{array}{l} p \to q \\ r \to s \\ p \vee r \\ \hline q \vee s \end{array}$$

Destructive dilemma is a valid argument form:

$$\begin{array}{l} p \to q \\ r \to s \\ (\sim q) \vee (\sim s) \\ \hline (\sim p) \vee (\sim r) \end{array}$$

Building Your Mathematical Vocabulary

The following terms and expressions were introduced or defined in this chapter; check your understanding of these words. The references in parentheses indicate the sections in which the terms and expressions were first used.[1]

Antecedent (1.8)
Argument (1.12)
Associative properties (1.6)
Commutative properties (1.6)
Compound statement (1.2)
Conclusion (1.8, 1.12)
Conjunction (1.2)
Consequent (1.8)
Consistent statements (1.6)
Constructive dilemma (1.14)
Contrapositive (1.10)
Converse (1.10)
DeMorgan's laws for statements (1.7)
Dependent statements (1.6)
Destructive dilemma (1.14)
Disjunction (1.3)
Disjunctive syllogism (1.12)
Distributive properties (1.6)
Dominant (1.6)
Double negation (1.4)
Exclusive disjunction (1.3)
Fallacious argument (1.13)
Fallacy (1.13)
Fallacy of affirming the consequent (1.13)
Fallacy of denying the antecedent (1.13)
Hypothesis (1.8)
Idempotent properties (1.6)
Implication (1.8)
Inclusive disjunction (1.3)
Inconsistent statements (1.6)
Independent statements (1.6)
Invalid disjunctive syllogism (1.13)
Inverse (1.10)
Law of contradiction (1.5)
Law of double negation (1.7)
Law of the excluded middle (1.5)
Law of the syllogism (1.12)
Logically equivalent statements (1.6)
Logically false statement (1.5)
Logically true statement (1.5)
Modus ponens (1.12)
Modus tollens (1.12)
Necessary condition (1.9)
Negation (1.4)
Open statement (1.6)
Premises (1.12)
Rule of detachment (1.12)
Rule of substitution (1.14)
Simple statement (1.1)
Sufficient condition (1.9)
Truth table (1.5)
Truth value (1.1)
Valid argument (1.12)

---

[1] This section will appear at the end of each chapter.

2

Sets

2.1

Sets and Their Descriptions

The concept of a set is fundamental to the study of mathematics. The properties of sets and relations that exist among sets are simple to understand, and furthermore, there is a close relationship between sets and mathematical logic. Intuitively, we consider a *set* as a collection of objects such as numbers, persons, places, or geometric figures. For example, the following are sets:

1. The numbers 1, 3, 5, 7, and 9
2. The Presidents of the United States
3. The baseball teams in the American League
4. The seven days of the week
5. All equilateral triangles
6. All lines in a plane
7. The prime numbers
8. The cities of Europe with populations over 1 million
9. The students in your mathematics class
10. The points on a line

A set is *well defined* if and only if it is always possible to determine whether or not a given object is in the collection. Each of the sets just listed is a well-defined set.

Capital letters are used to indicate particular sets to be discussed or studied. For example, we could write

A = The set of numbers 1, 3, 5, 7, and 9
B = The set of all equilateral triangles
C = The set of cities of Europe with populations over 1 million

If an object belongs to a set, the object is called a *member* or *element* of the set. Lowercase letters are used to indicate elements of a set. The relation of "membership" is the only important relation that exists between a given object and a given set.

If an object a is an element of a set A, we write $a \in A$. If an object a is not an element of a set A, we write $a \notin A$.

The symbol $\in$ (the Greek letter epsilon) is read "is an element of"; the symbol $\notin$ is read "is not an element of."

Example 1

Let P represent the set of prime numbers. Determine the truth values of statements (*a*) to (*d*):

(*a*) $3 \in P$
(*b*) $10 \notin P$
(*c*) $33 \in P$
(*d*) $17 \in P$

(*a*) True; 3 is a prime. (*b*) True; 10 is not a prime. (*c*) False; 33 is not a prime. (*d*) True; 17 is a prime.

Example 2

Let S represent the set of Presidents of the United States. Let g represent George Washington, h represent Alexander Hamilton, f represent Benjamin Franklin, and k represent John F. Kennedy. Replace ■ by $\in$ or $\notin$ to make each of the following statements true:

(*a*) g ■ S
(*b*) h ■ S
(*c*) f ■ S
(*d*) k ■ S

(*a*) $g \in S$ (*b*) $h \notin S$ (*c*) $f \notin S$ (*d*) $k \in S$

Sets may be defined in several ways; one way was used in the preceding material. A set may be defined also by means of a verbal statement that describes the objects which are elements of the set. For example,

A is the set of whole numbers less than 5;
B is the set of months of the year.

An object will be an element of the set if it fits the description given; otherwise, it will not be an element of the set. This method of defining a set is called the *description method*.

A set may be defined by listing each member of the set within a pair of braces. For example,

$A = \{0, 1, 2, 3, 4\}$
$B =$ {January, February, March, April, May, June, July, August, September, October, November, December}

An object will be an element of the set if it is contained in the list within the braces; otherwise, it will not be an element of the set. This method of defining a set is called the *listing* or *roster method*.

For many sets we may not want or be able to write a complete list of the particular members of the set. A convenient notation using the braces of the listing method and the descriptive technique of the description method exists and is called the *set-builder notation*. Thus, you can define a set by writing

$$\{x \mid \cdots\}$$

which is read "the set of all elements x such that . . . ." The common property possessed by each element x is described in the space . . . . For example,

$A = \{x \mid x \text{ is a whole number less than } 5\}$
$B = \{x \mid x \text{ is a month of the year}\}$

This method is called the *rule method* for defining a set.

It is often more convenient to use one of the three methods of defining a set than the others.

Example 3

Represent sets (*a*) to (*d*) by the listing method:

(*a*) $\{x \mid x \text{ is a vowel in the English alphabet}\}$
(*b*) $\{x \mid x \text{ is a day of the week}\}$
(*c*) $\{x \mid x \text{ is a prime number less than } 17\}$
(*d*) $\{x \mid x \text{ is a whole number greater than 3 and less than } 5\}$

(*a*) $\{a, e, i, o, u\}$ (*b*) {Sunday, Monday, Tuesday, Wednesday, Thursday, Friday, Saturday} (*c*) $\{2, 3, 5, 7, 11, 13\}$ (*d*) $\{4\}$

Exercises

In Exercises 1 through 4 let $A = \{1, 2, 3\}$ and $B = \{2, 3, 4\}$. Determine the truth value of each statement.

1. $4 \in A$
2. $2 \in A$ and $2 \in B$
3. $1 \in A$ or $1 \in B$
4. $1 \in A$ and $3 \notin B$

In Exercises 5 through 10 represent each set by the listing method.

5. $\{x|x$ is a country in North America that borders the United States$\}$
6. $\{x|x$ is a season of the year$\}$
7. $\{x|x$ is a consonant in the English alphabet$\}$
8. $\{x|x$ is a whole number less than 7$\}$
9. $\{x|x$ is a whole number such that $2x < 11\}$
10. $\{x|x$ is a vowel in the word "mathematics"$\}$

In Exercises 11 through 16 represent each set by using set-builder notation in the rule method.

11. {first baseman, second baseman, shortstop, third baseman}
12. $\{3, 4, 5, 6, 7, 8\}$
13. $\{1, 3, 5, 7, 9, 11, 13\}$
14. The set of months of the year containing the letter "r" in their names
15. The set of even prime numbers
16. {red, orange, yellow, green, blue, indigo, violet}

2.2

Finite and Infinite Sets

Sets may have a finite number of elements; that is, a set may have elements which can be counted, such as the set of vowels in the English alphabet,

$$\{a, e, i, o, u\}$$

which has five elements and the set of months of the year,

{January, February, March, April, May, June, July,
August, September, October, November, December}

which has twelve elements. Such sets are examples of *finite sets*. Other sets that have infinitely many elements exist; that is, a set may have elements which cannot be counted, such as the set of all equilateral triangles,

$$\{x \mid x \text{ is an equilateral triangle}\}$$

the set W of whole numbers,

$$W = \{0, 1, 2, 3, \ldots\}$$

the set N of natural numbers,

$$N = \{1, 2, 3, 4, \ldots\}$$

and the set of real numbers. Such sets are examples of *infinite sets*.

The three dots are used in the listing method for an infinite set whenever it is understood what the missing elements are. Note that the listing method is useful only for infinite sets in which a pattern relating the elements can be established. Three dots are also used in the listing method for finite sets with a large number of elements when a pattern can be established. For example,

$$\{a, b, c, \ldots, x, y, z\}$$

denotes the finite set of letters of the English alphabet.

Example 1

Represent sets (*a*) to (*d*) by the listing method:

(*a*) $\{x \mid x$ is a whole number less than $6\}$

(*b*) $\{x \mid x$ is an odd natural number$\}$

(*c*) $\{x \mid x$ is a natural number greater than $2\}$

(*d*) $\{x \mid x$ is a natural number greater than 4 and less than $7\}$

(*a*) $\{0, 1, 2, 3, 4, 5\}$ (*b*) $\{1, 3, 5, 7, \ldots\}$ (*c*) $\{3, 4, 5, 6, \ldots\}$
(*d*) $\{5, 6\}$

Example 2

Determine which of the following sets are finite and which are infinite:

(*a*) $\{3, 6, 9, 12, \ldots\}$

(*b*) $\{1, 4, 9, 16, 25\}$

(*c*) $\{x \mid x$ is a whole number that is a multiple of $5\}$

(*d*) $\{x \mid x$ is a natural number greater than $8\}$

(*e*) $\{x \mid x$ is a natural number less than $8\}$

(*f*) $\{x \mid x$ is a natural number equal to $8\}$

(*a*) Infinite set. (*b*) Finite set; the set has five elements. (*c*) Infinite set. (*d*) Infinite set. (*e*) Finite set; the set has seven elements. (*f*) Finite set; the set has one element.

A finite set may have no elements. For example, the following sets are finite sets with no elements:

The set of even primes greater than 2
The set of natural numbers less than 1
The set of whole numbers that are both odd and even
The set of quadrilaterals with exactly three sides

***A set that contains no elements is called the* null, *or* empty, set. *The null set is denoted by either* { } *or the symbol* $\varnothing$.**

Consider the sets

$$A = \{2, 3, 1, 7\} \qquad \text{and} \qquad B = \{3, 7, 2, 1\}$$

Every element of set A is also an element of set B; every element of set B is also an element of set A. In such a case, the two sets are said to be equal, and we write $A = B$.

Sets A and B are* equal *if and only if every element of A is an element of B and every element of B is an element of A. When A is equal to B, we write A = B; when A is not equal to B, we write A ≠ B.

Example 3

Replace ■ by $=$ or $\neq$ to make each of the following statements true:

(*a*) N ■ W

(*b*) $\{1, 3, 5, 7\}$ ■ $\{5, 7, 3, 1\}$

(*c*) {0, 1, 2,} ■ {1, 2}

(*d*) $\{x|x$ is a whole number but not a natural number$\}$ ■ {0}

(*a*) $N \neq W$ (*b*) $\{1, 3, 5, 7\} = \{5, 7, 3, 1\}$ (*c*) $\{0, 1, 2\} \neq \{1, 2\}$
(*d*) $\{x|x$ is a whole number but not a natural number$\} = \{0\}$

Exercises

In Exercises 1 through 4 represent each set by the listing method.

1. $\{x|x$ is an even natural number$\}$
2. $\{x|x$ is a whole number greater than $10\}$
3. $\{x|x$ is a natural number that is a multiple of $3\}$
4. $\{x|x$ is a natural number whose last digit is $5\}$

In Exercises 5 through 14 determine which of the sets are finite and which are infinite.

5. The set of baseball teams in the National League
6. The set of people in the world
7. The set of trees in Yellowstone National Park
8. The set of books in the Library of Congress
9. $\{x|x$ is a whole number less than $3\}$
10. $\{x|x$ is a whole number greater than $3\}$
11. $\{1, 2, 4, 8, 16, 32, \ldots\}$
12. $\{7, 14, 21, 28, \ldots\}$
13. $\{x|x$ is a natural number but not a whole number$\}$
14. $\{x|x$ is an even prime number$\}$

In Exercises 15 through 23 determine whether or not each set is the null set.

15. The set of states in the United States smaller than Rhode Island
16. The set of people born in China who live in Florida
17. The set of Presidents of the United States who were women

18. The set of Democrats who voted Republican in the 1972 Presidential Election
19. The set of prime numbers greater than 19 and less than 24
20. The set of odd natural numbers that belong to the set of whole numbers
21. The set of quadrilaterals that are rectangles
22. $\{0\}$
23. $\{\varnothing\}$
24. Let $x \in \{-1, 0, 1\}$. Replace ■ by $=$ or $\neq$ to make each of the following statements true:

    (*a*) $\{x \mid x^2 = 1\}$ ■ $\{1\}$
    (*b*) $\{x \mid x^2 = x\}$ ■ $\{0, 1\}$
    (*c*) $\{x \mid x^3 - 1 = 0\}$ ■ $\{-1, 1\}$
    (*d*) $\{x \mid x(x - 1)(x + 1) = 0\}$ ■ $\{-1, 0, 1\}$

2.3

Subsets

Every natural number is also a whole number. That is, every element of the set $\{1, 2, 3, 4, \ldots\}$ is also an element of the set $\{0, 1, 2, 3, \ldots\}$. In such a case the set N of natural numbers is said to be a *subset* of the set W of whole numbers, and we write $N \subseteq W$. Since the set W contains an element, namely 0, which is not an element of N, then the set W of whole numbers is not a subset of the set N of natural numbers, and we write $W \not\subseteq N$.

Set A is a* subset *of set B if and only if every element of A is also an element of B. When A is a subset of B, we write $A \subseteq B$; when A is not a subset of B, we write $A \not\subseteq B$.

For example,

$$\{2, 3, 5\} \subseteq \{1, 2, 3, 4, 5, 6\}$$
$$\{a, b, c, d\} \subseteq \{a, b, c, d\}$$
$$\{5\} \subseteq \{x \mid x \text{ is an odd natural number}\}$$
$$\{\text{Sunday, Monday, January}\} \not\subseteq \{x \mid x \text{ is a day of the week}\}$$

Note that by definition every set is a subset of itself. Furthermore, since the statement "Every element of A is also an element of B" is equivalent to "There is no element of A that is not an element of B," the null set $\{\ \}$ is a subset of every set.

$$A \subseteq A \qquad \text{for every set } A$$
$$\{\ \} \subseteq A \qquad \text{for every set } A$$

Example 1

Let N be the set of natural numbers, W the set of whole numbers, D the set of odd whole numbers, and E the set of even whole numbers. Determine the truth values of statements (a) to (d):

(a) $D \subseteq W$

(b) $D \subseteq N$

(c) $E \subseteq N$

(d) $E \nsubseteq W$

(a) True. (b) True. (c) False; every even whole number is not a natural number because $0 \in E$, but $0 \notin N$. (d) False; every even whole number is a whole number.

In the sets $R = \{1, 2, 3, 4, 5\}$, $S = \{1, 3, 4\}$, and $T = \{3, 5, 1, 2, 4\}$, note that

$$S \subseteq R \quad \text{but} \quad R \nsubseteq S,\ S \neq R$$
$$T \subseteq R \quad \text{and} \quad R \subseteq T,\ T = R$$

Although sets S and T are both subsets of R, set R is a subset only of T. Set R contains at least one element that is not an element of S. Since $T \subseteq R$ and $R \subseteq T$, every element of T is an element of R, and every element of R is an element of T. Thus $T = R$. We often wish to distinguish between the two types of subsets of R illustrated by S and T. The set S is said to be a *proper subset* of R, and we write $S \subset R$; set T is not a proper subset of R, and we write $T \not\subset R$.

Set A is a* proper subset *of set B if and only if $A \subseteq B$ and $A \neq B$. When A is a proper subset of B, we write $A \subset B$; when A is not a proper subset of B, we write $A \not\subset B$.

By definition, a set A is a proper subset of set B if and only if A is a subset of B, where B contains at least one element that is not an element of A. For example,

$$\{1, 2\} \subset \{1, 2, 3, 4, 5\}$$
$$\{a, e, i, o, u\} \subset \{x \mid x \text{ is a letter of the English alphabet}\}$$
$$\{1, 3, 5, 7\} \not\subset \{1, 3, 5, 7\}$$
$$\{x \mid x \text{ is an equilateral triangle}\} \subset \{x \mid x \text{ is an isosceles triangle}\}$$

As a result of the definitions of proper subset and null set, it follows that

$$A \not\subset A \quad \text{for every set } A$$
$$\{\ \} \subset A \quad \text{for every set } A \text{ except the null set}$$

No set is a proper subset of itself. The null set is a proper subset of every set other than itself.

Exercises

In Exercises 1 through 8 let $A = \{a, b, c, d, e\}$, $B = \{a, c, d\}$, and $C = \{c, d, e\}$. Determine the truth value of each statement.

1. $A \subseteq B$
2. $B \nsubseteq C$
3. $B \subseteq A$
4. $B \subset A$
5. $C \nsubseteq A$
6. $B \subset C$
7. $C \subset A$
8. $A \nsubseteq C$

In Exercises 9 through 16 let $x \in A$, $y \in B$, $A \subset B$, and $B \subset C$. Determine, if possible, the truth value of each statement.

9. $x \in B$
10. $x \notin C$
11. $y \in A$
12. $y \in C$
13. $B \subseteq C$
14. $A \neq C$
15. $A \subseteq C$
16. $A \subset C$

In Exercises 17 through 24 determine the truth value of each statement.

17. The set of squares is a subset of the set of rectangles.
18. The set of rectangles is not a subset of the set of quadrilaterals.
19. The set of triangles is a subset of the set of quadrilaterals.
20. The set of men is a subset of the set of teachers.
21. The set of teachers is not a subset of the set of men.
22. The set of horses is a subset of the set of animals.
23. The set of Catholics is a subset of the set of Christians.
24. The set of United States senators is not a subset of the set of Republicans.
25. Let A be the set of all triangles, B the set of all equilateral triangles, C the set of all isosceles triangles, and D the set of all equiangular triangles. State a relationship between sets (*a*) to (*f*):

    (*a*) A and B
    (*b*) A and C
    (*c*) B and C
    (*d*) A and D
    (*e*) B and D
    (*f*) C and D

2.4

Power Sets

The problem of determining all the sets that can be found by choosing elements from a given set arises frequently in mathematics. Consider the problem of determining all the subsets of the set $A = \{1, 2, 3, 4\}$. The only subset of A that contains no elements is the null set $\{\ \}$. There are four subsets of A that contain a single element:

$$\{1\}, \{2\}, \{3\}, \text{ and } \{4\}$$

If we pair the element 1 with each element that follows it in set A, pair the element 2 with each element that follows it in set A, and, finally, pair the element 3 with each element that follows it in set A, we can obtain the six subsets of A that contain exactly two elements:

$$\{1, 2\}, \{1, 3\}, \{1, 4\}, \{2, 3\}, \{2, 4\}, \text{ and } \{3, 4\}$$

To determine the subsets of A that contain exactly three elements, we can consider the elements that were left when the four subsets with single elements were formed. For example, when the set $\{1\}$ was formed, the set $\{2, 3, 4\}$ remained. Thus, there are four subsets of A that contain exactly three elements:

$$\{2, 3, 4\}, \{1, 3, 4\}, \{1, 2, 4\}, \text{ and } \{1, 2, 3\}$$

Finally, the only subset of A that contains four elements is the set A itself, that is, $\{1, 2, 3, 4\}$. Thus, the sixteen subsets of A are

$$\{\ \}, \{1\}, \{2\}, \{3\}, \{4\}, \{1, 2\}, \{1, 3\},$$
$$\{1, 4\}, \{2, 3\}, \{2, 4\}, \{3, 4\}, \{2, 3, 4\},$$
$$\{1, 3, 4\}, \{1, 2, 4\}, \{1, 2, 3\}, \{1, 2, 3, 4\}$$

The set that contains these sixteen "sets as elements" is called the power set of A. In general, the *power set* of any set S is the set of all subsets of S. Note that each element of a power set is itself a set.

Table 2.1 gives a list of the subsets of set A, where A has one, two, three, and four elements, respectively. Study the table carefully. Can you discover the rule for the number of subsets of a set A with n elements? There are two subsets of a set with one element, four (2×2) subsets of a set with two elements, eight $(2 \times 2 \times 2)$ subsets of a set with three elements, and sixteen $(2 \times 2 \times 2 \times 2)$ subsets of a set with four elements. The number of subsets of a set A with n elements is 2^n; n is called an *exponent* and means that we use 2 as a factor n times.

Table 2.1

Set A	Number of elements in A	Subsets of A	Number of subsets of A
$\{a\}$	1	$\{\ \}, \{a\}$	2
$\{a, b\}$	2	$\{\ \}, \{a\}, \{b\}, \{a, b\}$	4
$\{a, b, c\}$	3	$\{\ \}, \{a\}, \{b\}, \{c\}, \{a, b\}$ $\{a, c\}, \{b, c\}, \{a, b, c\}$	8
$\{a, b, c, d\}$	4	$\{\ \}, \{a\}, \{b\}, \{c\}, \{d\},$ $\{a, b\}, \{a, c\}, \{a, d\},$ $\{b, c\}, \{b, d\}, \{c, d\},$ $\{a, b, c\}, \{a, b, d\},$ $\{a, c, d\}, \{b, c, d\},$ $\{a, b, c, d\}$	16

Example 1

Determine how many elements are in the power set A if A has the following number of elements:
(*a*) zero
(*b*) five

(*a*) 1; there is only one subset of the null set, the null set itself.
(*b*) 2^5, that is, $2 \times 2 \times 2 \times 2 \times 2$. Thus, there are 32 subsets of a set with five elements.

Example 2

Steven plans to read some books during a summer vacation trip. He has the following six new books in his home library:

"A Treasury of Sherlock Holmes"
"The Call of the Wild"
"King Solomon's Mines"
"In Dubious Battle"
"Stories to Remember"
"The History of Baseball"

How many different nonempty sets of books can he take with him?

There are 2^6 subsets of a set with six elements. However, one of these sets is the null or empty set. Therefore, the number of different nonempty sets of books that Steven can take with him is $2^6 - 1$, that is, 63. (Remember that $2^6 = 2 \times 2 \times 2 \times 2 \times 2 \times 2 = 64$.)

Exercises

1. Consider the set $\{a, b, c, d, e\}$. Determine all the subsets with
   (*a*) two elements (*b*) one element
   (*c*) three elements (*d*) no elements
   (*e*) four elements (*f*) five elements
2. Determine the number of elements in the power set of A if A has the following number of elements: (*a*) seven; (*b*) ten.
3. Determine how many of the elements in the power set of A are proper subsets of A if A has the following number of elements: (*a*) two; (*b*) three; (*c*) four; (*d*) n.
4. Determine the number of nonempty subsets of A if A has the following number of elements: (*a*) two; (*b*) three; (*c*) five; (*d*) n.
5. List all the subsets of the set S, where $S = \{1, 2, 4, 8\}$. Find the sum of the elements in each subset; describe the sums.
6. Find the number of committees of three people that can be obtained from a set of five people.

2.6

Universal Sets and Complements

In any particular discussion of sets, it is usually necessary to limit the objects to be considered.

The set of all objects under study is called the* universal set *and is represented by the symbol U.

The universal set U serves as a reference set for the particular discussion. Every set A that is considered is a subset of U.

Different universal sets are used for different discussions. For example, if we wish to restrict a discussion to the set of days of the week, the universal set is

$$U = \{\text{Sunday, Monday, Tuesday, Wednesday, Thursday, Friday, Saturday}\}$$

If we wish to restrict our discussion to the set of whole numbers, the universal set is

$$U = \{0, 1, 2, 3, \ldots\}$$

Let the universal set U be the set of letters of the English alphabet:

$$U = \{x \mid x \text{ is a letter of the English alphabet}\}$$

The set U of letters of the English alphabet can be partitioned into two subsets: the set V of vowels,

$$V = \{a, e, i, o, u\}$$

and the set C of consonants,

$$C = \{b, c, d, f, g, h, j, k, l, m, n, p, q, r, s, t, v, w, x, y, z\}$$

Every element of the set U of letters is in either set V or set C. Furthermore, no letter is both a vowel and a consonant; that is, no element is in both V and C. Each of the sets V and C is the *complement* of the other with respect to set U.

> ***The* complement *of a set A is the set of elements in the universal set U that are not in set A. The complement of set A is denoted by $\bar{A}$.***

In this definition it is understood that A must be some subset of U. Thus, in symbolic notation,

$$\bar{A} = \{x \mid x \in U \text{ and } x \notin A\}$$

The symbol $\bar{A}$ is read "the complement of A."

Example 1

Let $U = \{1, 2, 3, 4, 5\}$ be the universal set. Find the complements of sets (*a*) to (*d*):

(*a*) $\{1, 2, 3\}$

(*b*) $\{4, 5\}$

(*c*) $\{1, 3, 4\}$

(*d*) $\{\ \}$

(*a*) $\{4, 5\}$ (*b*) $\{1, 2, 3\}$ (*c*) $\{2, 5\}$ (*d*) $\{1, 2, 3, 4, 5\}$. The universal set and the null set are complements of each other.

Example 2

Let the universal set be the set N of natural numbers. Determine the complements of sets (a) to (d):

(a) $\{1, 3, 5, 7, \ldots\}$, the set of odd natural numbers

(b) $\{2, 3, 5, 7, 11, 13, \ldots\}$, the set of prime numbers

(c) $\{x \mid x < 11\}$

(d) $\{x \mid x \text{ is less than 4 or greater than 6}\}$

(a) $\{2, 4, 6, 8, \ldots\}$, the set of even natural numbers (b) $\{x \mid x \text{ is a composite number or is equal to 1}\}$ (c) $\{11, 12, 13, 14, \ldots\}$ (d) $\{4, 5, 6\}$

Example 3

In the universal set $U = \{1, 2, 3, 4, 5, 6, 7, 8, 9, 10\}$, let A be the set of odd primes and B the set of odd numbers. Show that $A \subset B$ and $\bar{B} \subset \bar{A}$.

$A = \{3, 5, 7\}$ and $\bar{A} = \{1, 2, 4, 6, 8, 9, 10\}$
$B = \{1, 3, 5, 7, 9\}$ and $\bar{B} = \{2, 4, 6, 8, 10\}$

Because every element of A is also an element of B and $A \neq B$, then A is a proper subset of B; that is, $A \subset B$.

Similarly, because every element of $\bar{B}$ is also an element of $\bar{A}$ and $\bar{B} \neq \bar{A}$, then $\bar{B}$ is a proper subset of $\bar{A}$; that is, $\bar{B} \subset \bar{A}$.

There are occasions when we wish to determine the elements of one set that are not in another set. For example, if set P is the set of prime numbers

$$P = \{2, 3, 5, 7, 11, 13, 17, 19, 23, \ldots\}$$

and set E is the set of even natural numbers

$$E = \{2, 4, 6, 8, 10, 12, 14, 16, \ldots\}$$

then the elements of set P that are not in set E form the set of odd prime numbers

$$\{3, 5, 7, 11, 13, 17, 19, 23, \ldots\}$$

This set is called the *relative complement* of E in P and is denoted by $P - E$. That is,

$$P - E = \{3, 5, 7, 11, 13, 17, 19, 23, \ldots\}$$

Note that P and E are both subsets of the set N of natural numbers.

> ***If A and B are subsets of the universal set U, then the* relative complement *of A in B is the set of all elements in B that are not in A. The relative complement of A in B is denoted by $B - A$.***

The notation $B - A$ is read "the relative complement of A in B."

The concept of a relative complement is more general than the simple concept of a complement. Note that

$$\bar{A} = U - A$$

that is, "the complement of A" can be considered as "the relative complement of A in U."

Exercises

In Exercises 1 through 8 let $U = \{a, b, c, d, e\}$, $A = \{a, c, e\}$, $B = \{a, b, d\}$, and $C = \{b, c, d, e\}$. Find each set.

1. $\bar{A}$
2. $\bar{B}$
3. $U - C$
4. $B - A$
5. $C - B$
6. $C - (A - B)$
7. $A - \bar{C}$
8. $\overline{B - C}$

In Exercises 9 through 16 let the universal set be the set W of whole numbers. Let N be the set of natural numbers, P the set of prime numbers, E the set of even whole numbers, and D the set of odd whole numbers. Find each set.

9. $\bar{N}$
10. $W - N$
11. $\bar{E}$
12. $P - D$
13. $E - D$
14. $E - \bar{D}$
15. $\overline{\{0\}}$
16. $W - \bar{E}$
17. Find (*a*) $\bar{U}$; (*b*) $\bar{\varnothing}$.
18. Let A be any subset of some universal set U. Find (*a*) $A - A$; (*b*) $A - \bar{A}$.
19. Find $\overline{(\bar{A})}$, where $U = \{1, 2, 3, 4, 5\}$ and $A = \{1, 3, 4\}$.
20. Show that $\overline{(\bar{A})} = A$; that is, the complement of the complement of any set A is the set A itself.

2.6

Venn Diagrams

Sets and relations among sets may be represented geometrically by means of *Venn diagrams*. In a Venn diagram the universal set U is represented by the set of points of a rectangular region, as shown in Fig. 2.1. Any nonempty set A that is a proper subset of U is represented by the set of points of a circular region contained within the rectangular region, as shown in Fig. 2.2.

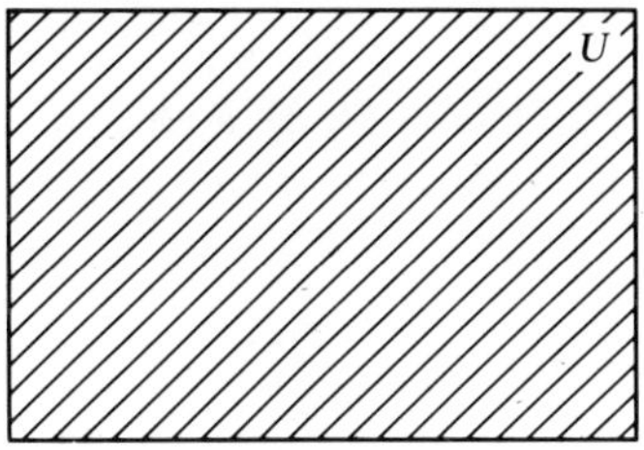

Figure 2.1

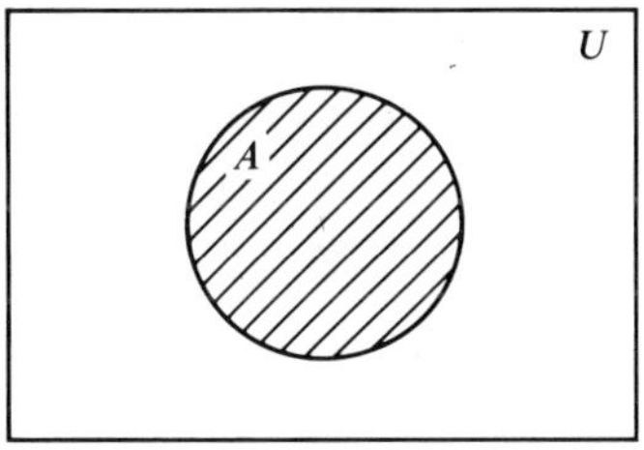

$A \subset U$ and $A \neq \{ \}$

Figure 2.2

The set $\bar{A}$, the complement of A, is represented by the set of points of the rectangular region that are not in the circular region representing A. The complement of set A is represented by the hatched region in Fig. 2.3. Note that the circular boundary of A is shown as a dashed line since the points of the circle represent elements of set A rather than elements of $\bar{A}$.

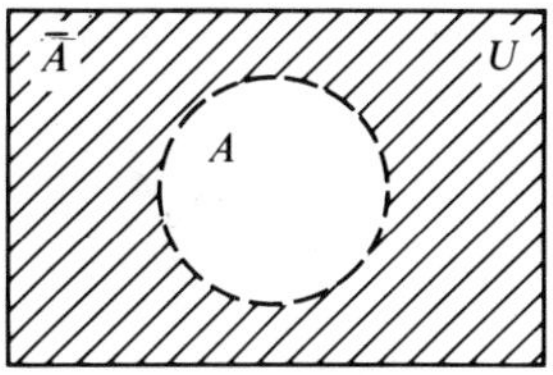

$\bar{A} = \{x | x \in U \text{ and } x \notin A\}$

Figure 2.3

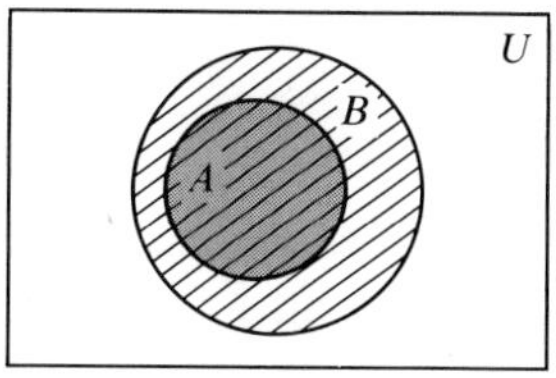

$A \subset B$

Figure 2.4

Venn diagrams are especially useful in illustrating relations that exist among sets. For example, if set A is a proper subset of set B, this relation may be represented by the Venn diagram shown in Fig. 2.4. The equality of sets A and B may be represented by the Venn diagram shown in Fig. 2.5.

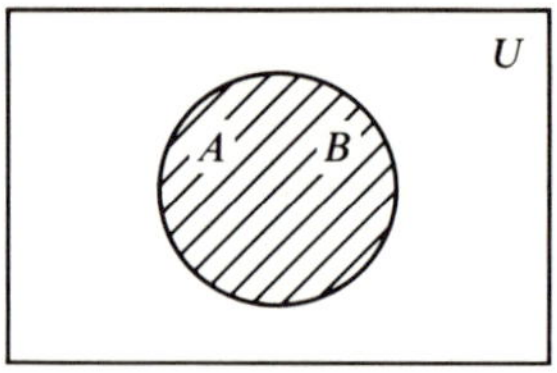

$A = B$

Figure 2.5

Example 1

Let $U = \{1, 2, 3, 4, 5, 6\}$, $A = \{1, 3\}$, $B = \{4, 6\}$, and $C = \{1, 2, 3\}$. Use Venn diagrams to represent the relations that exist among the four sets.

U is the universal set, so it is represented by the set of points in the rectangular region of Fig. 2.6. The sets A, B, and C are all proper

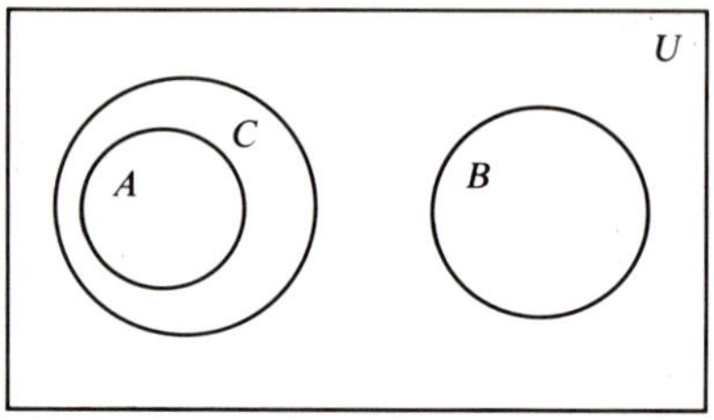

Figure 2.6

subsets of U; thus they are represented by circular regions contained within the rectangular region representing U. Since sets B and C have no elements in common, the circular regions representing them do not overlap. The circular region representing set A is contained entirely within the circular region representing set C because A is a proper subset of C.

Example 2

Consider Fig. 2.6. Shade the regions represented by sets (a) to (d):

(a) $\{5\}$

(*b*) {2, 4, 5, 6}
(*c*) {2}
(*d*) {1, 3, 5}

(*a*)

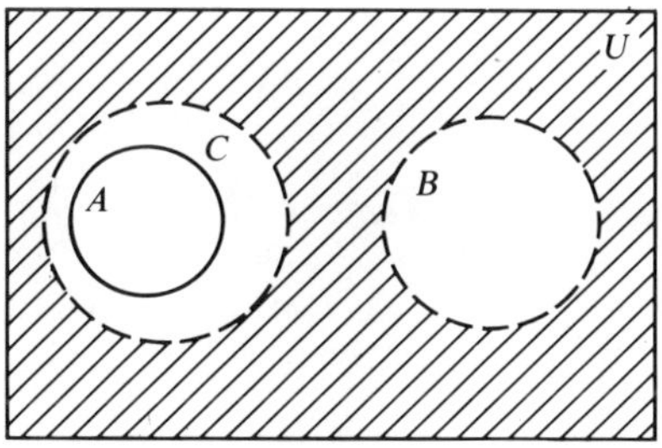

Figure 2.7

(*b*)

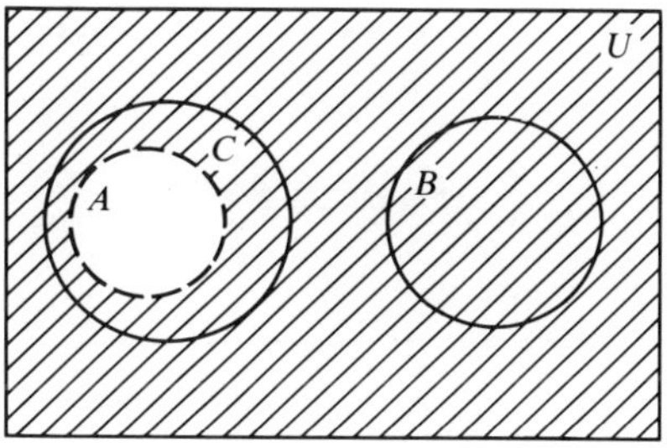

Figure 2.8

(*c*)

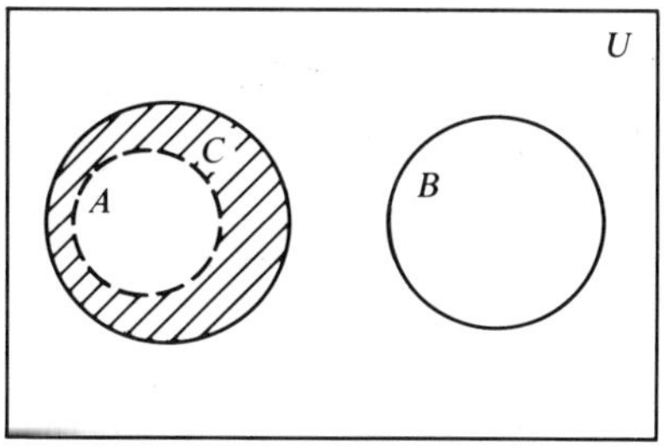

Figure 2.9

(*d*)

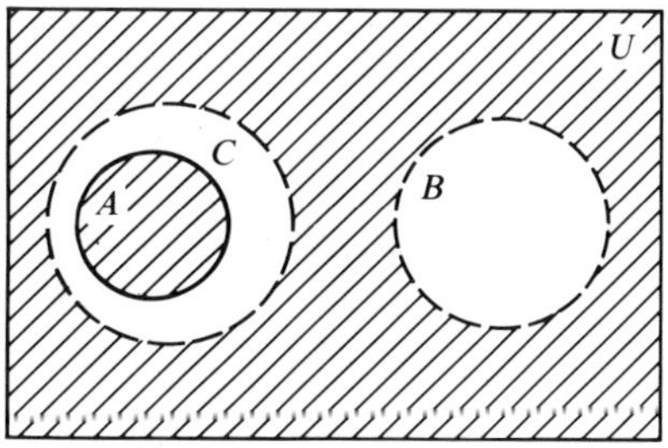

Figure 2.10

Note that in Fig. 2.8 the hatched region represents the complement of set A; that is,

$$\bar{A} = \{2, 4, 5, 6\}$$

Furthermore, in Fig. 2.9 the hatched region represents the relative complement of A in C; that is,

$$C - A = \{1, 2, 3\} - \{1, 3\} = \{2\}$$

Generally, if A and B are proper nonempty subsets of a universal set U, then the relative complement of A in B is represented by one of the four hatched regions in Figs. 2.11 through 2.14. The relative complement of A in B is represented by the hatched region in Fig. 2.11 if every element of A is also in B, by the hatched region in Fig. 2.12 if some but not all elements of A are in B, or by the hatched region in Figs. 2.13 or 2.14 if no element of A is in B.

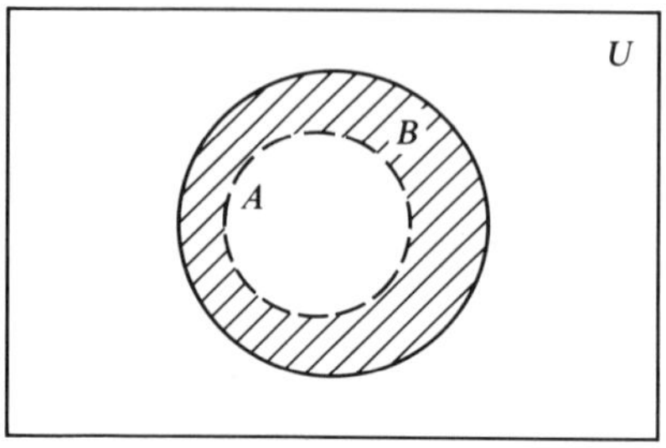

$A \subset B \qquad B - A$

Figure 2.11

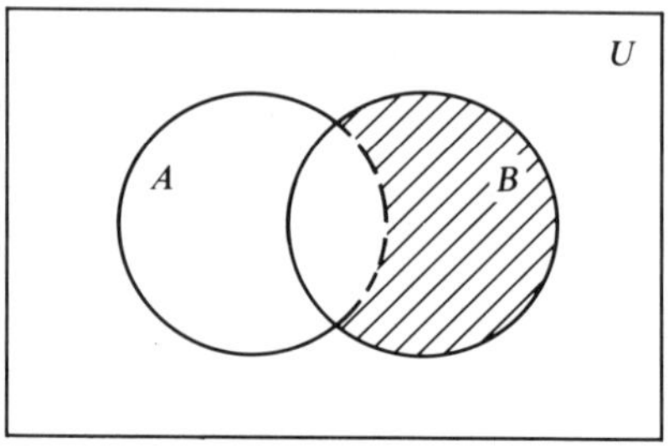

$B - A$

Figure 2.12

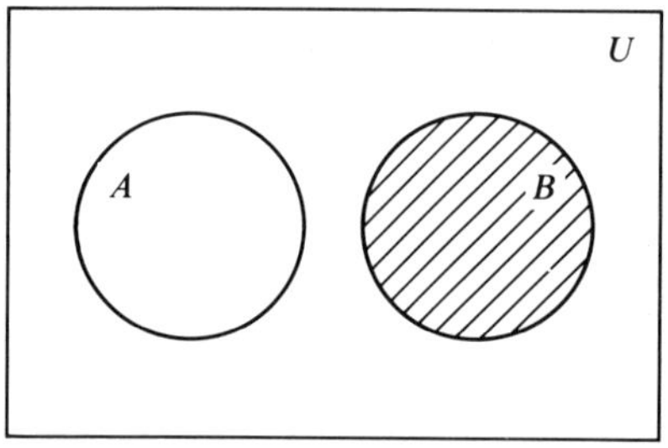

$B - A$

Figure 2.13

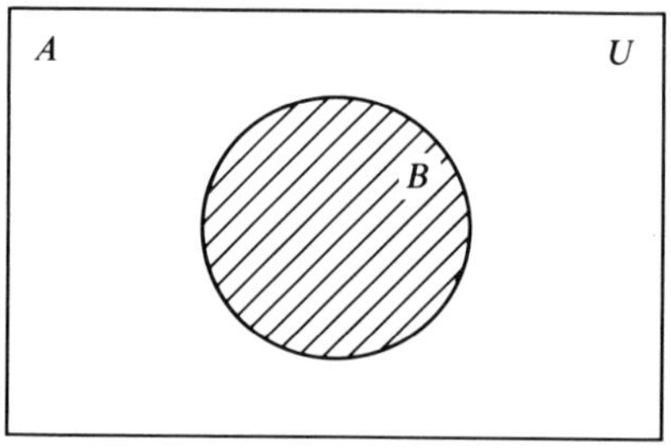

$A = \overline{B} \qquad B - A$

Figure 2.14

Example 3

Use Venn diagrams to show that if $A \subset B$, then $\overline{B} \subset \overline{A}$.

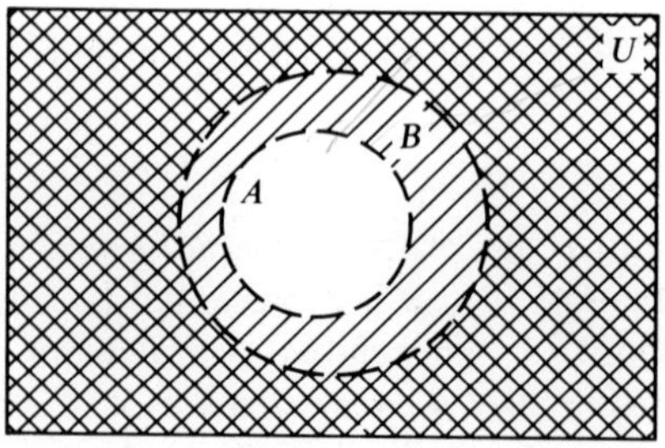

Figure 2.15

In Fig. 2.15 the universal set U is represented by the set of points in the rectangular region; sets A and B are represented by circular regions such that $A \subset B$. The complement of set A is represented by the region  ; the complement of set B is represented by the region  . Every point in the region 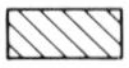 representing $\bar{B}$ is also a point in the region  representing $\bar{A}$. Thus, $\bar{B} \subseteq \bar{A}$. However, there exists at least one point in the region  representing set $\bar{A}$ that is not in the region  representing set $\bar{B}$. Hence, $\bar{B} \subseteq \bar{A}$; that is, $\bar{B}$ is a proper subset of $\bar{A}$.

Exercises

1. Let $U = \{a, b, c, d, e\}$, $A = \{a, b, d, e\}$, $B = \{a, e\}$, and $C = \{d, e\}$. Use Venn diagrams to represent the relations that exist among the four sets.

In Exercises 2 through 9 consider the Venn diagram in the solution of Exercise 1. Shade the region represented by each set.

2. $\{b, c, d\}$
3. $\{a, d, e\}$
4. $\{e\}$
5. $\{b, d\}$
6. $\{a, b, c\}$
7. $\{d\}$
8. $\{a, b\}$
9. $\{c\}$
10. Use Venn diagrams to show that if $A \subset B$ and $B \subset C$, then $A \subset C$.

In Exercises 11 through 18 consider the following Venn diagram:

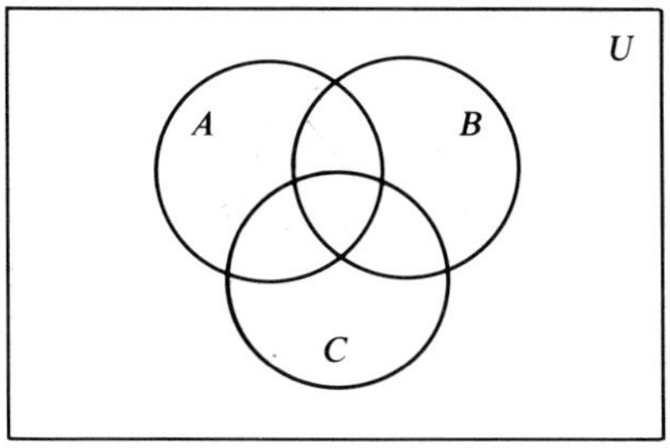

Shade each region.

11. $(A - B) - C$
12. $A - \bar{B}$
13. $A - (B - C)$
14. $B - (A - \bar{B})$

15. $\overline{A - B}$ 16. $A - (\overline{B - C})$

17. $U - (A - C)$ 18. $(A - B) - (A - C)$

19. On a certain day, three sports writers picked the following teams to win:

{New York, Chicago, Boston, Cleveland}
{New York, Boston, Detroit, Oakland}
{Cleveland, New York, Baltimore, Oakland}

Minnesota was not picked to win by any of the writers. Use Venn diagrams to determine the teams that played each other on that day.

2.7

Intersection of Sets

In Sec. 2.6 we noted that it is possible for two sets to have elements in common. For example, the sets

$$A = \{a, b, c, e\} \qquad \text{and} \qquad B = \{a, c, d\}$$

have the elements a and c in common. The set $\{a, c\}$ is called the *intersection* of A and B.

The* intersection *of two sets A and B is the set of all elements that belong to both A and B. The intersection of sets A and B is denoted by $A \cap B$.

In this definition it is understood that A and B are subsets of the same universal set U. Symbolically,

$$A \cap B = \{x \mid x \in A \text{ and } x \in B\}$$

The notation $A \cap B$ is read "A intersection B."

Example 1

Let A be the set of multiples of 2 in the set N of natural numbers, and let B be the set of multiples of 3 in the set N of natural numbers. Determine $A \cap B$.

$A = \{2, 4, \mathbf{6}, 8, 10, \mathbf{12}, 14, 16, \mathbf{18}, 20, 22, \mathbf{24}, \ldots\}$
$B = \{3, \mathbf{6}, 9, \mathbf{12}, 15, \mathbf{18}, 21, \mathbf{24}, 27, \ldots\}$

(Some of the elements that belong to both A and B are shown in dark type.) Since

$$A \cap B = \{x \mid x \in A \text{ and } x \in B\}$$

then

$$A \cap B = \{6, 12, 18, 24, \ldots\}$$

The intersection of sets A and B consists of those elements of the universal set N that are multiples of 6.

Example 2

Let A be a subset of some universal set U. Simplify expressions (a) and (b):

(a) $A \cap A$
(b) $A \cap \bar{A}$

(a) $A \cap A = A$ (b) $A \cap \bar{A} = \varnothing$

The statement of equality expressed in Example 2(a) is called the *idempotent property of the intersection of sets.*

Example 3

Let $A = \{1, 2, 4\}$, $B = \{1, 4, 5\}$, and $C = \{1, 4, 7\}$. Show that although $A \cap B = A \cap C$, it does not follow that $B = C$.

Since $A \cap B = \{1, 4\}$ and $A \cap C = \{1, 4\}$, then $A \cap B = A \cap C$. However, $B \neq C$ since $\{1, 4, 5\} \neq \{1, 4, 7\}$.

Example 4

Consider $A = \{a, b, c, d, e\}$, $B = \{a, b, d, g, h, j\}$, and $C = \{a, c, d, j\}$. Show that

$$(A \cap B) \cap C = A \cap (B \cap C)$$

$$A \cap B = \{a, b, c, d, e\} \cap \{a, b, d, g, h, j\} = \{a, b, d\}$$
$$(A \cap B) \cap C = \{a, b, d\} \cap \{a, c, d, j\} = \{a, d\}$$
$$B \cap C = \{a, b, d, g, h, j\} \cap (a, c, d, j\} = \{a, d, j\}$$
$$A \cap (B \cap C) = \{a, b, c, d, e\} \cap \{a, d, j\} = \{a, d\}$$

Since

$$\{a, d\} = \{a, d\}$$

then

$$(A \cap B) \cap C = A \cap (B \cap C)$$

The statement of equality expressed in Example 4 is called the *associative property of the intersection of sets.*

The Venn diagrams in Figs. 2.16 through 2.21 illustrate six possible situations for the intersection of A and B, where A and B are nonempty proper subsets of U. In each figure $A \cap B$ is represented by the hatched region. Note that in all six cases $A \cap B \subseteq A$ and $A \cap B \subseteq B$.

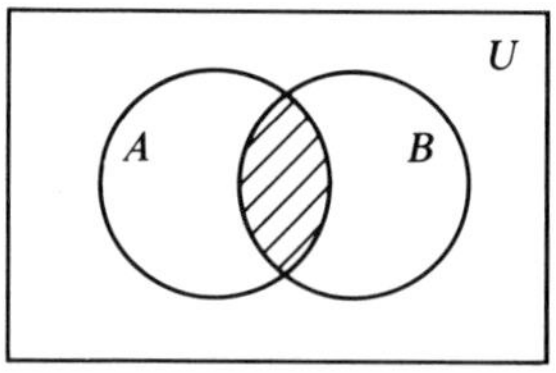

$A \cap B$

Figure 2.16

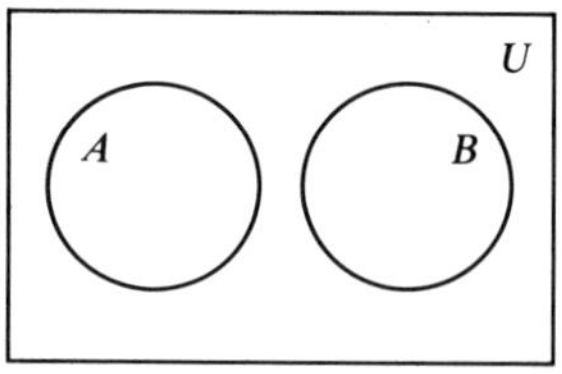

$A \cap B = \{\ \}$

Figure 2.17

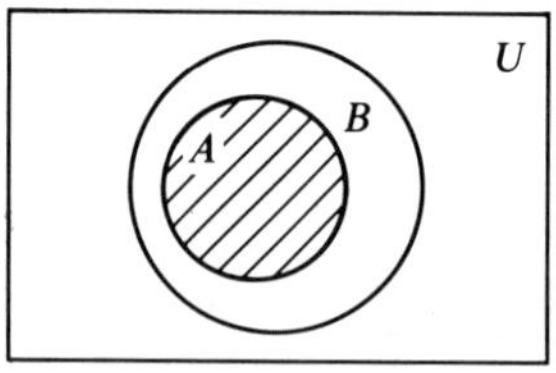

$A \subset B \quad A \cap B = A$

Figure 2.18

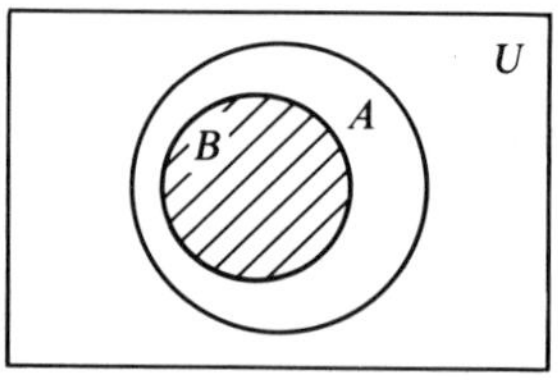

$B \subset A \quad A \cap B = B$

Figure 2.19

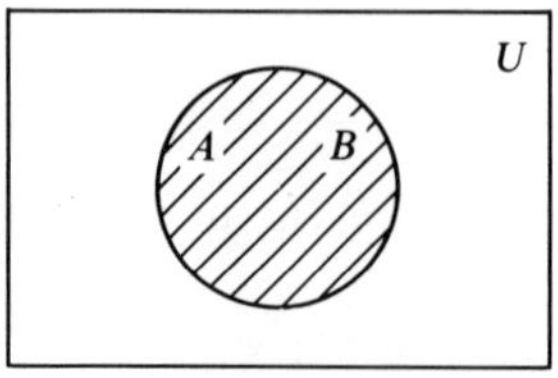

$A \cap B = A = B$

Figure 2.20

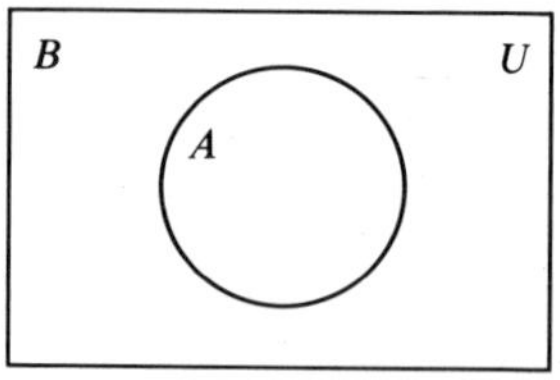

$B = \bar{A} \quad A \cap B = \{\ \}$

Figure 2.21

If the intersection of two nonempty sets A and B is the null set, then A and B are called *disjoint sets*. Venn diagrams for two disjoint sets are shown in Figs. 2.17 and 2.21. Note that a set A and its complement $\bar{A}$ are disjoint sets.

Example 5

Use Venn diagrams to show that

$$(A \cap B) \cap C = A \cap (B \cap C)$$

The universal set U is represented by the set of points in the rectangular regions. The three circular regions represent the sets A, B, and C, respectively. We consider the most general situation possible; that is, we consider it possible for some elements to be simultaneously in any two sets or in all three sets.

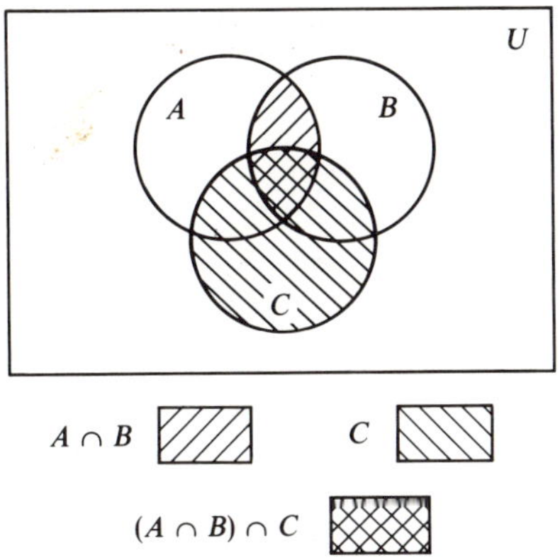

Figure 2.22

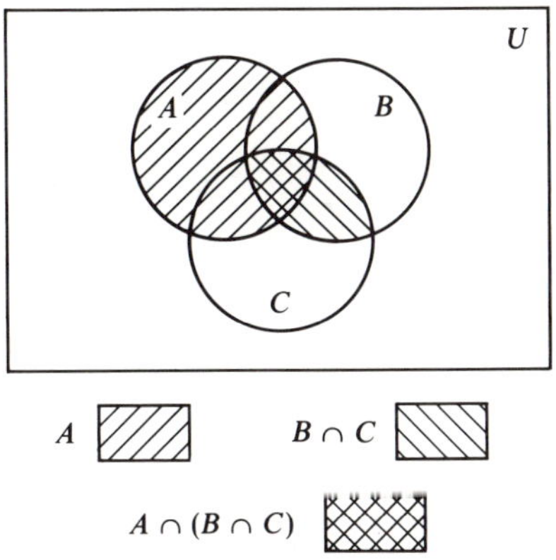

Figure 2.23

Since the cross-hatched regions of $(A \cap B) \cap C$ and $A \cap (B \cap C)$ in Figs. 2.22 and 2.23 are identical, then

$$(A \cap B) \cap C = A \cap (B \cap C)$$

Exercises

In Exercises 1 through 10 let $U = \{1, 2, 3, 4, 5, 6, 7, 8, 9, 10\}$, $A = \{1, 2, 3, 4, 5, 6, 7\}$, $B = \{1, 4, 7, 10\}$, and $C = \{2, 4, 5, 7, 8, 10\}$. Find each set.

1. $A \cap B$
2. $A \cap C$
3. $B \cap C$
4. $(A \cap B) \cap C$

5. $A \cap (B \cap C)$
6. $A \cap \overline{B}$
7. $A \cap (C - B)$
8. $\overline{A \cap B}$
9. $\overline{A} \cap \overline{B}$
10. $(A - B) \cap B$
11. Consider $A = \{1, 2, 3, 4, 5\}$, $B = \{1, 3, 5, 7, 9\}$, and $C = \{2, 3, 5, 7\}$. Show that

$$(A \cap B) \cap C = A \cap (B \cap C)$$

12. Let A be the set of multiples of 3 in the set N of natural numbers, and let B be the set of multiples of 5 in the set N of natural numbers. Determine $A \cap B$. Describe the set.
13. Let A be the set of factors of 24 in the set N of natural numbers, and let B be the set of factors of 30 in the set N of natural numbers. Determine $A \cap B$, and describe the set.
14. Let A be a subset of some universal set U. Simplify the following expressions: (*a*) $A \cap \varnothing$; (*b*) $A \cap U$.
15. Find (*a*) $U \cap U$; (*b*) $U \cap \varnothing$.

In Exercises 16 through 19 shade the required region in each Venn diagram.

16.

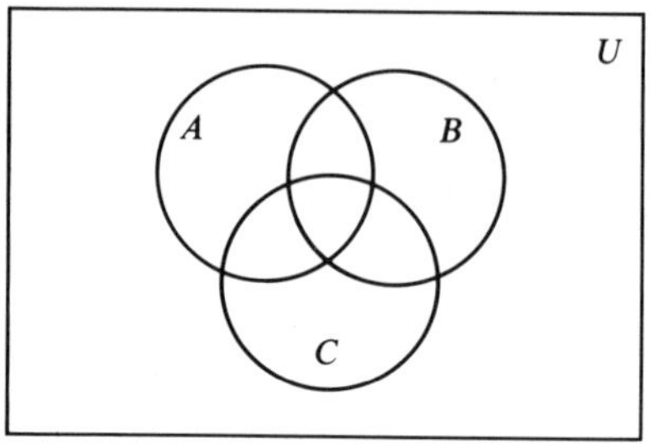

$A \cap (\overline{B} \cap C)$

17.

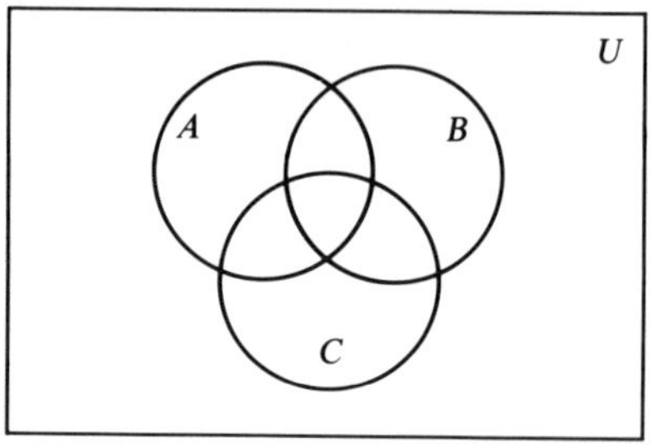

$(\overline{A} \cap \overline{B}) \cap \overline{C}$

18.

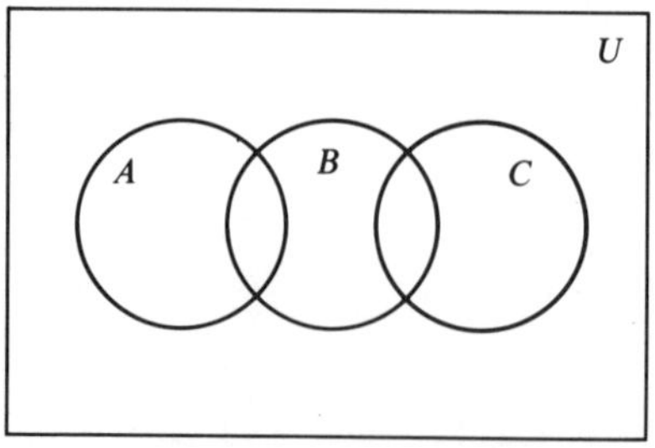

$\overline{A} \cap (B \cap \overline{C})$

19.

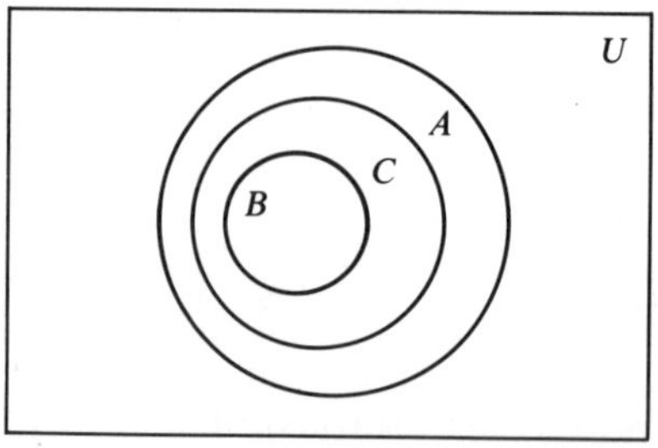

$A - (\overline{B} \cap C)$

In Exercises 20 through 23 label each shaded region in terms of A, B, and C and their complements.

20.
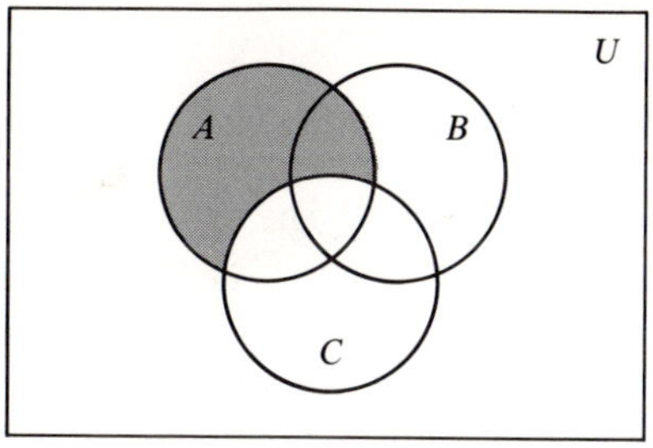

21.
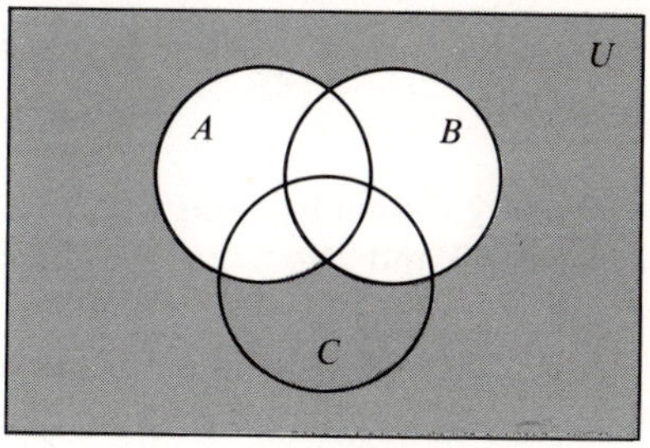

22.
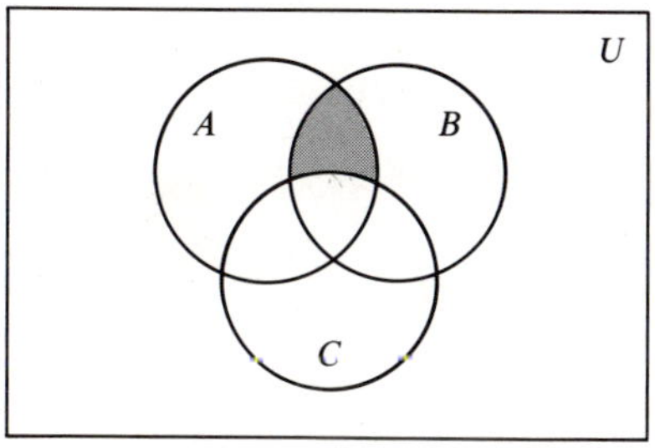

23.
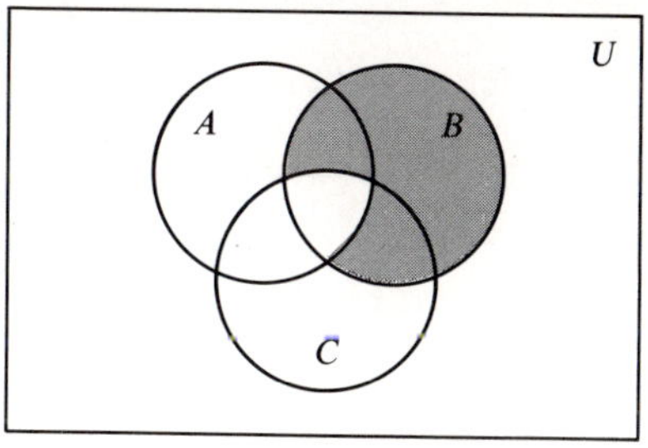

24. Show by definition that $A \cap \bar{B} = A - B$.
25. Use Venn diagrams to show that $A \cap \bar{B} = A - B$.
26. Show by definition that $A \cap B = B \cap A$ (*commutative property of the intersection of sets*).
27. Let U be the set of members of the United States Congress, S the set of senators, and R the set of members of the House of Representatives. Describe (*a*) $S \cap R$; (*b*) $\bar{S}$; (*c*) $\bar{R}$.
28. Let S be the set of all people. Determine the complements of sets (*a*) to (*d*):
    (*a*) The set of citizens of Canada
    (*b*) The set of people who are not women
    (*c*) The set of people who are teachers and live in Canada
    (*d*) The set of people who are male and not Canadian citizens
29. Let A be the set of human beings, B the set of United States citizens, C the set of people who live in Chicago, and D the set of all men. Describe each of the following sets: (*a*) $B \cap C$; (*b*) $A \cap D$; (*c*) $C \cap D$; (*d*) $A \cap B$; (*e*) $\bar{B} \cap C$; (*f*) $B \cap \bar{D}$; (*g*) $(B \cap \bar{C}) \cap D$; (*h*) $B - (\bar{C} \cap D)$.

2.8

Union of Sets

Consider the set of odd natural numbers less than 10:

$$D = \{1, 3, 5, 7, 9\}$$

and the set of primes less than 10:

$$P = \{2, 3, 5, 7\}$$

Both these sets are subsets of the set

$$S = \{1, 2, 3, 5, 7, 9\}$$

and every element of this set S is either in set D or set P. The set S is called the *union* of D and P.

The* union *of two sets A and B is the set of all elements that belong to either A or B. The union of sets A and B is denoted by A ∪ B.

In this definition it is understood that A and B are subsets of the same universal set U. Symbolically,

$$A \cup B = \{x | x \in A \text{ or } x \in B\}$$

The notation $A \cup B$ is read "A union B."

Example 1

Let N be the set of natural numbers, W the set of whole numbers, E the set of even whole numbers, and D the set of odd whole numbers. Determine the truth values of statements (*a*) to (*d*):

(*a*) $D \cup E = W$

(*b*) $N \cup D = N$

(*c*) $N \cap W = N$

(*d*) $E \cup N = N$

(*a*) True; the union of the set of odd whole numbers and the set of even whole numbers is the set of whole numbers. (*b*) True. (*c*) True. (*d*) False; 0 is an even whole number that is not a natural number.

Example 2

Let A be a subset of some universal set U. Simplify expressions (*a*) and (*b*):

(*a*) $A \cup A$

(*b*) $A \cup \bar{A}$

$$(a)\ A \cup A = A \quad (b)\ A \cup \bar{A} = U$$

The statement of equality expressed in Example 2(*a*) is called the *idempotent property of the union of sets.*

Example 3

Let $A = \{2, 4, 5, 8\}$, $B = \{1, 2, 5\}$, and $C = \{2, 4, 6\}$. Show that

$$(A \cup B) \cup C = A \cup (B \cup C)$$

$$A \cup B = \{2, 4, 5, 8\} \cup \{1, 2, 5\} = \{1, 2, 4, 5, 8\}$$
$$(A \cup B) \cup C = \{1, 2, 4, 5, 8\} \cup \{2, 4, 6\} = \{1, 2, 4, 5, 6, 8\}$$
$$B \cup C = \{1, 2, 5\} \cup \{2, 4, 6\} = \{1, 2, 4, 5, 6\}$$
$$A \cup (B \cup C) = \{2, 4, 5, 8\} \cup \{1, 2, 4, 5, 6\} = \{1, 2, 4, 5, 6, 8\}$$

Since

$$\{1, 2, 4, 5, 6, 8\} = \{1, 2, 4, 5, 6, 8\}$$

then

$$(A \cup B) \cup C = A \cup (B \cup C)$$

The statement of equality expressed in Example 3 is called the *associative property of the union of sets.*

Example 4

Use the sets in Example 1 and show that

$$A \cap (B \cup C) = (A \cap B) \cup (A \cap C)$$

$$B \cup C = \{1, 2, 5\} \cup \{2, 4, 6\} = \{1, 2, 4, 5, 6\}$$
$$A \cap (B \cup C) = \{2, 4, 5, 8\} \cap \{1, 2, 4, 5, 6\} = \{2, 4, 5\}$$
$$A \cap B = \{2, 4, 5, 8\} \cap \{1, 2, 5\} = \{2, 5\}$$
$$A \cap C = \{2, 4, 5, 8\} \cap \{2, 4, 6\} = \{2, 4\}$$
$$(A \cap B) \cup (A \cap C) = \{2, 5\} \cup \{2, 4\} = \{2, 4, 5\}$$

Since

$$\{2, 4, 5\} = \{2, 4, 5\}$$

then

$$A \cap (B \cup C) = (A \cap B) \cup (A \cap C)$$

The statement of equality expressed in Example 4 is called the *distributive property of intersection over union.*

The Venn diagrams in Figs. 2.24 through 2.29 illustrate six possible situations for the union of A and B, where A and B are nonempty proper subsets of U. In each figure $A \cup B$ is represented by the hatched region. Note especially that in all six cases $A \subseteq A \cup B$ and $B \subseteq A \cup B$.

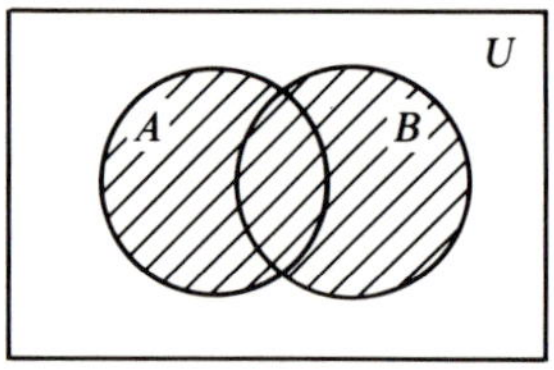

$A \cup B$

Figure 2.24

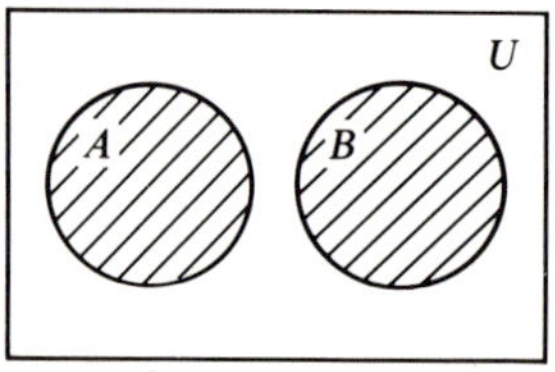

$A \cup B$

Figure 2.25

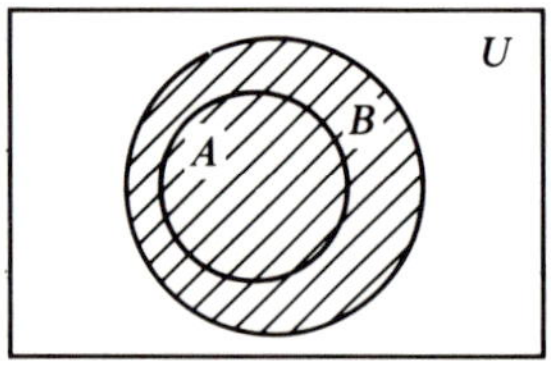

$A \subset B \quad A \cup B = B$

Figure 2.26

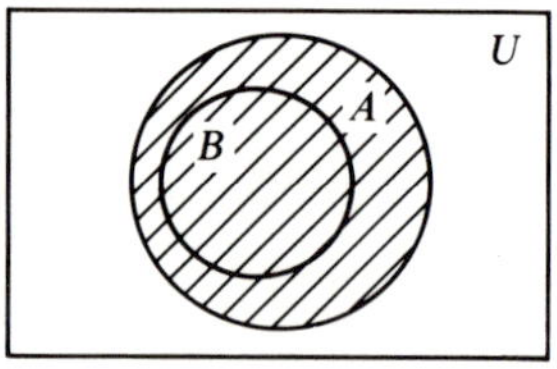

$B \subset A \quad A \cup B = A$

Figure 2.27

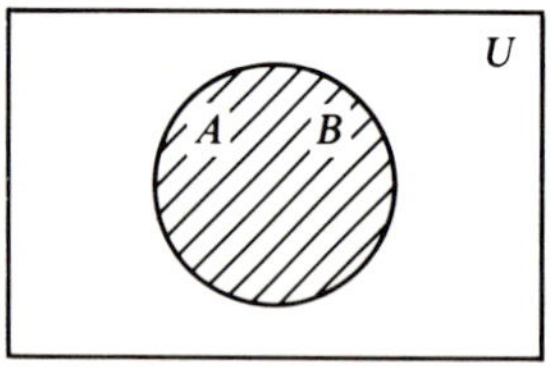

$A \cup B = A = B$

Figure 2.28

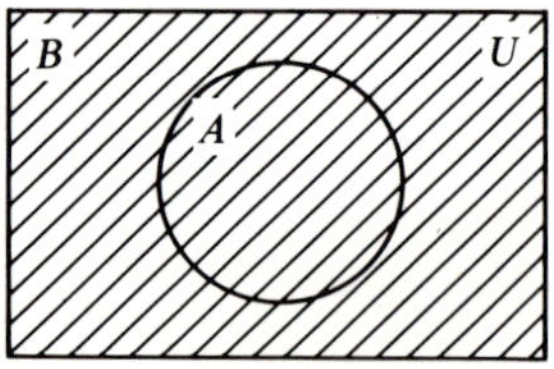

$B = \bar{A} \quad A \cup B = U$

Figure 2.29

Example 5

Use Venn diagrams to show that

$$\overline{(A \cup B)} = \bar{A} \cap \bar{B}$$

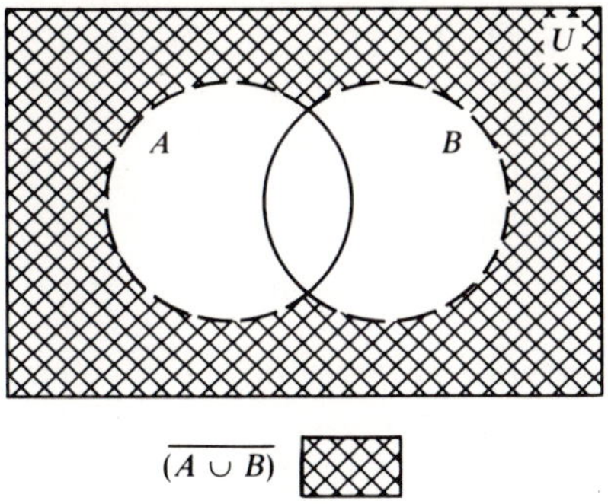

Figure 2.30

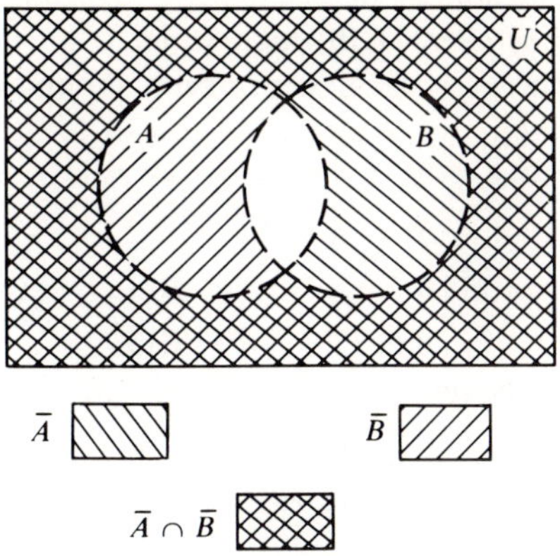

Figure 2.31

The cross-hatched regions in Figs. 2.30 and 2.31 are identical; hence

$$\overline{(A \cup B)} = \bar{A} \cap \bar{B}$$

In Example 5 we showed that the complement of the union of two sets is equal to the intersection of their complements. It can also be shown by the use of Venn diagrams that

$$\overline{(A \cap B)} = \bar{A} \cup \bar{B}$$

That is, the complement of the intersection of two sets is equal to the union of their complements. These two equalities of sets are called *DeMorgan's laws for sets.*

Exercises

In Exercises 1 through 10 let $U = \{a, b, c, d, e, f, g, h\}$, $A = \{a, b, c, d\}$, $B = \{a, b, d, f, g\}$, and $C = \{b, c, d, e, h\}$. Find each set.

1. $A \cup B$
2. $A \cup C$
3. $B \cup \bar{C}$
4. $A \cap (B \cup C)$
5. $\overline{A \cup B}$
6. $\bar{A} \cup B$
7. $B \cup (\bar{A} \cap C)$
8. $C - (A \cup B)$
9. $A \cup (U - C)$
10. $\overline{(A \cap B)} \cup C$

In Exercises 11 through 15 let $U = \{1, 2, 3, 4, 5, 6\}$, $A = \{1, 2, 5\}$, $B = \{1, 3, 5, 6\}$, and $C = \{1, 4, 6\}$. Verify each statement.

11. $\overline{(A \cup B)} = \bar{A} \cap \bar{B}$
12. $\overline{(A \cap B)} = \bar{A} \cup \bar{B}$
13. $A \cap (B \cup C) = (A \cap B) \cup (A \cap C)$
14. $A \cup (B \cap C) = (A \cup B) \cap (A \cup C)$ (*distributive property of union over intersection*)
15. $(A \cup B) \cup C = A \cup (B \cup C)$
16. Let A be a subset of some universal set U. Simplify the following expressions: (*a*) $A \cup \varnothing$; (*b*) $A \cup U$.
17. Let A and B be subsets of some universal set U. Simplify the following expressions:

    (*a*) $A \cup (A \cap \bar{A})$
    (*b*) $A \cap (A \cup \bar{A})$
    (*c*) $A \cap (A \cup B)$
    (*d*) $A \cup (A \cap B)$
18. Find (*a*) $U \cup U$; (*b*) $U \cup \varnothing$.

In Exercises 19 through 24 shade the required region in each Venn diagram.

19.

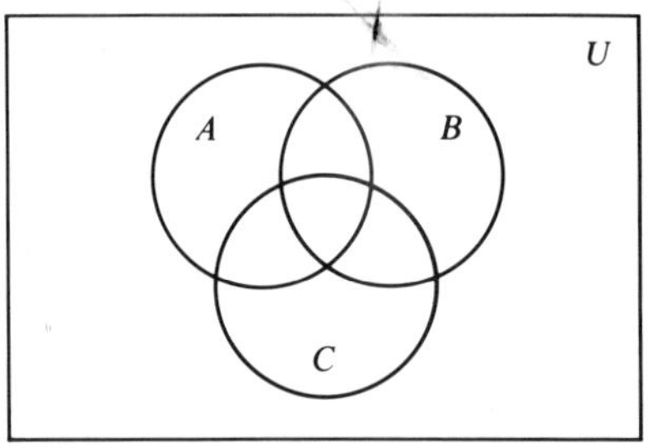

$(A \cup B) \cup \bar{C}$

20.

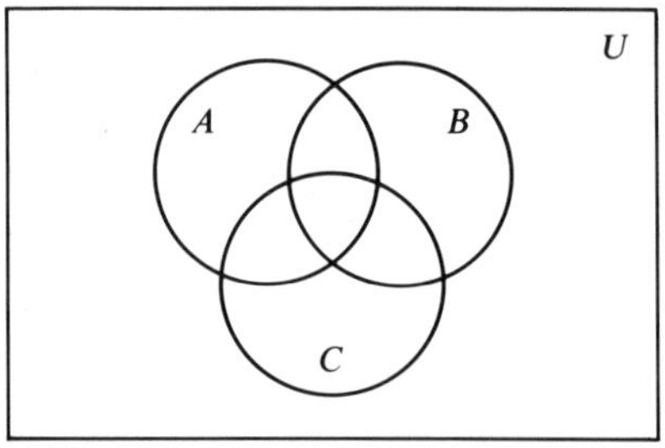

$A \cup (B \cap \bar{C})$

21.

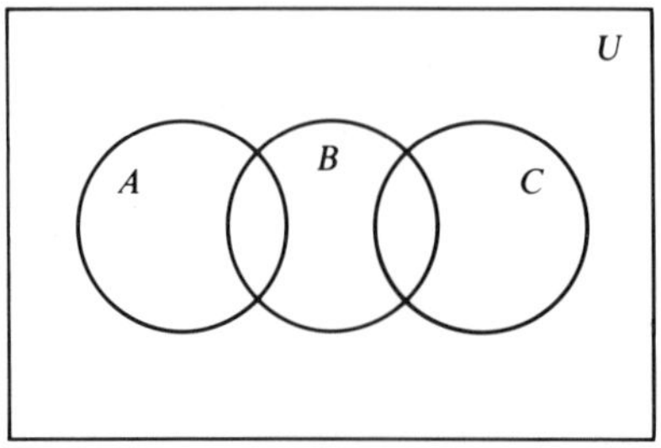

$A \cap (\bar{B} \cup C)$

22.

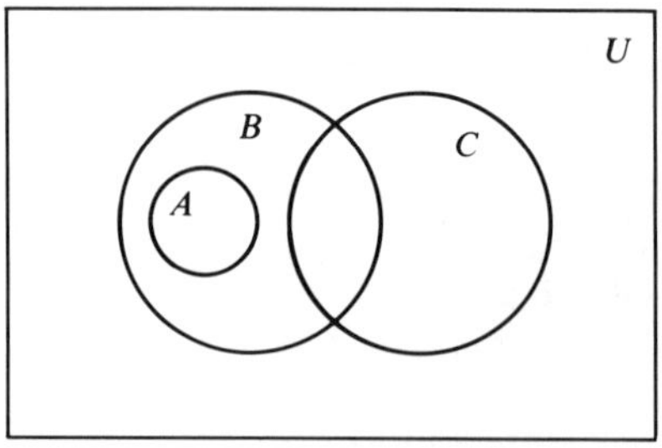

$\bar{A} \cup \overline{(B \cap C)}$

23.

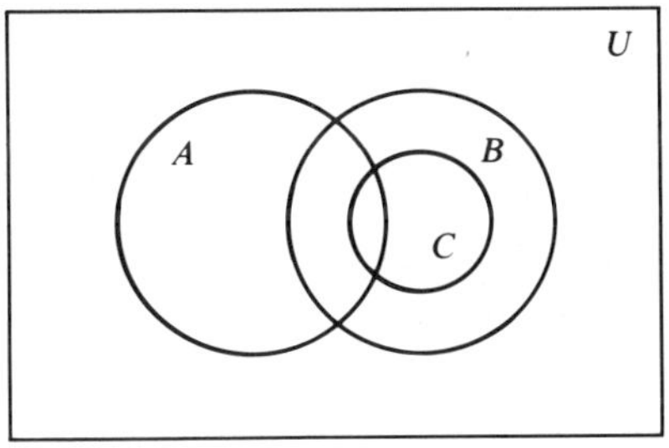

$A \cap (\bar{B} \cup C)$

24.

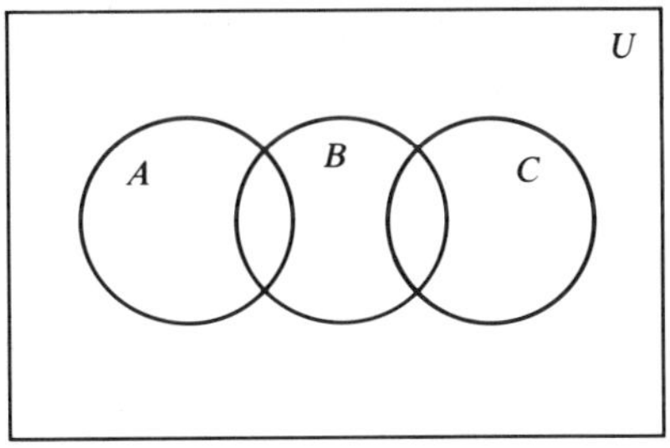

$(A \cup B) \cap (A \cup C)$

In Exercises 25 through 28 label each shaded region in terms of A, B, and C and their complements:

25.

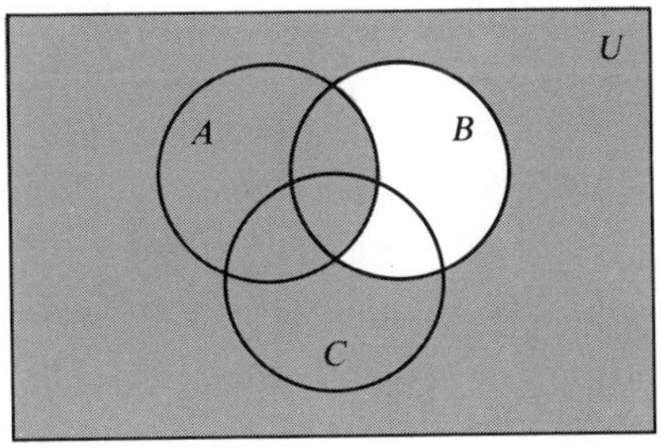

26.

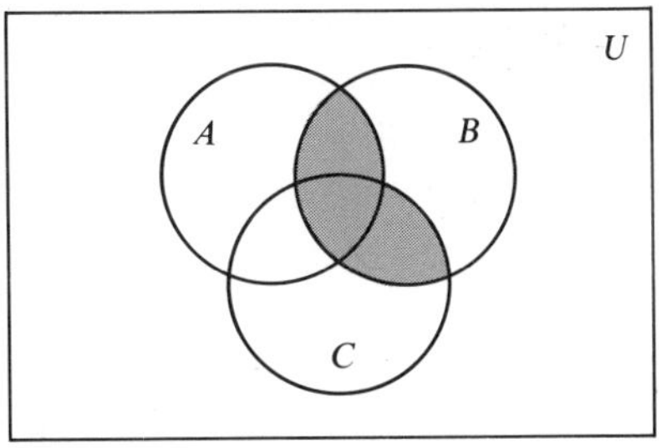

27.

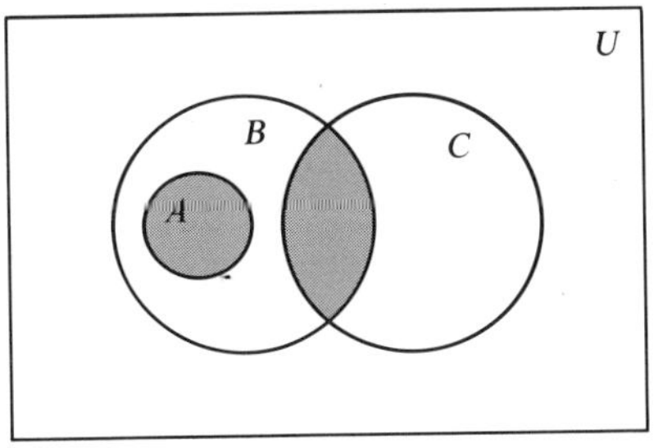

28.

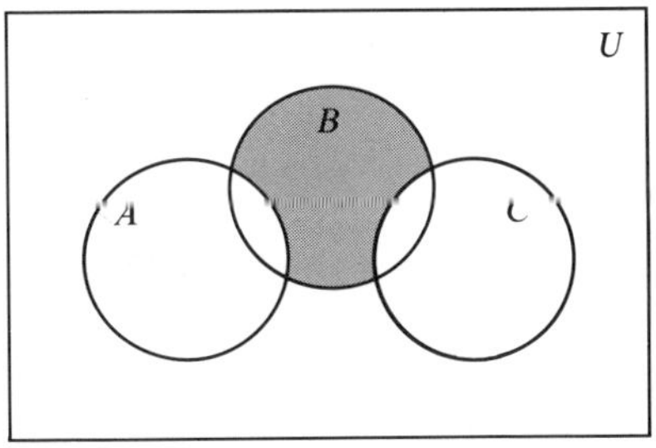

29. Use Venn diagrams to show that

$$\overline{(A \cap B)} = \bar{A} \cup \bar{B}$$

30. Use Venn diagrams to show that

$$(A \cup B) \cup C = A \cup (B \cup C)$$

31. Use Venn diagrams to show that

$$A \cap (B \cup C) = (A \cap B) \cup (A \cap C)$$

32. Use Venn diagrams to show that

$$A \cup (B \cap C) = (A \cup B) \cap (A \cup C)$$

33. Use Venn diagrams to show that

$$(A \cap B) \cup C \neq A \cap (B \cup C)$$

34. Let U be the set of all college students. Consider the following subsets:

$P =$ The set of all football players
$H =$ The set of all honor students
$M =$ The set of all males
$F =$ The set of all females
$S =$ The set of all seniors

Describe sets (a) to (f) in symbolic notation:
(a) The set of all senior football players
(b) The set of all female honor students
(c) The set of all football players who are not honor students
(d) The set of all honor students who are either football players or females
(e) The set of all males who are either football players or seniors
(f) The set of all nonseniors who are male football players or female honor students

35. Show by definition that $A \cup B = B \cup A$ (*commutative property of the union of sets*).

36. Let U be the set of shoppers in the local market. Consider the following subsets:

$A =$ Those who purchased item x
$B =$ Those who purchased item y
$C =$ Those who purchased item z

Draw a Venn diagram representing the shoppers. Shade in the region representing those shoppers who purchased:
(a) Items x and y, but not item z
(b) Item x, but not items y or z
(c) Items x, y, and z
(d) Item x, or items y and z

2.9

Number of Elements in a Set

Associated with every finite set A is a number, denoted by $n(A)$, which represents the number of elements in set A. The number $n(A)$ is called the *cardinal number* of set A.

Consider the following problem. In a certain high school, there are 35 students enrolled in mathematics, 26 in biology, and 18 in French. Included among these students are 15 students studying both mathematics and biology, 9 studying both mathematics and French, 12 studying both biology and French, and only 4 students studying all three subjects. How many different students are there in the three classes? The question can be answered by constructing the Venn diagram shown in Fig. 2.32.

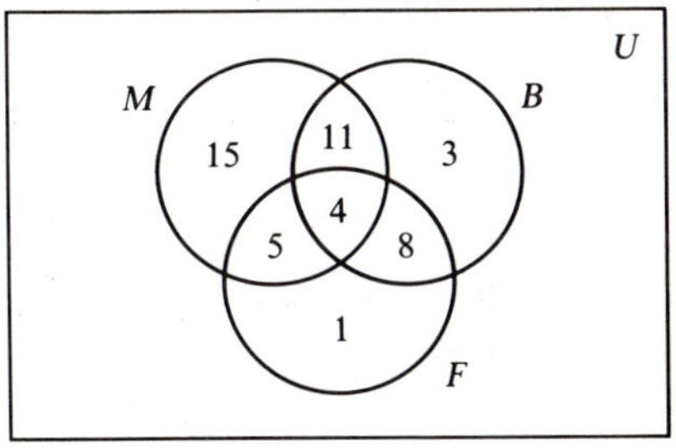

Figure 2.32

Let the set of points in circular region M represent the set of students studying mathematics, in circular region B the set of students studying biology, and in circular region F the set of students studying French. Now,

$$n(M \cap B \cap F) = \mathbf{4}$$

and we indicate this in the appropriate region of the diagram. The region representing those students studying both biology and French has associated with it the number 12; that is, $n(B \cap F) = 12$. Hence, there are eight students studying both biology and French but not mathematics; that is,

$$\begin{aligned} n(B \cap F \cap \overline{M}) &= n(B \cap F) - n(B \cap F \cap M) \\ &= 12 - 4 = \mathbf{8} \end{aligned}$$

The number of students studying both mathematics and French but not biology is given by

$$\begin{aligned} n(M \cap F \cap \overline{B}) &= n(M \cap F) - n(M \cap F \cap B) \\ &= 9 - 4 = \mathbf{5} \end{aligned}$$

Similarly, the number of students studying both mathematics and biology but not French is given by

$$\begin{aligned} n(M \cap B \cap \overline{F}) &= n(M \cap B) - n(M \cap B \cap F) \\ &= 15 - 4 = \mathbf{11} \end{aligned}$$

The number of students studying French only is given by

$$\begin{aligned} n(F \cap \overline{M} \cap \overline{B}) &= n(F) - [n(F \cap M \cap \overline{B}) + n(F \cap M \cap B) \\ &\quad + n(F \cap \overline{M} \cap B)] \\ &= 18 - [5 + 4 + 8] = \mathbf{1} \end{aligned}$$

The number of students studying biology only is given by

$$\begin{aligned} n(B \cap \overline{M} \cap \overline{F}) &= n(B) - [n(B \cap M \cap \overline{F}) + n(B \cap M \cap F) \\ &\quad + n(B \cap \overline{M} \cap F)] \\ &= 26 - [11 + 4 + 8] = \mathbf{3} \end{aligned}$$

Finally, the number of students studying mathematics only is given by

$$\begin{aligned} n(M \cap \overline{B} \cap \overline{F}) &= n(M) - [n(M \cap B \cap \overline{F}) + n(M \cap B \cap F) \\ &\quad + n(M \cap \overline{B} \cap F)] \\ &= 35 - [11 + 4 + 5] = \mathbf{15} \end{aligned}$$

The total number of different students in the three classes may now be obtained by summing the numbers appearing in the various regions of the Venn diagram in Fig. 2.32. Thus, there are **4** + **8** + **5** + **11** + **1** + **3** + **15**, that is, 47, different students in the three classes.

Exercises

1. The Varsity Lettermen's Club of State University consists of: 34 men in football, 9 in basketball, 22 in baseball, 3 in football and basketball, 4 in basketball and baseball, and 12 in football and baseball. No man lettered in all three sports. (*a*) How many members are there in the club? (*b*) How many members lettered in one sport only?
2. Two hundred people participated in a survey of the magazines they read. The results were:

   141 read magazine *A*
   56 read magazine *B*
   74 read magazine *C*
   17 read magazines *A* and *B*
   40 read magazines *A* and *C*
   36 read magazines *B* and *C*
   15 read magazines *A*, *B*, and *C*

   Find how many people read (*a*) at least one of the magazines; (*b*) none of the magazines; (*c*) exactly one of the magazines; (*d*) exactly two of the magazines.
3. A college has an enrollment of 4,325 students. On one weekend 2,755 students attended a football game and 750 students attended a dance. Determine the number of students that attended either the game or the dance if 475 did both.
4. Every student of a small language-oriented liberal arts college must take at least one language course each term. During the last term 135 students studied French, 160 studied German, 110 studied Spanish, 105 studied Russian, 83

studied French and German, 46 studied French and Spanish, 29 studied French and Russian, 50 studied German and Spanish, 56 studied German and Russian, 36 studied Spanish and Russian, 24 studied French, German, and Spanish, 19 studied French, German, and Russian, 9 studied French, Spanish, and Russian, 10 studied German, Spanish, and Russian, and 2 studied all four languages. (*a*) How many students are enrolled at the college? (*b*) How many students take exactly one language course? (*c*) How many students take French and German only? (*d*) How many students take Russian and one other language?

5. Use a Venn diagram to show that

$$n(A \cup B) = n(A) + n(B) - n(A \cap B)$$

6. Use a Venn diagram to show that

$$n(A \cup B \cup C) = n(A) + n(B) + n(C) - n(A \cap B) - n(A \cap C) - n(B \cap C) + n(A \cap B \cap C)$$

7. Determine the conditions under which

$$n(A \cup B) = n(A) + n(B)$$

2.10

Cartesian Sets

The order is of no importance when listing the elements of a set; thus we may write $\{a, b, c\} = \{c, a, b\}$. However, we often wish to consider a set of elements that consist of pairs of numbers in which the order of appearance of the numbers is important. For example, consider using the elements of the two sets

$$A = \{a, b\} \qquad \text{and} \qquad B = \{c, d, e\}$$

to form a third set whose elements are the pairs

$$(a, c), (a, d), (a, e), (b, c), (b, d), \text{and } (b, e)$$

We denote this third set as $A \times B$; that is,

$$A \times B = \{(a, c), (a, d), (a, e), (b, c), (b, d), (b, e)\}$$

The number of elements belonging to set $A \times B$ equals the product of the number of elements of A and the number of elements of B. Notice that an element of the set $A \times B$ consists of a pair of numbers whose first member belongs to set A and whose second member belongs to B. Such pairs of numbers are called *ordered pairs*.

The set of all ordered pairs of the form (x, y), where $x \in A$ and $y \in B$, is called the* Cartesian set *$A \times B$ of sets A and B.

The Cartesian set is named after the famous French mathematician René Déscartes and is of fundamental importance in the development of analytic geometry. The Cartesian set $A \times B$ is also often called the *product set* of A and B.

Example 1

Determine the Cartesian set $A \times B$, where

$$A = \{0, 1, 2\} \qquad \text{and} \qquad B = \{3, 4\}$$

By definition of the Cartesian set, $A \times B$ is the set of all ordered pairs (x, y), where $x \in \{0, 1, 2\}$ and $y \in \{3, 4\}$. Hence,

$$A \times B = \{(0, 3), (0, 4), (1, 3), (1, 4), (2, 3), (2, 4)\}$$

Example 2

Determine the Cartesian set $A \times A$, where $A = \{1, 2\}$.

$$A \times A = \{(1, 1), (1, 2), (2, 1), (2, 2)\}$$

Example 3

Let $A = \{m, n\}$ and $B = \{r, s\}$. Show that $A \times B \neq B \times A$.

By definition,

$$A \times B = \{(m, r), (m, s), (n, r), (n, s)\}$$
$$B \times A = \{(r, m), (r, n), (s, m), (s, n)\}$$

Then

$$A \times B \neq B \times A$$

since, for example, $(m, s) \in A \times B$, but $(m, s) \notin B \times A$.

Exercises

In Exercises 1 through 6 find $A \times B$ for each pair of sets A and B.

1. $A = \{1, 2\}$ and $B = \{3, 4\}$
2. $A = \{0, 1\}$ and $B = \{0, 1\}$
3. $A = \{m\}$ and $B = \{m, n, r\}$
4. $A = \{3, 4, 5\}$ and $B = \{0, 1, 2, 3, \ldots\}$
5. $A = \{$apple, pear, lemon, orange$\}$ and $B = \{$blue, green, yellow$\}$
6. $A = \{\ \}$ and $B = \{5, 10, 15\}$
7. Find $B \times A$ for sets A and B in Exercise 1. Is $A \times B = B \times A$ in this case?
8. Find $B \times A$ for sets A and B in Exercise 2. Is $A \times B = B \times A$ in this case?
9. Find $A \times A$, where $A = \{a, b, c\}$.
10. Determine the number of elements in $A \times B$, where A has 5 members and B has 12 members.
11. If $A = \{1, 2, 3, 4\}$, find the subset of $A \times A$ with ordered pairs (x, y), where
    (*a*) $x = y$
    (*b*) $x \neq y$
    (*c*) $x = 3$
    (*d*) $x > y$
12. The concept of a Cartesian set can be extended to the set $A \times B \times C$ whose elements are ordered triples (x, y, z), where $x \in A$, $y \in B$, and $z \in C$. Determine $A \times B \times C$, where $A = \{0, 1\}$, $B = \{2, 5\}$, and $C = \{3\}$.
13. Find $A \times A \times A$, where $A = \{r, s\}$ (see Exercise 12).
14. Determine the number of elements in $A \times B \times C$, where A has 3 members, B has 12, and C has 5 (see Exercise 12).

2.11

Relations

The relationships that exist among the elements of a set help to describe a structure for that set. The concept of a "relation" is one with which you are familiar; a relation is a form of connective between two elements of a set. For example, in Chap. 1 we used the following relations with respect to the set of statements:

$\Rightarrow$ "implies" $\Leftrightarrow$ "is equivalent to"

The following relations exist with respect to the set of real numbers:

$=$ "equals" $>$ "is greater than" $<$ "is less than"

In this chapter the relations used with respect to the set of subsets of some universal set included:

$=$ "equals" $\subseteq$ "is a subset of" $\subset$ "is a proper subset of"

The following examples illustrate relations defined on other sets:

Set	Relation
Of people	is the father of
	has the same weight as
Of people in the United States	lives in the same state as
Of lines in a plane	is parallel to
	is perpendicular to
Of points on a line	coincides with
	is to the right of
Of professional baseball players	belongs to the same team as
Of real numbers	is 2 more than
Of triangles	has the same area as

In each example it is always possible to determine whether or not a given element of the set is in the relation stated to another element of the set.

Let S be some set. A* relation *$\mathcal{R}$ is defined on S if and only if for a and b any two elements of S the truth value of the statement

a is in the relation $\mathcal{R}$ to b

can be determined.

Consider the "less than" and "greater than" relations with respect to the set S, where $S = \{1, 2, 3\}$. Note that whenever $a < b$, then $b > a$ for any two elements a and b of the set S. For example,

$$1 < 2 \quad \text{and} \quad 2 > 1$$
$$1 < 3 \quad \text{and} \quad 3 > 1$$
$$2 < 3 \quad \text{and} \quad 3 > 2$$

Each of the relations "less than" and "greater than" is the inverse of the other relation.

Let $\mathcal{R}'$ and $\mathcal{R}$ be two relations defined on a set S. The relation $\mathcal{R}'$ is the* inverse *of $\mathcal{R}$ if and only if b is in the relation $\mathcal{R}'$ to a whenever a is in the relation $\mathcal{R}$ to b.

Example 1

Determine inverse relations for the following sets and the relations defined on them:

Set	Relation
(*a*) Of people	is a daughter of
(*b*) Of lines in a plane	is perpendicular to
(*c*) Of points on a line	is to the left of
(*d*) Of real numbers	is a factor of

(*a*) is a parent of the girl (*b*) is perpendicular to (*c*) is to the right of (*d*) is a multiple of

A relation may be considered as a rule that associates with each element of some set A one or more elements of some set B. For example, consider the sets

$$A = \{3, 6, 7, 14, 25\} \qquad \text{and} \qquad B = \{2, 3, 5, 7\}$$

Let $\mathscr{R}$ be a relation which associates with each element of A the elements in set B that are factors of that element. The relation $\mathscr{R}$ associates with 3 the element 3 because 3 "has the factor" 3; associates with 6 the elements 2 and 3 because 6 "has the factor" 2 and 3; associates with 7 the element 7 because 7 "has the factor" 7; associates with 14 the elements 2 and 7 because 14 "has the factor" 2 and 7; and associates with 25 the element 5 because 25 "has the factor" 5.

The relation above may be illustrated by means of a diagram, as shown in Fig. 2.33. The elements of sets A and B are listed in column form. An arrow is drawn from each element of A to those elements of B that are its factors.

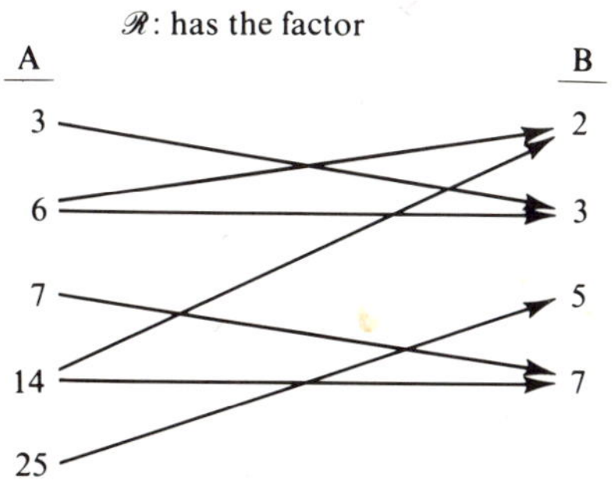

Figure 2.33

Given any relation $\mathscr{R}$ defined on a set A, we can determine the set

$$\{(a, b) \mid a \text{ is in the relation } \mathscr{R} \text{ to } b\}$$

Since this set consists of all ordered pairs (a, b) for which "a is in the relation $\mathscr{R}$ to b," we can define this set as the relation. Note that since $a \in A$ and $b \in A$, each element (a, b) belongs to the Cartesian set $A \times A$. Hence the relation

$$\{(a, b) \mid a \text{ is in the relation } \mathscr{R} \text{ to } b\}$$

is a subset of $A \times A$.

> ***A* relation *defined on a set A is some subset of the Cartesian set A × A.***

For example, let $A = \{1, 2, 3, 4\}$. Then the "less than" relation defined on A can be considered as the set of ordered pairs

$$\{(1, 2), (1, 3), (1, 4), (2, 3), (2, 4), (3, 4)\}$$

That is,

$$\{(x, y) \mid x < y\} = \{(1, 2), (1, 3), (1, 4), (2, 3), (2, 4), (3, 4)\}$$

where $x \in A$ and $y \in A$.

The first member x of the ordered pair (x, y) of a relation is called the *independent variable*; the second member y of the ordered pair (x, y) of a relation is called the *dependent variable*. The set of values of x used in the relation is called the *domain* of the relation; the set of values of y used in the relation is called the *range* of the relation. In the preceding example the domain of the "less than" relation

$$\{(1, 2), (1, 3), (1, 4), (2, 3), (2, 4), (3, 4)\}$$

is

$$\{1, 2, 3\}$$

and the range of the relation is

$$\{2, 3, 4\}$$

Note that both the domain and the range are subsets of A.

In general, a relation can exist between elements of two distinct sets A and B. Therefore, a more general definition of a relation as a set of ordered pairs is possible.

> ***A* relation *between the members of two sets A and B is some subset of A × B.***

A relation that is a subset of $A \times B$ shall be referred to as being defined on the Cartesian set $A \times B$.

Example 2

Determine the elements of the relation

$$\mathscr{R} = \{(x, y) \mid y = 2x + 1\}$$

defined on the Cartesian set $A \times B$, where $A = \{1, 2, 3, 4\}$ and $B = \{1, 2, 3, 4, 5, 6, 7\}$. State the domain and range of the relation.

Since

$$2(\mathbf{1}) + 1 = \mathbf{3}$$
$$2(\mathbf{2}) + 1 = \mathbf{5}$$

and

$$2(\mathbf{3}) + 1 = \mathbf{7}$$

then

$$\mathscr{R} = \{(1, 3), (2, 5), (3, 7)\}$$

The domain of $\mathscr{R}$ is the set

$$\{1, 2, 3\}$$

and the range of $\mathscr{R}$ is the set

$$\{3, 5, 7\}$$

Exercises

1. Determine inverse relations for the following sets and the relations defined on them:

Set	Relation
(*a*) Of people	is taller than
(*b*) Of people	is the father of
(*c*) Of real numbers	is 1 more than
(*d*) Of lines in a plane	is parallel to
(*e*) Of triangles	is congruent to
(*f*) Of children in a family	is older than

In Exercises 2 through 4 consider the relation $\mathscr{R}$, which associates with each element of A the elements of B. Illustrate each relation by means of a diagram.

2. $\mathscr{R}$: is less than, $A = \{1, 2, 3, 4\}$, and $B = \{2, 3, 4, 5\}$
3. $\mathscr{R}$: is equal to the square of, $A = \{1, 4, 9, 16\}$, and $B = \{1, 2, 3, 4\}$
4. $\mathscr{R}$: has a common factor with, $A = \{2, 3, 6, 8\}$, and $B = \{2, 9, 15\}$

In Exercises 5 through 7 what relation is suggested by each diagram?

5. 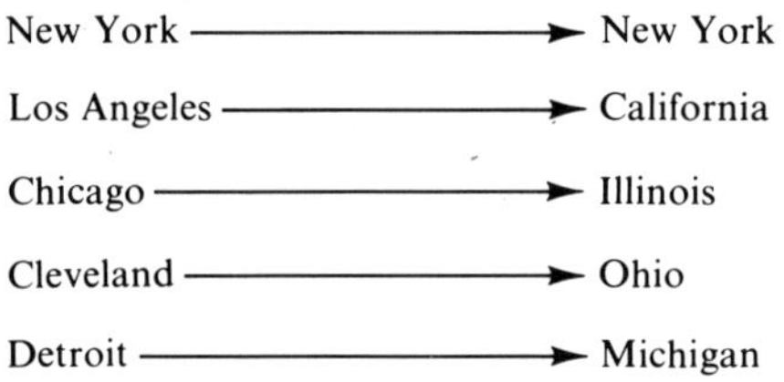

6. 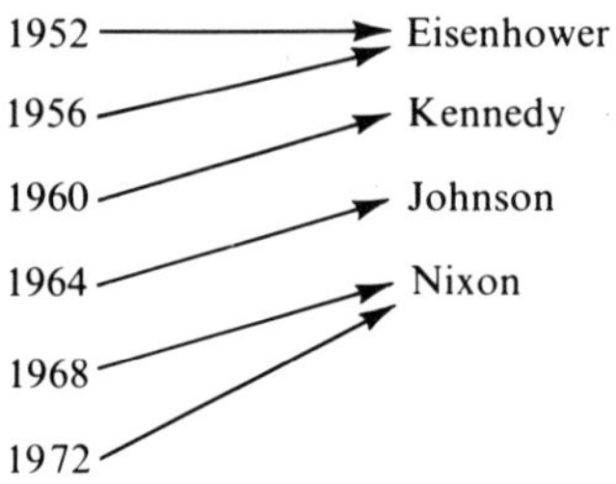

7. 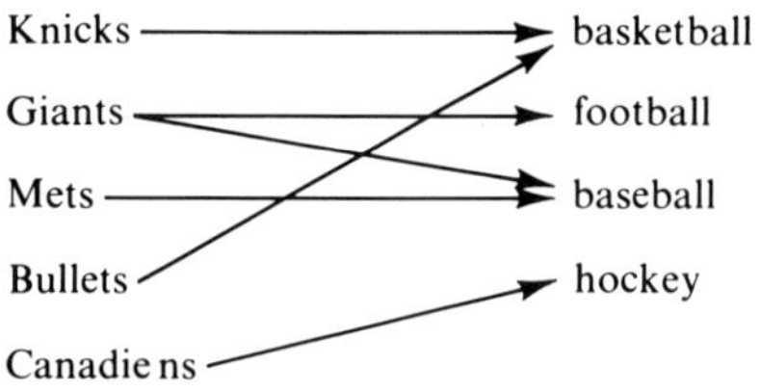

8. Illustrate by means of a diagram the inverse of the following relation:

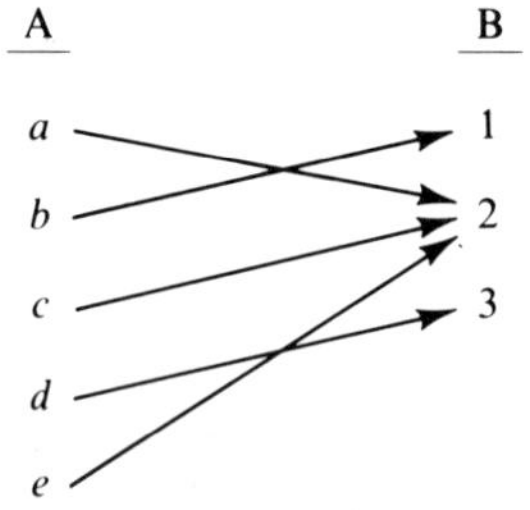

In Exercises 9 through 14 state the domain and range of each relation.

9. $\{(1, 5), (3, 2), (4, 6), (7, 6)\}$
10. $\{(2, 3), (5, 2), (1, 6), (5, 10)\}$
11. $\{(0, 1), (0, 2), (0, 3), (0, 4), (0, 5)\}$
12. $\{(x, y) \mid y = 2x\}$, where $(x, y) \in A \times A$ and $A = \{1, 2, 3, 4, 5, 6, 7, 8\}$
13. $\{(x, y) \mid y = x^2\}$, where $(x, y) \in N \times N$ and N is the set of natural numbers
14. The set of ordered pairs (x, y), where if x is a negative number, then y is equal to 0, and if x is a nonnegative number, then y is equal to 1

In Exercises 15 through 20 determine the elements of each relation defined on the given Cartesian set $A \times B$. State the domain and range of each relation.

15. $\{(x, y) \mid y = x\}$; $A = \{1, 2, 3\}$ and $B = \{1, 3, 5\}$
16. $\{(x, y) \mid y \geq x\}$; $A = \{1, 2, 3, 4\}$ and $B = \{2, 3, 5\}$
17. $\{(x, y) \mid x + y \text{ is odd}\}$; $A = B = \{1, 2, 3, 4, 5, 6\}$
18. $\{(x, y) \mid xy = 0\}$; $A = B = \{0, 1, 2\}$
19. $\{(x, y) \mid y = x^3\}$; $A = \{1, 2, 3, 4\}$ and $B = \{1, 2, 3, 4, \ldots\}$
20. $\{(x, y) \mid y = 2x + 1\}$; $A = \{1, 2, 3, 4\}$ and $B = \{1, 2, 3, 4, \ldots\}$
21. Find the inverse $\mathscr{R}'$ of the relation $\mathscr{R}$, where

$$\mathscr{R} = \{(2, 3), (5, 1), (7, 2)\}$$

22. Find the inverse $\mathscr{R}'$ of the relation $\mathscr{R}$, where

$$\mathscr{R} = \{(x, y) \mid y = 2x + 1\}$$

23. List the elements of $\mathscr{R}$ and $\mathscr{R}'$ in Exercise 22 if $\mathscr{R}$ is defined on the set A, where $A = \{1, 2, 3, 4, \ldots, 10\}$.
24. Determine how many relations can be defined on a set with (*a*) two, (*b*) three, (*c*) four, and (*d*) n elements.

2.12

Functions

Consider the relations illustrated in Figs. 2.34 and 2.35:

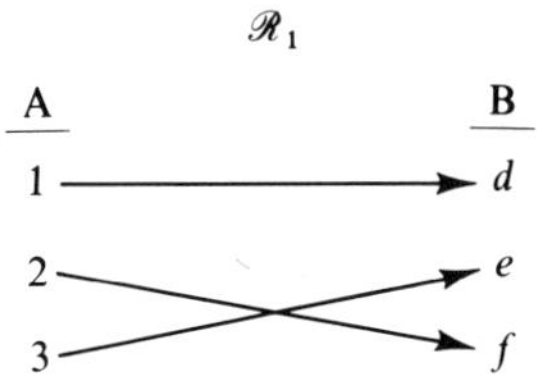

Figure 2.34

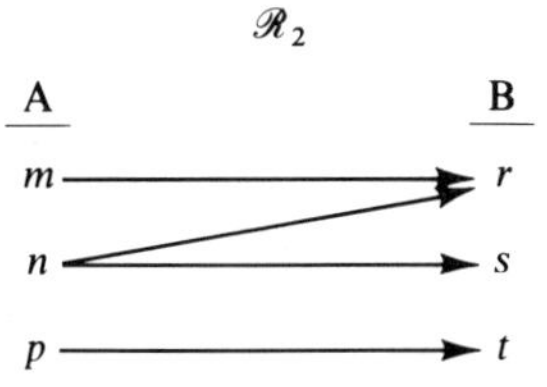

Figure 2.35

The relations differ in one very important property. In the relation $\mathscr{R}_1$ each element in the domain A has associated with it a unique element in the range B. However, in the relation $\mathscr{R}_2$ this is not the case. The element n in the domain has associated with it two elements in the range, namely r and s. Relation $\mathscr{R}_1$ is a special form of relation called a *function*; the relation $\mathscr{R}_2$ is not a function.

> ***A* function *is a relation for which there is exactly one value of the dependent variable for each value of the independent variable.***

If each of the relations $\mathscr{R}_1$ and $\mathscr{R}_2$ are written as a set of ordered pairs, we have

$$\mathscr{R}_1 = \{(1, d), (2, f), (3, e)\}$$
$$\mathscr{R}_2 = \{(m, r), (n, r), (n, s), (p, t)\}$$

Note that $\mathscr{R}_2$ contains two ordered pairs with identical first members: (n, r) and (n, s). No two ordered pairs of $\mathscr{R}_1$ have the same first member. Thus, a function is a set of ordered pairs for which no two pairs have the same first member.

Example 1

Determine whether or not each relation is a function:

(*a*) $\{(a, b), (b, d), (c, e), (d, g)\}$

(*b*) $\{(1, 3), (2, 5), (0, 6), (4, 1), (2, 7)\}$

(*a*) The relation is clearly a function since only one value y exists for each value x. (*b*) The relation is not a function since y has two values when $x = 2$.

It is important to note that although a function is a set of ordered pairs (x, y), mathematicians often call the dependent variable y a function of the independent variable x. For example, in the case of the relation

$$\{(x, y) \mid y = x + 1\}$$

we often speak of y, or $x + 1$, as being a function of x.

Consider the expression $x^2 + x + 3$. For each value of x, this expression has a corresponding value. If $x = 0$, the value of the expression is 3; if $x = 1$, the value of the expression is 5; if $x = 2$, the value of the expression is 9; and so forth. We abbreviate these statements by denoting the "function of x" by "$f(x)$." The symbol $f(x)$ is read "f of x." Then

$$f(x) = x^2 + x + 3$$
$$f(0) = 0^2 + 0 + 3 = 3$$
$$f(1) = 1^2 + 1 + 3 = 5$$
$$f(2) = 2^2 + 2 + 3 = 9$$
$$\cdots\cdots\cdots\cdots\cdots\cdots$$

Thus, $f(0)$ is the value of $f(x)$ when $x = 0$; $f(1)$ is the value of $f(x)$ when $x = 1$; $f(2)$ is the value of $f(x)$ when $x = 2$; and so forth. In general, $f(a)$ is the value of $f(x)$ when $x = a$.

Different letters may be used to denote different functions. For example, we may write $f(x)$ for the function $x + 2$, $g(x)$ for the function x^2, and $h(x)$ for the function $2x + 1$.

Example 2

Given $f(x) = 3x^2 + x - 1$, find

(*a*) $f(1)$
(*b*) $f(2)$
(*c*) $f(a)$
(*d*) $f[f(1)]$

(*a*) $f(1) = 3 \cdot 1^2 + 1 - 1 = 3$ (*b*) $f(2) = 3 \cdot 2^2 + 2 - 1 = 13$
(*c*) $f(a) = 3a^2 + a - 1$ (*d*) $f[f(1)] = f(3) = 3 \cdot 3^2 + 3 - 1 = 29$

The functions discussed thus far have been functions of one independent variable. In Chap. 3 we shall have occasion to consider functions of two or more independent variables. For example, consider expressions such as

$$x^2 + 2xy + y \qquad x + y + 1 \qquad x + y^2$$

These expressions are functions of the two independent variables x and y since their values are uniquely determined when values are assigned to x and y, respectively. We may denote such functional expressions by the symbol $f(x, y)$, which is read "f of x and y." Thus, if

$$f(x, y) = x^2 + 2xy + y$$

then

$$f(0, 1) = 0^2 + 2 \cdot 0 \cdot 1 + 1 = 1$$
$$f(2, 1) = 2^2 + 2 \cdot 2 \cdot 1 + 1 = 9$$
$$f(1, 3) = 1^2 + 2 \cdot 1 \cdot 3 + 3 = 10$$

and so forth.

Example 3

Determine

(*a*) $f(1, 2)$, where $f(x, y) = xy$
(*b*) $g(2, 0)$, where $g(x, y) = x^2 + y$
(*c*) $h(3, 5)$, where $h(x, y) = 7x + y^2 - xy$

(*a*) $f(1, 2) = 1 \cdot 2 = 2$ (*b*) $g(2, 0) = 2^2 + 0 = 4$ (*c*) $h(3, 5) = 7 \cdot 3 + 5^2 - 3 \cdot 5 = 31$

Exercises

In Exercises 1 through 4 state whether or not each relation is a function.

1. $\{(1, 3), (2, 5), (3, 1), (4, 3)\}$
2. $\{(1, 0), (2, 1), (1, 3), (3, 4)\}$
3. $\{(x, y) \mid y > x\}$, where $x, y \in A$ and $A = \{1, 2, 3\}$
4. $\{(x, y) \mid y = 2x\}$, where $x, y \in A$ and $A = \{0, 1, 2, 3, 4\}$
5. Determine if the relation expressed by the following table of values is a function:

x	3	2	1	2	3	4	5
y	9	4	1	1	1	4	9

6. Determine how many functions can be defined on a set with (*a*) two; (b) three; (*c*) four; (*d*) *n* elements.
7. Given $f(x) = 5x^2 + 2x + 1$, find (*a*) $f(0)$; (*b*) $f(2)$; (c) $f(a)$; (*d*) $f(a + 2)$; (*e*) $f(a) + f(2)$; (*f*) $f[f(2)]$.
8. For each relation in Exercises 1 through 4, (*a*) find the inverse; (*b*) state whether or not the inverse relation is a function.
9. Is the inverse of a function always a function?
10. A function whose range is a set with a single element k is a *constant function*. Find an expression for $f(x)$ for such a function.
11. The function $\{(x, f(x)) \mid f(x) = ax + b\}$, where a and b are real numbers and $a \neq 0$, is called a *linear function*. Determine (*a*) $f(0)$; (*b*) $f(1)$; (*c*) $f(2)$; (*d*) $f(3)$; (*e*) $f(1) - f(0)$; (*f*) $f(2) - f(1)$; (*g*) $f(3) - f(2)$.
12. Given $f(x, y) = x + 3y$, find (*a*) $f(0, 0)$; (*b*) $f(1, 0)$; (*c*) $f(0, 1)$; (*d*) $f(2, 3)$.
13. Given $f(x, y) = x^2 + xy + y^2$, find (*a*) $f(1, 2)$; (*b*) $f(2, 1)$; (*c*) $f(a, b)$; (*d*) $f(b, a)$.
14. Given $f(x, y, z) = x^2 + y - z$, find (*a*) $f(1, 0, 1)$; (*b*) $f(1, 2, 3)$; (*c*) $f(1, 1, 1)$; (*d*) $f(3, 2, 1)$.

Keyed Summary

In this chapter we studied sets and their representation by Venn diagrams. The concepts and methods presented with respect to sets are part of the body of knowledge called the *algebra of sets*.

2.1 *Sets and Their Descriptions*

A set is a collection of objects.

A set is well defined if and only if it is possible to determine whether or not a given object is in the collection.

If an object belongs to a set, the object is called a member or element of the set. If an object a is an element of a set A, we write $a \in A$; if an object a is not an element of a set A, we write $a \notin A$.

Sets may be defined by the description, listing, or roster methods, and by the rule method using set-builder notation.

2.2 *Finite and Infinite Sets*

Sets that have a finite number of elements are classified as finite; sets that have infinitely many elements are classified as infinite.

A set that contains no elements is called the null or empty set. The null set is denoted by either { } or the symbol $\varnothing$.

Sets A and B are equal if and only if every element of A is an element of B and every element of B is an element of A. When A is equal to B, we write $A = B$; when A is not equal to B, we write $A \neq B$.

2.3 *Subsets*

Set A is a subset of set B if and only if every element of A is also an element of B. When A is a subset of B, we write $A \subseteq B$; when A is not a subset of B, we write $A \not\subseteq B$.

Set A is a proper subset of set B if and only if $A \subseteq B$ and $A \neq B$. When A is a proper subset of B, we write $A \subset B$; when A is not a proper subset of B, we write $A \not\subset B$.

2.4 *Power Sets*

The power set of any set S is the set of all subsets of S.

2.5 *Universal Sets and Complements*

The set of all objects under study is called the universal set and is represented by the symbol U.

The complement of a set A is the set of elements in the universal set U that are not in set A. The complement of set A is denoted by $\bar{A}$.

$$\bar{A} = \{x \mid x \in U \text{ and } x \notin A\}$$

If A and B are subsets of the universal set U, then the relative complement of A in B is the set of all elements in B that are not in A. The relative complement of A in B is denoted by $B - A$.

$$B - A = \{x \mid x \in B \text{ and } x \notin A\}$$

2.6 Venn Diagrams

Sets and relations among sets may be represented geometrically by means of Venn diagrams.

The Venn diagrams for A and $\bar{A}$, where A is a nonempty proper subset of U, are shown by the hatched regions in Figs. 2.2 and 2.3, respectively.

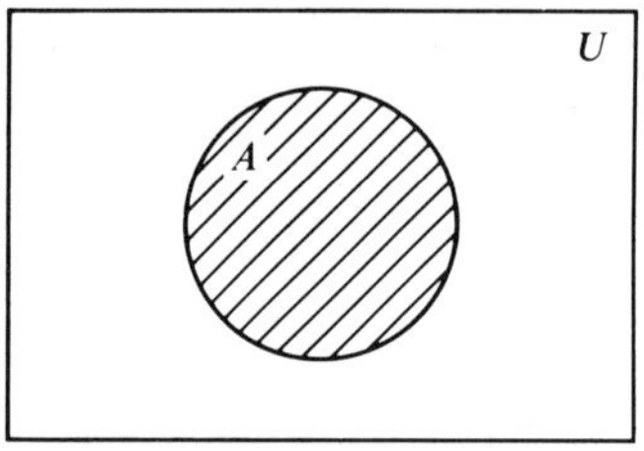

$A \subset U$ and $A \neq \{\ \}$

Figure 2.2

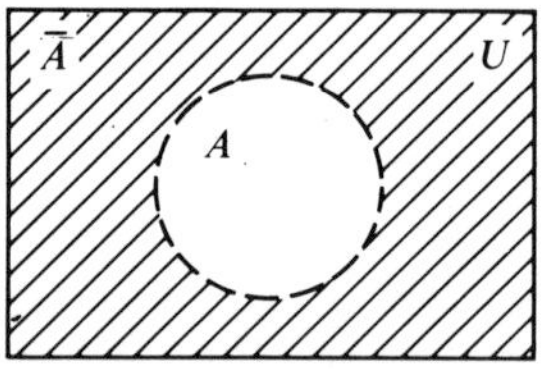

$\bar{A} = \{x \mid x \in U \text{ and } x \notin A\}$

Figure 2.3

2.7 Intersection of Sets

The intersection of two sets A and B is the set of all elements that belong to both A and B. The intersection of sets A and B is denoted by $A \cap B$.

$$A \cap B = \{x \mid x \in A \text{ and } x \in B\}$$

If the intersection of two nonempty sets A and B is the null set, then A and B are called disjoint sets.

In general, the Venn diagram for the intersection of sets A and B, where A and B are nonempty proper subsets of U, is shown by the hatched region in Fig. 2.16.

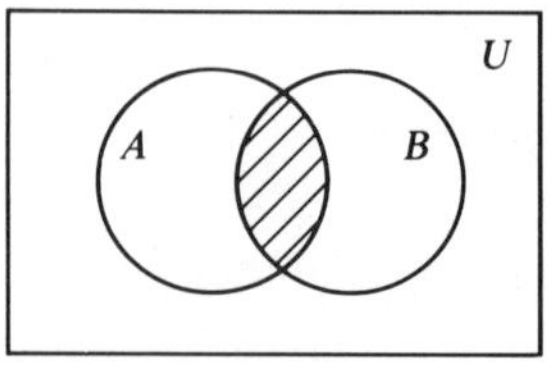

$A \cap B$

Figure 2.16

2.8 Union of Sets

The union of two sets A and B is the set of all elements that belong to either A or B. The union of sets A and B is denoted by $A \cup B$.

$$A \cup B = \{x \mid x \in A \text{ or } x \in B\}$$

The equality of sets

$$\overline{(A \cup B)} = \bar{A} \cap \bar{B}$$

and

$$\overline{(A \cap B)} = \bar{A} \cup \bar{B}$$

are called DeMorgan's laws for sets.

In general, the Venn diagram for the union of sets A and B, where A and B are nonempty proper subsets of U, is shown by the hatched region in Fig. 2.24.

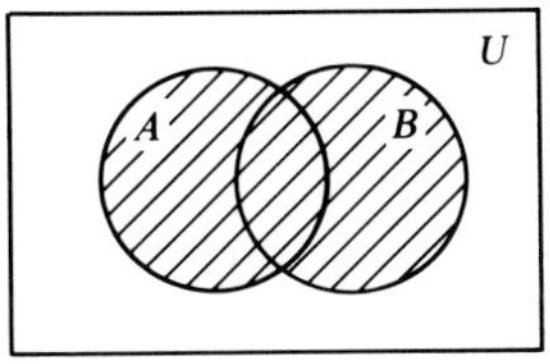

$A \cup B$

Figure 2.24

2.9 Number of Elements in a Set

Associated with every finite set A is a number, denoted by $n(A)$, which represents the number of elements in set A. The number $n(A)$ is called the cardinal number of set A.

2.10 Cartesian Sets

The set of all ordered pairs of the form (x, y), where $x \in A$ and $y \in B$, is called the Cartesian set $A \times B$ of sets A and B.

2.11 Relations

Let S be some set; let a and b be any two elements of S. A relation $\mathscr{R}$ is defined on S if and only if the truth value of the statement

$$a \text{ is in the relation } \mathscr{R} \text{ to } b$$

can be determined.

Let $\mathscr{R}$ and $\mathscr{R}'$ be two relations defined on a set S. The relation $\mathscr{R}'$ is the inverse of $\mathscr{R}$ if and only if b is in the relation $\mathscr{R}'$ to a whenever a is in the relation $\mathscr{R}$ to b.

A relation may be considered as a rule that associates with each element of some set A one or more elements of some set B.

A relation defined on a set A is some subset of the Cartesian set $A \times A$. A relation between the members of two sets A and B is some subset of $A \times B$.

The first member x of the ordered pair (x, y) of a relation is called the independent variable; the second member y of the ordered pair (x, y) of a relation is called the dependent variable. The set of values of x used in the relation is called the domain of the relation; the set of values of y used in the relation is called the range of the relation.

2.12 Functions

A function is a relation for which there is exactly one value of the dependent variable for each value of the independent variable.

The symbol $f(x)$, which is read "f of x," is used to denote a function of x. The symbol $f(x, y)$, which is read "f of x and y," is used to denote a function of the two independent variables x and y.

Building Your Mathematical Vocabulary

Associative properties (2.7, 2.8)
Cardinal number (2.9)
Cartesian set (2.10)
Commutative properties (2.7, 2.8)
Complement (2.5)
Constant function (2.12)
DeMorgan's laws (2.8)
Dependent variable (2.11)
Description method (2.1)
Disjoint sets (2.7)
Distributive properties (2.8)
Domain (2.11)
Element (2.1)
Empty set (2.2)
Equal (2.2)
Exponent (2.4)
Finite set (2.2)
Function (2.12)
Idempotent properties (2.7, 2.8)
Independent variable (2.11)
Infinite set (2.2)
Intersection (2.7)
Inverse of a relation (2.11)
Linear function (2.12)

Listing method (2.1)
Member (2.1)
Natural numbers (2.2)
Null set (2.2)
Ordered pair (2.10)
Power set (2.4)
Product set (2.10)
Proper subset (2.3)
Range (2.11)
Relation (2.11)
Relative complement (2.5)
Roster method (2.1)
Rule method (2.1)
Set (2.1)
Set-builder notation (2.1)
Subset (2.3)
Union (2.8)
Universal set (2.5)
Venn diagram (2.6)
Well defined (2.1)
Whole numbers (2.2)

Page 107

elements relations

2 - 16

3 - 512

4

n elements — 2^{n^2}, n is the number of elements in set.

Page 110

functions

2 — 9

3 — 64

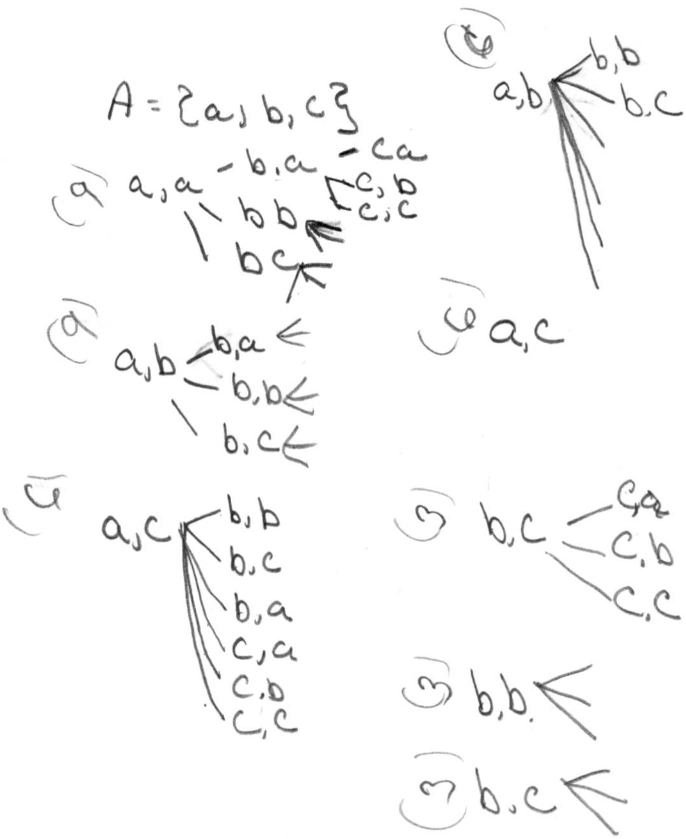

a, a
a, b
a, c
b, a
b, b
b. c
c, a
c, b
c c
∅

3

Boolean Algebra and Applications

3.1

Isomorphisms

In Chaps. 1 and 2 we noted that some of the basic properties of the algebra of statements and the algebra of sets are very similar. In the study and comparison of mathematical systems in general, we often can identify two structures which are so similar that they differ only in the notation used. For example, consider the operation of "addition" $\oplus$ defined on the set {0, 1, 2, 3} and the operation of "multiplication" $\otimes$ defined on the set {1, 5, 8, 12} as illustrated in the following tables:

Addition $\oplus$

$\oplus$	0	1	2	3
0	0	1	2	3
1	1	2	3	0
2	2	3	0	1
3	3	0	1	2

Multiplication $\otimes$

$\otimes$	1	5	8	12
1	1	5	8	12
5	5	12	1	8
8	8	1	12	5
12	12	8	5	1

These two tables are *abstractly identical*; that is, they differ only in the notation or symbols used for the elements of each set and the interpretation placed upon the operations used. To show this, consider a *one-to-one correspondence* between the elements of the set {0, 1, 2, 3} and the set {1, 5, 8, 12} described as

$$
\begin{aligned}
0 &\leftrightarrow 1\\
1 &\leftrightarrow 5\\
2 &\leftrightarrow 12\\
3 &\leftrightarrow 8
\end{aligned}
$$

If each of the elements 1, 5, 8, and 12 is replaced by the elements 0, 1, 3, and 2, respectively, in the second table, and ⊗ is replaced by ⊕, then the tables can be made identical:

⊗	1	5	8	12
1	1	5	8	12
5	5	12	1	8
8	8	1	12	5
12	12	8	5	1

becomes

⊕	0	1	3	2
0	0	1	3	2
1	1	2	0	3
3	3	0	2	1
2	2	3	1	0

which can be written as

⊕	0	1	2	3
0	0	1	2	3
1	1	2	3	0
2	2	3	0	1
3	3	0	1	2

by rearranging rows three and four and columns three and four. Thus we say that the two sets with the operations defined on them are abstractly identical or have the same structure. This relationship between the two structures is called an *isomorphism*.

> ***An* isomorphism *is a one-to-one correspondence between sets that preserves the operations defined on them.***

The idea of an isomorphism between two sets A and B can be exhibited by means of the diagram in Fig. 3.1. To each pair $a, b \in A$, the operation $*$

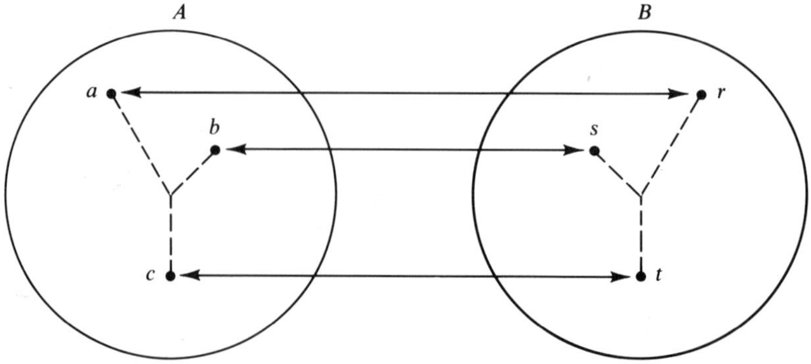

Figure 3.1

assigns a third element c; to each pair $r, s \in B$, the operation $\cdot$ assigns a third element t. Consider a one-to-one correspondence that associates a and b in set A with r and s in set B, respectively. Then an isomorphism exists between the two sets provided c (that is, $a * b$) is associated with t (that is, $r \cdot s$) under the one-to-one correspondence.

Example 1

Determine the one-to-one correspondence between the elements of $S = \{a, b, c\}$ and $T = \{p, q, r\}$ that preserves the operations defined on the sets by the following tables:

$*$	a	b	c
a	c	a	b
b	a	b	c
c	b	c	a

$\cdot$	p	q	r
p	p	q	r
q	q	r	p
r	r	p	q

The identity elements for S and T are b and p, respectively. This suggests that they should correspond. Consider the following one-to-one correspondence:

$$\begin{gathered} a \leftrightarrow q \\ b \leftrightarrow p \\ c \leftrightarrow r \end{gathered}$$

If we replace p, q, and r by b, a, and c, respectively, and replace the operation $\cdot$ by $*$ in the second table, then the given tables can be made identical:

$\cdot$	p	q	r
p	p	q	r
q	q	r	p
r	r	p	q

becomes

$*$	b	a	c
b	b	a	c
a	a	c	b
c	c	b	a

which can be written as

*	a	b	c
a	c	a	b
b	a	b	c
c	b	c	a

by rearranging rows one and two and columns one and two.

Example 2

Show that there is an isomorphism between the set of whole numbers $\{0, 1, 2, 3, \ldots\}$ under addition and the set $\{2^0, 2^1, 2^2, 2^3, \ldots\}$ under multiplication, using the one-to-one correspondence $0 \leftrightarrow 2^0$, $1 \leftrightarrow 2^1$, $2 \leftrightarrow 2^2$, $3 \leftrightarrow 2^3$, ...

Consider m and n as any two whole numbers. Using the one-to-one correspondence described,

$$m \leftrightarrow 2^m \qquad \text{and} \qquad n \leftrightarrow 2^n$$

Furthermore,

$$m + n \leftrightarrow 2^{m+n}$$

Since $2^m \cdot 2^n = 2^{m+n}$, then

$$m + n \leftrightarrow 2^m \cdot 2^n$$

and an isomorphism exists between the two sets.

The concept of an isomorphism between two sets will be useful in describing an application of the algebra to be studied in this chapter.

Exercises

In Exercises 1 through 6 determine a one-to-one correspondence between the elements of the two given sets that preserves the operations defined on each.

1. The set $\{1, 2, 3, 4\}$ and the set $\{a, b, c, d\}$ under the operations defined on the sets by the following tables:

$*$	1	2	3	4
1	1	2	3	4
2	2	1	4	3
3	3	4	1	2
4	4	3	2	1

$\cdot$	a	b	c	d
a	a	b	c	d
b	b	a	d	c
c	c	d	a	b
d	d	c	b	a

2. The set of odd natural numbers under addition and the set of even natural numbers under addition
3. The set $\{a, b, c\}$ and the set $\{x, y, z\}$ under the operations defined on the sets by the following tables:

$*$	a	b	c
a	b	a	b
b	c	b	a
c	a	c	b

$\cdot$	x	y	z
x	z	y	x
y	z	z	y
z	y	x	z

4. The set of natural numbers under addition and the set of positive integral powers of $\frac{1}{2}$ under multiplication
5. The sets $\{1, 2, 3, 4\}$ and $\{2, 4, 6, 8\}$ under the operations defined on the sets by the following tables:

$*$	1	2	3	4
1	1	2	3	4
2	2	4	1	3
3	3	1	4	2
4	4	3	2	1

$\cdot$	2	4	6	8
2	4	8	2	6
4	8	6	4	2
6	2	4	6	8
8	6	2	8	4

6. The sets $\{1, -1, i, -i\}$ and $\{a, b, c, d\}$ under the operations defined on the sets by the following tables:

$\cdot$	1	-1	i	$-i$
1	1	-1	i	$-i$
-1	-1	1	$-i$	i
i	i	$-i$	-1	1
$-i$	$-i$	i	1	-1

$*$	a	b	c	d
a	b	c	d	a
b	c	d	a	b
c	d	a	b	c
d	a	b	c	d

3.2

Binary Boolean Arithmetic

In this section we shall describe a very simple arithmetic called *binary Boolean arithmetic*, which has many applications. Later in this chapter we shall show the relationship of this arithmetic to the algebra of statements and the algebra of sets.

The set of elements in a binary Boolean arithmetic consists only of the real numbers 0 and 1. Two operations, addition (+) and multiplication (·), are defined on the set {0, 1} by means of Tables 3.1 and 3.2:

Table 3.1 Binary Boolean Addition Table

+	0	1
0	0	1
1	1	1

Table 3.2 Binary Boolean Multiplication Table

·	0	1
0	0	0
1	0	1

Note that the results of these operations are the same as those for the ordinary addition and multiplication of the real numbers 0 and 1, except that $1 + 1 = 1$.

Example 1

Find the values of expressions (*a*) and (*b*):

(*a*) $[(1 + 0) \cdot 1] + (1 \cdot 0)$

(*b*) $[(1 \cdot 0) + 0] \cdot (1 + 1)$

(*a*) $[(1 + 0) \cdot 1] + (1 \cdot 0) = (1 \cdot 1) + (1 \cdot 0) = 1 + 0 = 1$

(*b*) $[(1 \cdot 0) + 0] \cdot (1 + 1) = (0 + 0) \cdot (1 + 1) = 0 \cdot 1 = 0$

In mathematics, if one statement is obtained from another by an interchange of certain terms, then each statement is the *dual* of the other. For example, the statement

Two points determine a line

is the dual of the statement

Two lines determine a point

under the interchange of the words "point" and "line." Conversely, the second statement is the dual of the first. In certain geometries both these statements are true. In Euclidean geometry, the second statement is not necessarily true because the lines may be parallel.

If the operations $<$ and $>$ are interchanged in the algebra of real numbers, then the dual of the statement

If $a < b$ and $b < c$, then $a < c$

becomes

If $a > b$ and $b > c$, then $a > c$

Each of these statements is true.

Example 2

Write the dual of the statement

$$A \cap (B \cup C) = (A \cap B) \cup (A \cap C)$$

under the interchange of the operations $\cap$ and $\cup$.

$$A \cup (B \cap C) = (A \cup B) \cap (A \cup C)$$

Note that the given statement and its dual are both true. That is, (*a*) the operation of intersection is distributive over the union of sets, and (*b*) the operation of union is distributive over the intersection of sets.

In binary Boolean arithmetic, the dual of every statement obtained by interchanging the numbers 0 and 1 while also interchanging the operations + and · is true if and only if the original statement is true.

For example, using the addition and multiplication tables, each of the following statements and its dual are both true:

Statement	Dual of the Statement
$0 + 0 = 0$	$1 \cdot 1 = 1$
$0 + 1 = 1$	$1 \cdot 0 = 0$
$1 + 0 = 1$	$0 \cdot 1 = 0$
$1 + 1 = 1$	$0 \cdot 0 = 0$

This property of binary Boolean arithmetic is called the *principle of duality*; it will be extremely useful in discussing properties of the operations in Secs. 3.3 and 3.4.

Example 3

Find the value of

$$(1 \cdot 1) + [1 + (0 + 1)]$$

Write the dual of this statement and find its value.

$$\begin{aligned}(1 \cdot 1) + [1 + (0 + 1)] &= (1 \cdot 1) + (1 + 1) \\ &= 1 + 1 = 1\end{aligned}$$

The dual of the given statement is

$$(0 + 0) \cdot [0 \cdot (1 \cdot 0)]$$

Its value is 0 since the value of the given statement is 1.

Exercises

In Exercises 1 through 4 find the value of each expression in the binary Boolean arithmetic system.

1. $(1 \cdot 1) + [0 \cdot (1 + 0)]$
2. $[(0 \cdot 1) + (1 \cdot 0)] \cdot (1 + 1)$
3. $[(1 + 0) + 1] \cdot (0 + 1)$
4. $[(1 \cdot 1) + 1] \cdot (1 \cdot 0)$

In Exercises 5 through 8 write the dual of the statement in the indicated exercise and find its value.

5. Exercise 1
6. Exercise 2
7. Exercise 3
8. Exercise 4

In Exercises 9 through 12 write the dual of each statement about sets under the interchange of the operations $\cap$ and $\cup$. Determine whether or not each statement and its dual have the same truth value.

9. $A \cap A = A$
10. $A \cap B = B \cap A$
11. $(A \cup B) \cup C = A \cup (B \cup C)$
12. $A \cap (A \cup B) = A$

In Exercises 13 through 17 write the dual of each statement of logical equivalence under the interchange of the operations $\wedge$ and $\vee$. Determine whether or not each dual is also a valid statement of logical equivalence.

13. $p \wedge p \Leftrightarrow p$
14. $p \wedge q \Leftrightarrow q \wedge p$
15. $(p \vee q) \vee r \Leftrightarrow p \vee (q \vee r)$
16. $p \wedge (q \vee r) \Leftrightarrow (p \wedge q) \vee (p \wedge r)$
17. $\sim (p \wedge q) \Leftrightarrow (\sim p) \vee (\sim q)$

3.3

Binary Boolean Functions

In the algebra of real numbers, the values of a function such as that expressed by the equation

$$f(x) = x^2 + 3x - 5$$

cannot be listed for all replacement values of x. Infinitely many replacement values exist. However, in binary Boolean algebra a function of a finite number of variables can be evaluated for all replacement values. For example, a function of two variables $B(x, y)$ has only four possible pairs of replacement values for x and y. That is, the only replacement values for x and y are 0 and 0, 0 and 1, 1 and 0, and 1 and 1. Thus, it is possible to list all values of $B(x, y)$.

Example 1

Find all the values of $B(x, y)$, where

$$B(x, y) = x \cdot (x + y)$$

$$B(0, 0) = 0 \cdot (0 + 0) = 0 \cdot 0 = 0$$
$$B(0, 1) = 0 \cdot (0 + 1) = 0 \cdot 1 = 0$$
$$B(1, 0) = 1 \cdot (1 + 0) = 1 \cdot 1 = 1$$
$$B(1, 1) = 1 \cdot (1 + 1) = 1 \cdot 1 = 1$$

Two binary Boolean functions are* equivalent *if and only if the values of the functions are equal for each possible set of replacement values of the variables.

Example 2

If possible, simplify the binary Boolean function in Example 1.

In the following table the four possible pairs of replacement values for x and y are considered; the resulting values of $B(x, y)$ are listed in the third column.

x	y	$B(x, y)$
0	0	**0**
0	1	**0**
1	0	**1**
1	1	**1**

Since the values of $B(x, y)$ are identical to those for x,

$$B(x, y) = x$$

that is,

$$x \cdot (x + y) = x$$

in binary Boolean algebra.

The results of Example 2 express a fundamental theorem or law of binary Boolean algebra. Using the results of Example 2 and the principle of duality, we may obtain another theorem:

$$x + (x \cdot y) = x$$

This theorem and its dual are called the *absorption properties*. The principle of duality will save us considerable time and effort in the development of binary Boolean algebra—every theorem proved will have a corresponding valid dual.

Example 3

Show that

$$(x + x) \cdot y = x \cdot y$$

---

x	y	$x + x$	$(x + x) \cdot y$	$x \cdot y$
0	0	0	**0**	**0**
0	1	0	**0**	**0**
1	0	1	**0**	**0**
1	1	1	**1**	**1**

Columns one and two contain the possible pairs of replacement values for x and y. Column three may be obtained by using the values in column one and the binary Boolean addition table; column four may be obtained by using the values in columns three and two and the binary Boolean multiplication table. Finally, column five may be obtained by using the values in columns one and two and the binary Boolean multiplication table. Since columns four and five are identical,

$$(x + x) \cdot y = x \cdot y$$

Exercises

1. Find $B(1)$, where $B(x) = (x \cdot x) + x$
2. Find $B(0, 1)$, where $B(x, y) = [(x + x) \cdot y] + y$
3. Find $B(1, 1)$, where $B(x, y) = [(x \cdot y) + x] \cdot y$
4. Find $B(1, 0, 1)$, where $B(x, y, z) = (x + y) \cdot (y + z)$
5. Find $B(0, 1, 1)$, where $B(x, y, z) = [(x \cdot x) + (y \cdot z)] + y$

In Exercises 6 through 10 use a table of values for x and y to prove each statement of equivalence.

6. $x \cdot (x + x) = x$
7. $x + (x \cdot y) = x$

8. $(x \cdot x) + y = x + y$
9. $y \cdot [x + (x \cdot y)] = x \cdot y$
10. $(x + y) \cdot (x + z) = x + (y \cdot z)$
11. The Short Life Insurance Company agrees to pay the $25,000 value of a policy if both John and Mary die, or if John dies. What binary Boolean algebra theorem is illustrated by this situation?

3.4

Properties of Binary Boolean Algebra

In this section we shall examine some basic properties of a binary Boolean algebra. The proofs of a representative number of these properties will be presented; proofs of the remaining properties are left to the student as exercises. The properties will be listed in pairs of dual statements.

Idempotent properties

$$x + x = x$$
$$x \cdot x = x$$

To prove that $x + x = x$, only the two cases $x = 0$ and $x = 1$ need to be considered:

x	$x + x$
0	**0**
1	**1**

Because both columns are identical, $x + x = x$. By the principle of duality, it follows that $x \cdot x = x$.

Commutative properties

$$x + y = y + x$$
$$x \cdot y = y \cdot x$$

To prove that $x \cdot y = y \cdot x$, the following four cases need to be considered:

x	y	$x \cdot y$	$y \cdot x$
0	0	**0**	**0**
0	1	**0**	**0**
1	0	**0**	**0**
1	1	**1**	**1**

Since columns three and four are identical, $x \cdot y = y \cdot x$. Once again, by the principle of duality, $x + y = y + x$.

Associative properties

$$(x + y) + z = x + (y + z)$$
$$(x \cdot y) \cdot z = x \cdot (y \cdot z)$$

Note that as a result of these properties and the idempotent properties, coefficients and exponents other than 0 and 1 are unnecessary in binary Boolean algebra. For example, we do not write $x + x + x = 3x$ because

$$\begin{aligned} x + x + x &= (x + x) + x \\ &= x + x \\ &= x \end{aligned}$$

Indeed, in general,

$$\underbrace{x + x + x + \cdots + x}_{n \text{ terms}} = x$$

Similarly, we do not write $x \cdot x \cdot x = x^3$ because

$$\begin{aligned} x \cdot x \cdot x &= (x \cdot x) \cdot x \\ &= x \cdot x \\ &= x \end{aligned}$$

In general,

$$\underbrace{x \cdot x \cdot x \cdot \cdots \cdot x}_{n \text{ factors}} = x$$

Distributive properties

$$x \cdot (y + z) = (x \cdot y) + (x \cdot z)$$
$$x + (y \cdot z) = (x + y) \cdot (x + z)$$

To prove that $x \cdot (y + z) = (x \cdot y) + (x \cdot z)$, eight cases need to be considered because each variable may be replaced by one of two values:

x	y	z	$y + z$	$x \cdot (y + z)$	$(x \cdot y)$	$(x \cdot z)$	$(x \cdot y) + (x \cdot z)$
0	0	0	0	**0**	0	0	**0**
0	0	1	1	**0**	0	0	**0**
0	1	0	1	**0**	0	0	**0**
1	0	0	0	**0**	0	0	**0**
0	1	1	1	**0**	0	0	**0**
1	0	1	1	**1**	0	1	**1**
1	1	0	1	**1**	1	0	**1**
1	1	1	1	**1**	1	1	**1**

Column four is obtained by using the values in columns two and three and the binary Boolean addition table; column five is obtained by using the values in columns one and four and the binary Boolean multiplication table. Column six is obtained by using the values in columns one and two and the binary Boolean multiplication table; column seven is obtained by using the values in columns one and three and the binary Boolean multiplication table. Column eight is obtained by using the values in columns six and seven and the binary Boolean addition table. Since columns five and eight are identical,

$$x \cdot (y + z) = (x \cdot y) + (x \cdot z)$$

Example 1

Find

(*a*) $x + 1$
(*b*) $x + 0$
(*c*) $1 \cdot x$
(*d*) $0 \cdot x$

(*a*)

x	$x + 1$
0	**1**
1	**1**

$x + 1 = 1$

(*b*)

x	$x + 0$
0	**0**
1	**1**

$x + 0 = x$

(c)

x	$1 \cdot x$
0	**0**
1	**1**

$$1 \cdot x = x$$

(d)

x	$0 \cdot x$
0	**0**
1	**0**

$$0 \cdot x = 0$$

In the algebra of real numbers, an expression such as $4 + 2 \cdot 7$ has only one interpretation:

$$4 + 2 \cdot 7 \qquad \text{means} \qquad 4 + (2 \cdot 7)$$

That is, when the parentheses are deleted, multiplication takes precedence over addition. Hereafter we shall adopt the same convention in binary Boolean algebra. For example, we may write

$$\begin{array}{rcl} x + x \cdot y & \text{to mean} & x + (x \cdot y) \\ x \cdot y + x \cdot z & \text{to mean} & (x \cdot y) + (x \cdot z) \\ x + x \cdot x + x \cdot y & \text{to mean} & x + (x \cdot x) + (x \cdot y) \end{array}$$

Be careful to note that if you wish to write an expression such as $(x + x) \cdot y$, the parentheses are necessary because

$$(x + x) \cdot y = x \cdot y$$

while

$$x + x \cdot y = x$$

Example 2

Simplify the binary Boolean function $B(x, y)$, where

$$B(x, y) = x + x \cdot x + x \cdot y$$

$x + x \cdot x + x \cdot y = x + x + x \cdot y$	idempotent property $(\cdot)$
$= (x + x) + x \cdot y$	associative property $(+)$
$= x + x \cdot y$	idempotent property $(+)$
$= x$	absorption property $(+)$

Exercises

1. Use a table of values for x to prove that $x \cdot x = x$.
2. Use a table of values for x and y to prove that $x + y = y + x$.
3. Use a table of values for x, y, and z to prove the associative properties.
4. Use a table of values for x, y, and z to prove the distributive property

$$x + (y \cdot z) = (x + y) \cdot (x + z)$$

5. Simplify (*a*) $x + x + x + x$; (*b*) $x \cdot x \cdot x$.

In Exercises 6 through 14 simplify each binary Boolean function.

6. $x \cdot (x + 1)$
7. $x \cdot (x + 0) + x$
8. $x \cdot x + y$
9. $y \cdot (x + x \cdot y)$
10. $(x + y) \cdot (x + z)$
11. $x + x \cdot x + x \cdot y$
12. $(x \cdot x) + (y \cdot z) + y$
13. $x + y \cdot (x \cdot x + x \cdot y)$
14. $x \cdot (y + x \cdot z) \cdot (y + z)$

3.5

Switching Circuits

A mechanical or electrical system (network) consists of (1) a circuit, or path, along which a fluid, gas, or electrical energy flows, and (2) switches that control the flow. Electronic digital computers, dial-telephone switching systems, and networks of pumping stations and pipes for natural gas or oil are examples of such systems. Simply stated, mechanical or electrical switches are *two-state (bistable) devices*; that is, such switches are either "closed" or "open." When closed, the fluid, gas, or electrical current is allowed to pass, and when open, the circuit is interrupted and no flow exists along that path.

Binary Boolean algebra may be used to study the algebra of switching circuits because the elements 0 and 1 can be used to represent open and closed switches, respectively. The operations of binary Boolean addition and multiplication can be used to represent whether the switches are connected in *parallel* or *series form*, respectively.

Consider two electrical switches, denoted by x and y, connected in *parallel form*. Four possible conditions exist, as illustrated in Figs. 3.2 to 3.5.

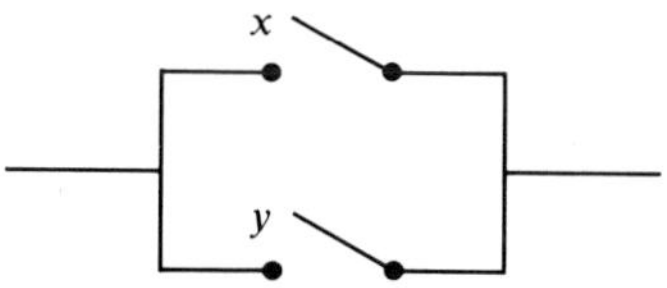

Switch x is open
Switch y is open

Figure 3.2

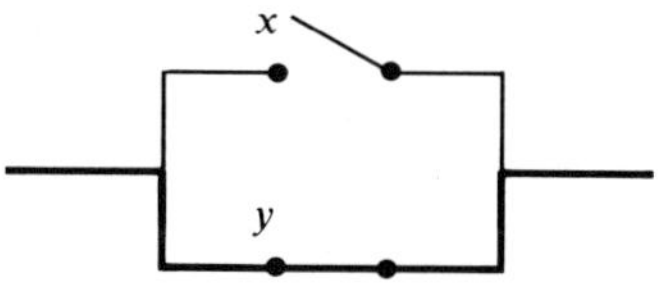

Switch x is open
Switch y is closed

Figure 3.3

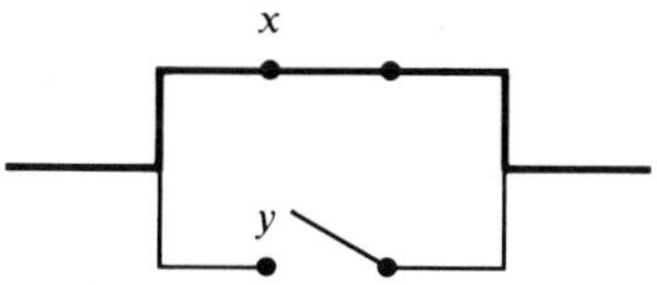

Switch x is closed
Switch y is open

Figure 3.4

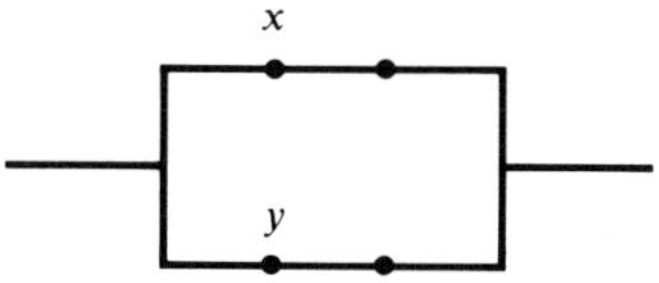

Switch x is closed
Switch y is closed

Figure 3.5

It is assumed that the ends of each circuit are connected to appropriate power sources and output devices. A circuit can be open (broken or interrupted) or closed only at a switch. Both switches x and y are shown open in Fig. 3.2; hence the circuit is open and no electricity flows. If either switch x or switch y is closed, as in Figs. 3.3 to 3.5, then the circuit is closed and electricity flows. Table 3.3 summarizes the properties of the parallel switches x and y.

Table 3.3

x	y	Parallel circuit
Open	Open	Open
Open	Closed	Closed
Closed	Open	Closed
Closed	Closed	Closed

Note that Table 3.3 is similar in structure to the binary Boolean addition table in Sec. 3.2, where

0	is replaced by	open
1	is replaced by	closed

Consider now the two electrical switches x and y connected in *series form*. Once again, four possible conditions exist, as illustrated in Figs. 3.6 to 3.9.

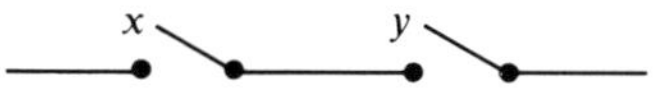

Switch x is open
Switch y is open

Figure 3.6

Switch x is open
Switch y is closed

Figure 3.7

Switch x is closed
Switch y is open

Figure 3.8

Switch x is closed
Switch y is closed

Figure 3.9

When two switches x and y are connected in series form, the circuit will be open and no electricity flows if x or y is open. Hence the circuits in Figs. 3.6 to 3.8 are open. If both switches x and y are closed, as in Fig. 3.9, the circuit is closed and electricity flows. Table 3.4 summarizes the properties of the series switches x and y.

Table 3.4

x	y	Series circuit
Open	Open	Open
Open	Closed	Open
Closed	Open	Open
Closed	Closed	Closed

Note that Table 3.4 is similar in structure to the binary Boolean multiplication table in Sec. 3.2 where, again, 0 is replaced by open and 1 is replaced by closed.

Exercises

Consider the following networks of electrical switches:

(*a*)

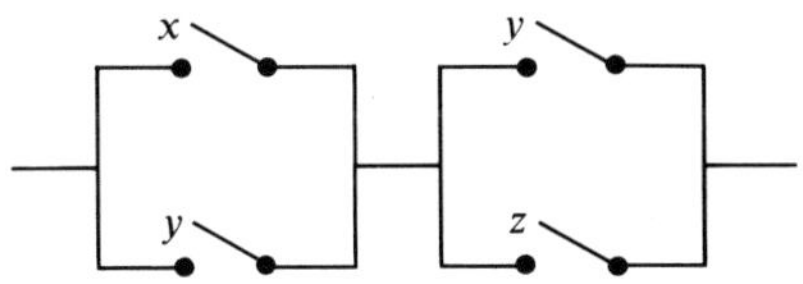

(*b*)

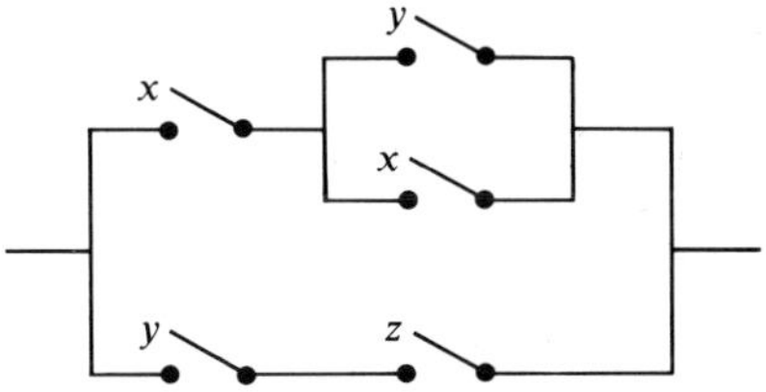

Note that if two switches operate so as to be open or closed simultaneously, we shall denote them by the same letter. In Exercises 1 through 5 determine whether or not there is a flow of electricity in each network under the following conditions:

1. Switch x is closed; all other switches are open.
2. Switch y is closed; all other switches are open.
3. Switch z is closed; all other switches are open.
4. Switches x and z are closed; switch y is open.
5. Switches y and z are closed; switch x is open.
6. Draw a network of switching circuits in which the three switches x, y, and z are all in parallel form.
7. Draw a network of switching circuits in which the three switches x, y, and z are all in series form.
8. Draw a network of switching circuits in which switches x and y are in parallel form and both switches are in series form with switch z.

3.6

The Algebra of Switching Circuits

The binary Boolean arithmetic table and the switching circuit tables are abstractly identical. That is, there is an isomorphism between binary Boolean algebra and the algebra of switching circuits, where

$$0 \leftrightarrow \text{open}$$
$$1 \leftrightarrow \text{closed}$$
$$\text{Addition } (+) \leftrightarrow \text{parallel circuit}$$
$$\text{Multiplication } (\cdot) \leftrightarrow \text{series circuit}$$

Hereafter, when discussing the algebra of switching circuits, we shall denote a switch by a single lowercase letter such as w, x, y, or z. If two switches operate so as to be open or closed simultaneously, we shall denote them by the same letter.

A switch has the* value 0 *when open and the* value 1 *when closed.

A switch that is always open is represented by 0; a switch that is always closed is represented by 1.

A* parallel circuit *(that is, a circuit consisting of two switches* x *and* y *connected in parallel form) will be denoted by* $x + y$*. A* series circuit *will be denoted by* $x \cdot y$*.

A circuit has the value 0 when open and the value 1 when closed.

Because each binary Boolean function corresponds to a combination of parallel and series circuits, each function is said to *represent* a switching circuit. Conversely, because each switching circuit corresponds to a binary Boolean function, each switching circuit is said to *actualize* a function.

Example 1

Represent the switching circuit in Fig. 3.10 by a binary Boolean function.

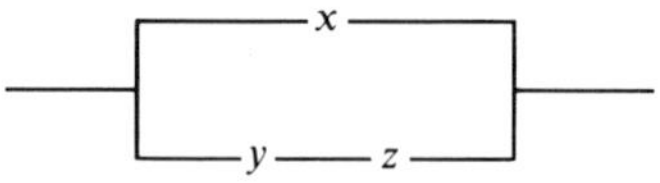

Figure 3.10

Since switch x is in parallel form with the series of switches y and z,

$$x + y \cdot z$$

is the binary Boolean function representing the switching circuit.

Switches are generally the more expensive items of equipment in most switching networks, so if possible it is usually desirable to reduce the number of switches used. Furthermore, because mechanical failure of switches is frequently the cause of circuit breakdown or interruption, a reduction in the number of switches lessens the chance of such an occurrence. For these reasons it is desirable to simplify the switching of networks by reducing the number of switches used.

Two switching circuits are said to be* equivalent *if and only if the values of the two circuits are equal for all possible values of the switches.

Examples 2 and 3 illustrate the use of binary Boolean algebra in finding simpler equivalent switching circuits.

Example 2

Simplify the switching network shown in Fig. 3.11.

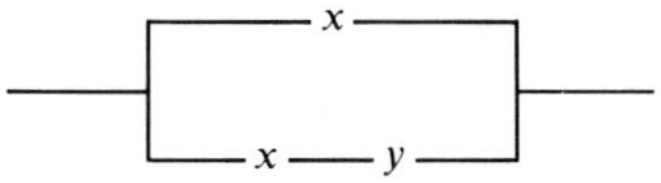

Figure 3.11

The binary Boolean function that represents the switching network is

$$x + x \cdot y$$

By the absorption property of addition, this function is equal to the simpler function x. Therefore the switching network in Fig. 3.12 is equivalent to that shown in Fig. 3.11.

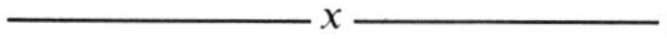

Figure 3.12

Note that in Fig. 3.11 the parallel circuit is closed if and only if switch x is closed. Whether or not switch y is closed is not important with respect to the closure conditions of the circuit.

Example 3

Use the distributive property of binary Boolean algebra to simplify the switching circuit shown in Fig. 3.13.

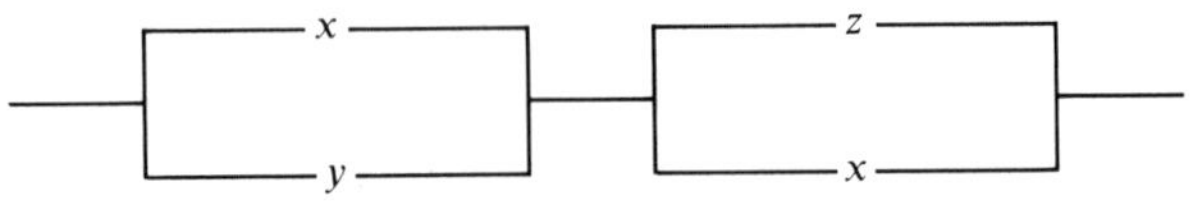

Figure 3.13

The binary Boolean function that represents the switching circuit is

$$(x + y) \cdot (x + z)$$

By the distributive property,

$$x + (y \cdot z) = (x + y) \cdot (x + z)$$

The switching circuit that actualizes the function $x + (y \cdot z)$ is shown in Fig. 3.14 and is equivalent to the given switching circuit. Note that only three switches instead of four are required.

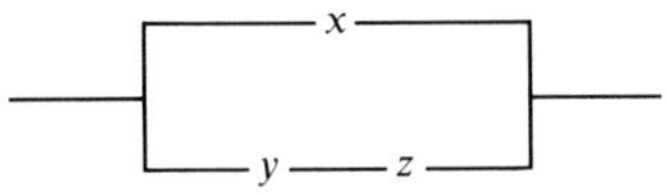

Figure 3.14

Example 4

Simplify the switching network shown in Fig. 3.15.

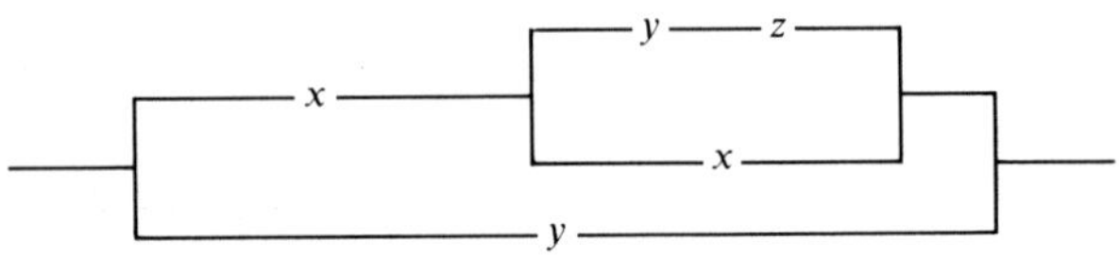

Figure 3.15

We first represent the switching network by a binary Boolean function, as shown in Fig. 3.16.

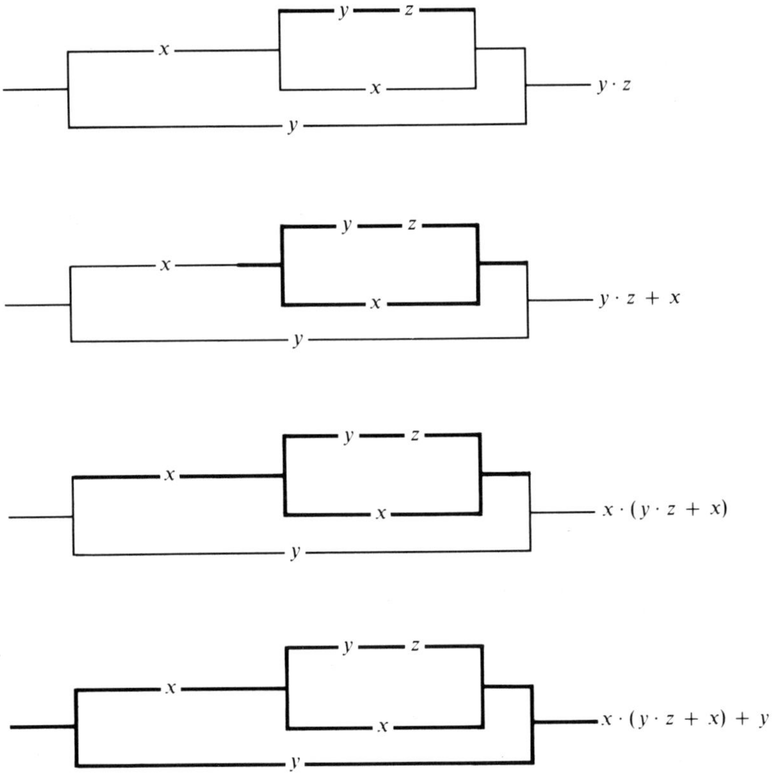

Figure 3.16

Thus, $x \cdot (y \cdot z + x) + y$ represents the switching network. Now,

$$x \cdot (y \cdot z + x) + y = x + y$$

by the absorption property of multiplication. Since the switching network in Fig. 3.17 actualizes the function $x + y$, it represents a simplified equivalent network to that shown in Fig. 3.15.

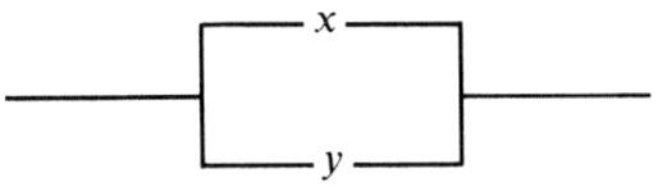

Figure 3.17

Exercises

In Exercises 1 through 6 represent each switching circuit by a binary Boolean function.

1. 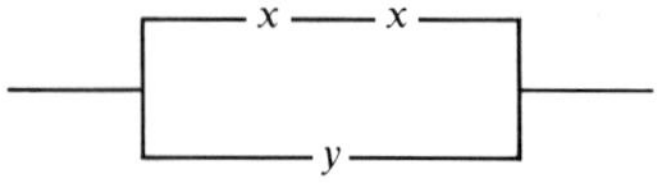

2. 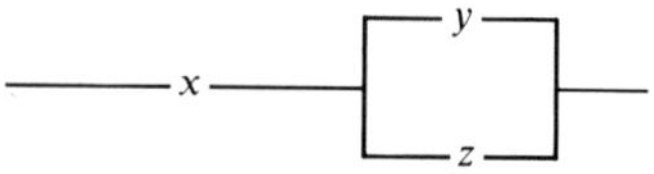

3. 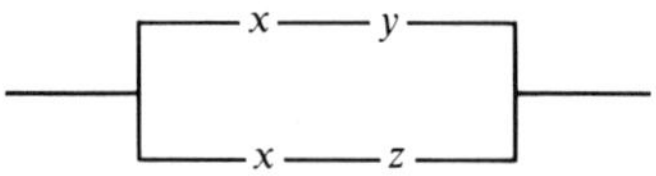

4. 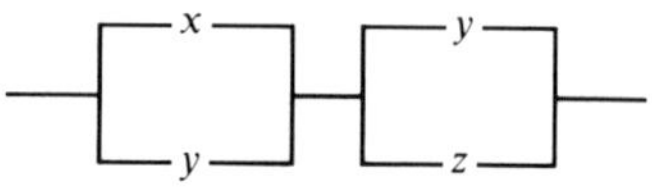

5. 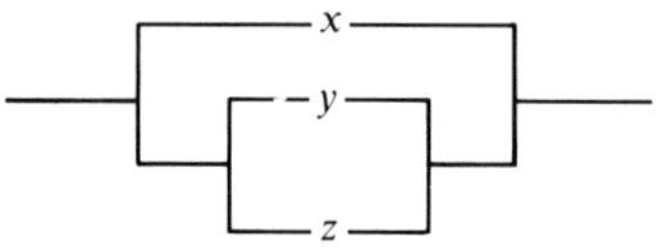

6. 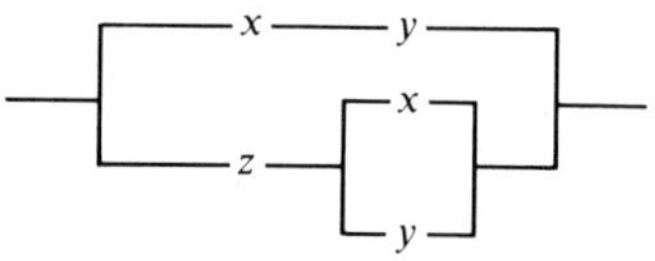

In Exercises 7 through 15 simplify each switching network.

7. The network shown in Exercise 1
8. The network shown in Exercise 3
9. The network shown in Exercise 4

10. 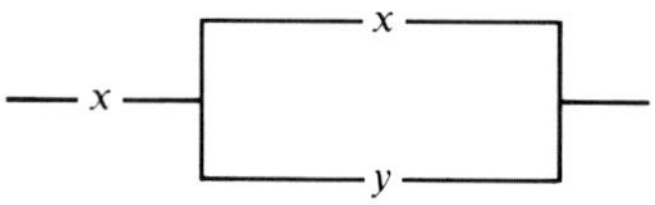

11. 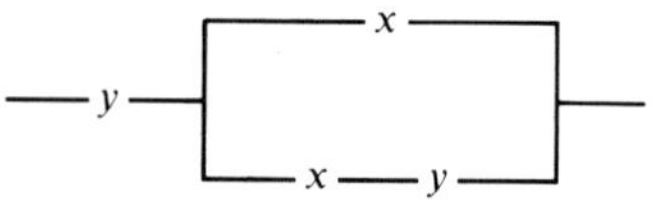

12. 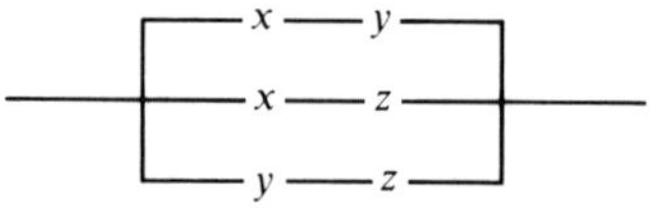

13. 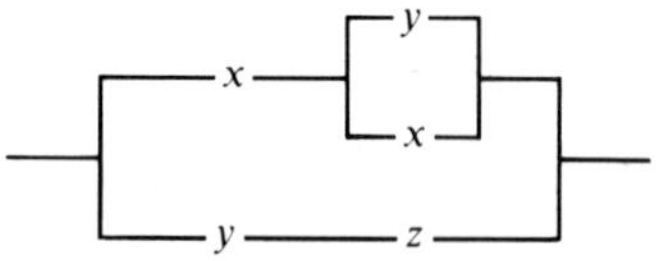

14. 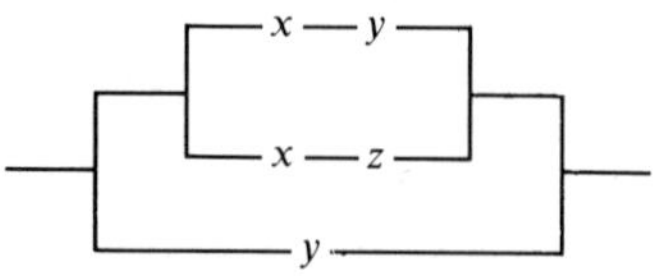

15. 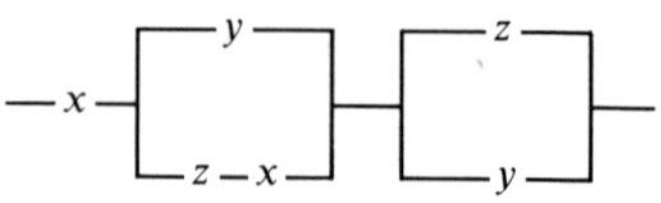

3.7

Complements in Binary Boolean Algebra

Consider the variable x in binary Boolean algebra; the only possible values of x are 0 and 1. It is convenient to consider a new variable, denoted by x', such that the value of x' is 1 when the value of x is 0; the value of x' is 0 when the value of x is 1. Each of the variables x and x' is the *complement* of the other. Note that generally the complement of the complement of a variable x is simply x; that is,

$$(x')' = x$$

Example 1

Show that

(*a*) $x + x' = 1$
(*b*) $x \cdot x' = 0$

(a)

x	x'	$x + x'$
0	1	**1**
1	0	**1**

$x + x' = 1$

(b)

x	x'	$x \cdot x'$
0	1	**0**
1	0	**0**

$x \cdot x' = 0$

Since $x + x' = 1$ for all possible values of x, we could have shown that $x \cdot x' = 0$ by using the principle of duality instead of examining a table of possible values of x.

Example 2

Show that

$$x \cdot (x' + y) = x \cdot y$$

$x \cdot (x' + y) = x \cdot x' + x \cdot y$	Distributive property
$= 0 + x \cdot y$	Example 1
$= x \cdot y$	Example 1(*b*) of Sec. 3.4.

Alternate approach:

x	y	x'	$x' + y$	$x \cdot (x' + y)$	$x \cdot y$
0	0	1	1	**0**	**0**
0	1	1	1	**0**	**0**
1	0	0	0	**0**	**0**
1	1	0	1	**1**	**1**

The four possible combinations of replacement values for x and y are considered. Because columns five and six are identical,

$$x \cdot (x' + y) = x \cdot y$$

By the principle of duality and the results of Example 2, it follows that

$$x + x' \cdot y = x + y$$

Example 3

Show that

$$(x + y)' = x' \cdot y'$$

that is, the complement of the sum of two terms in a binary Boolean algebra is the product of the complements of each term.

x	y	$x + y$	$(x + y)'$	x'	y'	$x' \cdot y'$
0	0	0	**1**	1	1	**1**
0	1	1	**0**	1	0	**0**
1	0	1	**0**	0	1	**0**
1	1	1	**0**	0	0	**0**

By the principle of duality and the results of Example 3, it follows that

$$(x \cdot y)' = x' + y'$$

that is, the complement of the product of two factors in a binary Boolean algebra is the sum of the complements of each factor.

The results of Example 3 suggest a general rule for finding the complement of any binary Boolean function, which we now state without proof:

In binary Boolean algebra, the complement of a function can be obtained by interchanging the operations of + and · throughout while also interchanging each variable and its complement.

Example 4

Find the complement of the binary Boolean function

$$x + y \cdot z'$$

$(x + y \cdot z')' = x' \cdot (y \cdot z')'$	The complement of the sum of two terms is the product of the complements of each term.
$= x' \cdot [y' + (z')']$	The complement of the product of two factors is the sum of the complements of each factor.
$= x' \cdot (y' + z)$	The complement of the complement of a variable is that variable.

Exercises

In Exercises 1 through 4 find the value of each binary Boolean function when (*a*) $x = 0$, $y = 1$, $z = 0$; (*b*) $x = 1$, $y = 0$, $z = 1$; (*c*) $x = 1$, $y = 1$, $z = 0$; (*d*) $x = 1$, $y = 1$, $z = 1$.

1. $(x' + 1) \cdot y'$
2. $x' \cdot y' + z$

3. $x \cdot (x + y)' + z$
4. $[(x + y) \cdot z' + 1] \cdot (x' + y)$

In Exercises 5 through 9 use a table of values to demonstrate each equivalence.

5. $(x + y) \cdot (x' + z) = x \cdot z + x' \cdot y$
6. $x \cdot (x + y)' = 0$
7. $x \cdot y + x' \cdot z + y \cdot z = x \cdot y + x' \cdot z$
8. $(x + y) \cdot (x' + z) \cdot (y + z) = (x + y) \cdot (x' + z)$
9. $x \cdot y + y \cdot z + z \cdot x = (x + y) \cdot (y + z) \cdot (z + x)$
10. Use the results of Exercise 7 and the principle of duality to prove the results of Exercise 8.
11. Prove that if $a + x = b + x$ and $a + x' = b + x'$, then $a = b$.
12. Prove that if $a \cdot x = b \cdot x$ and $a \cdot x' = b \cdot x'$, then $a = b$.

In Exercises 13 through 17 simplify each binary Boolean function.

13. $x \cdot (x + x')$
14. $(x + y) \cdot (x' + y')$
15. $(x + y) \cdot (x + y')$
16. $x \cdot y' + x \cdot z + x \cdot y$
17. $(x \cdot y \cdot z) + (x \cdot y' \cdot z) + (x' \cdot y' \cdot z)$

In Exercises 18 through 24 find the complement of each binary Boolean function.

18. $x' + y'$
19. $x' \cdot y'$
20. $w + x + y + z$
21. $w \cdot x \cdot y \cdot z$
22. $x \cdot (y + z')$
23. $x \cdot (y + 1)$
24. $x \cdot y + 1$
25. An isomorphism between binary Boolean algebra and statements involving conjunctions and disjunctions exists:

$0 \leftrightarrow$ open $\leftrightarrow$ false statement
$1 \leftrightarrow$ closed $\leftrightarrow$ true statement
Addition $(+) \leftrightarrow$ parallel circuit $\leftrightarrow$ "or"
Multiplication $(\cdot) \leftrightarrow$ series circuit $\leftrightarrow$ "and"

What connective in logic is represented by the binary Boolean function $(x + y) \cdot (x \cdot y)'$?

3.8

More About Switching Circuits

The operation of finding the complement of an element in binary Boolean algebra has useful applications to switching circuit theory. A pair of switches can be designed so that when one switch is open, the second switch is closed; when the first switch is closed, the second switch is open. Each switch is called the *negative* of the other. For example, the parallel circuit in Fig. 3.18 is always closed (electrical current flows) because either switch x or switch x' must be closed. The switching network in Fig. 3.18 demonstrates that the value of the binary Boolean function $x + x'$ is always 1, as noted in Example 1 of Sec. 3.7.

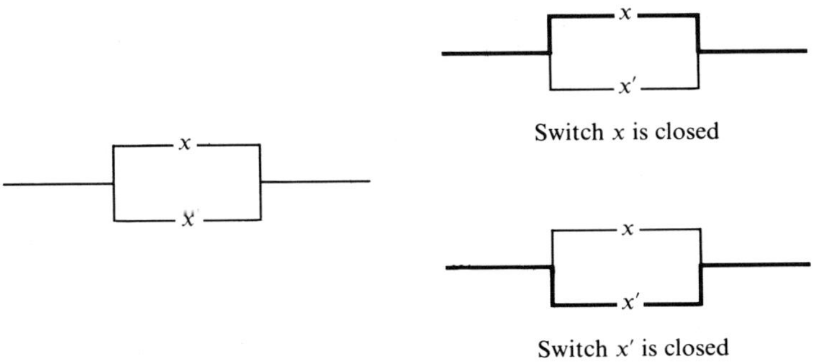

Figure 3.18

The series circuit in Fig. 3.19 is always open (electrical current does not flow) because either switch x or switch x' must be open. The switching network in Fig. 3.19 demonstrates that the value of the binary Boolean function $x \cdot x'$ is always 0, as noted in Example 1 of Sec. 3.7.

Figure 3.19

Example 1

Find the binary Boolean function that represents the switching network in Fig. 3.20.

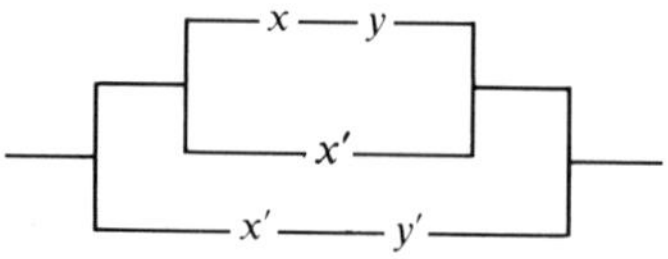

Figure 3.20

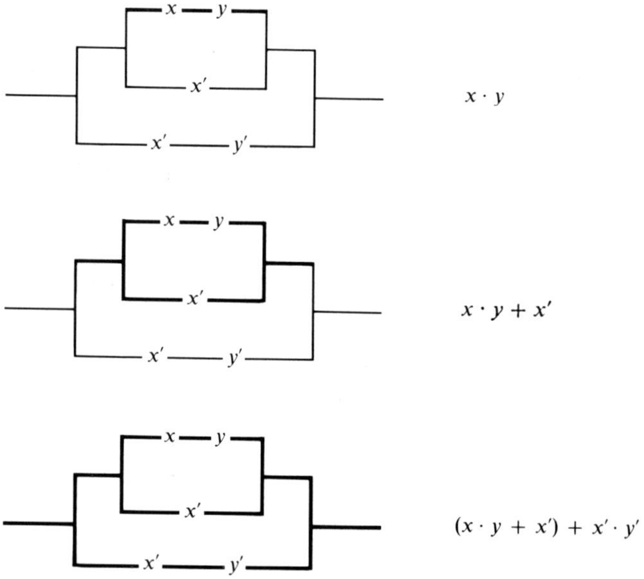

Figure 3.21

Thus, $(x \cdot y + x') + x' \cdot y'$ represents the switching network in Fig. 3.20.

Example 2

Simplify the switching network in Fig. 3.20 of Example 1.

$(x \cdot y + x') + x' \cdot y' = x \cdot y + (x' + x' \cdot y')$	Associative property (+)
$= x \cdot y + x'$	Absorption property (+)
$= x' + x \cdot y$	Commutative property (+)
$= x' + y$	Results of Example 2 of Sec. 3.7 and the principle of duality

The switching circuit that actualizes the function $x' + y$ is shown in Fig. 3.22 and is equivalent to the switching circuit in Fig. 3.20.

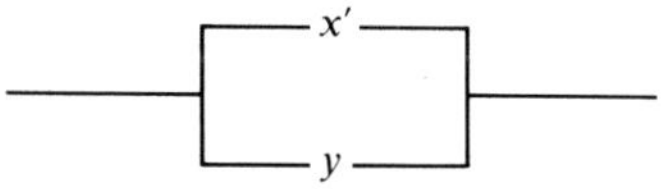

Figure 3.22

Example 3

Simplify the switching network in Fig. 3.23.

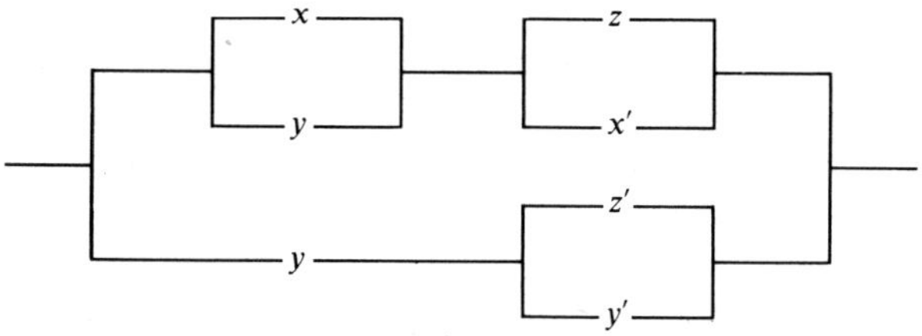

Figure 3.23

The binary Boolean function that represents the switching network is

$$(x + y) \cdot (z + x') + y \cdot (z' + y')$$

This function is equivalent to

$$(x + y) \cdot (z + x') + y \cdot z' + y \cdot y'$$
$$(x + y) \cdot (z + x') + y \cdot z'$$
$$(x \cdot z + x \cdot x' + y \cdot z + y \cdot x') + y \cdot z'$$
$$x \cdot z + y \cdot z + y \cdot x' + y \cdot z'$$
$$x \cdot z + y \cdot x' + y \cdot (z + z')$$
$$x \cdot z + y \cdot x' + y$$
$$x \cdot z + y$$

The switching network that actualizes the binary Boolean function $x \cdot z + y$ is a simpler equivalent network, as shown in Fig. 3.24.

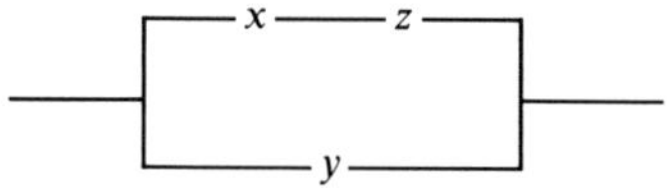

Figure 3.24

Exercises

In Exercises 1 through 4 find the binary Boolean function that represents each switching circuit.

1.

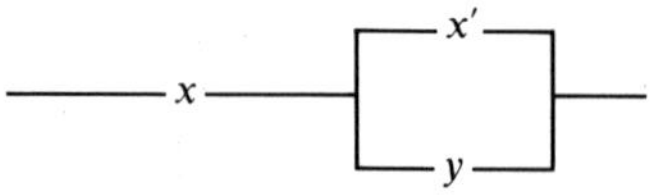

2.

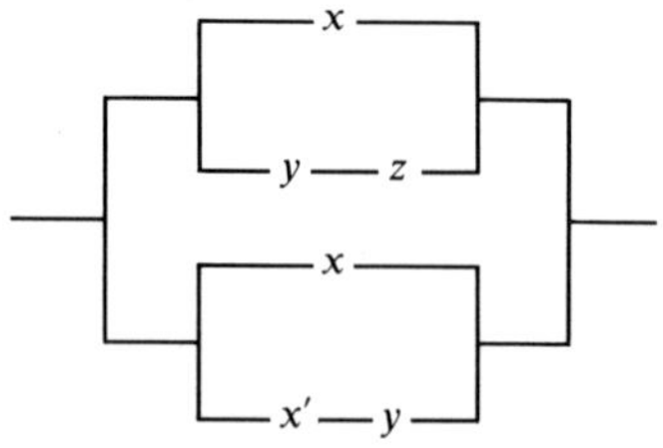

3.

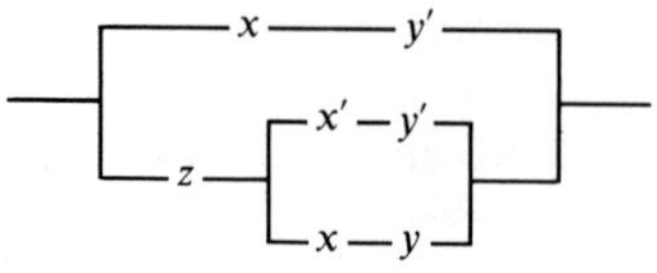

4.

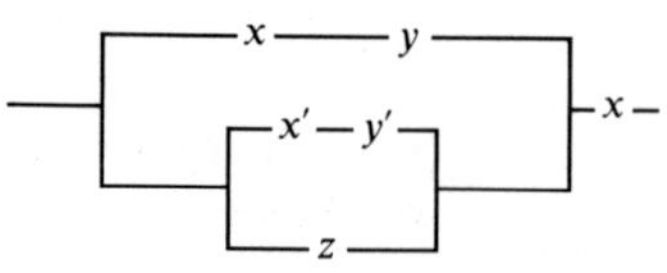

In Exercises 5 through 12 simplify each switching circuit.

5. The network shown in Exercise 1
6. The network shown in Exercise 2
7. The network shown in Exercise 3
8. The network shown in Exercise 4

9.

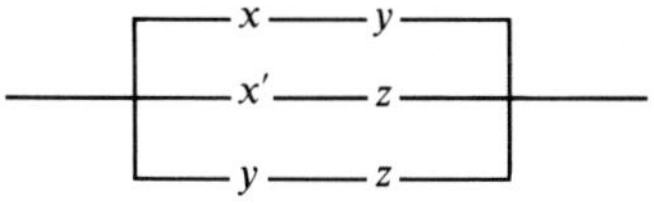

10.

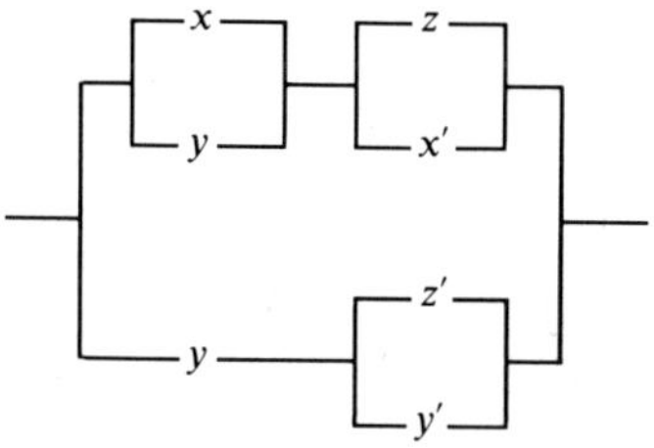

11.

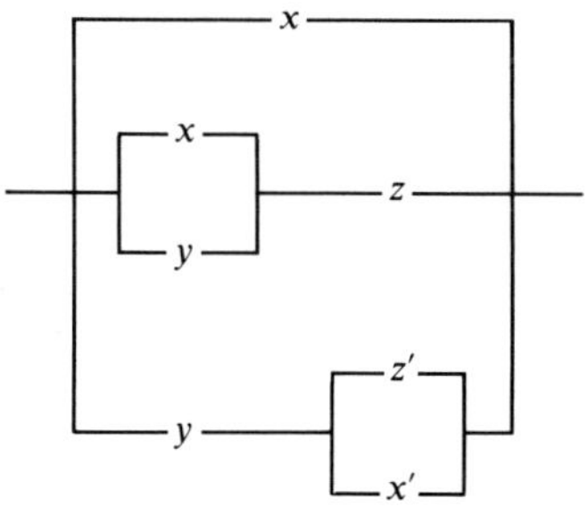

12.

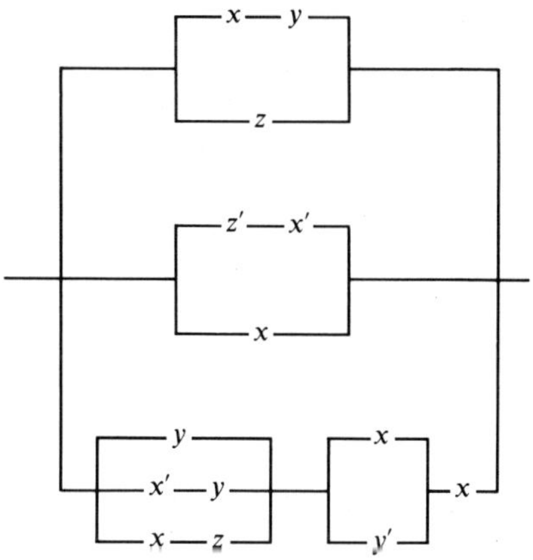

13. Demonstrate that the following two networks are equivalent:

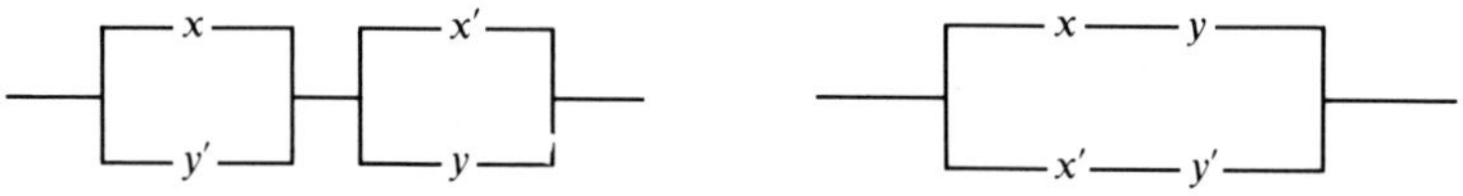

3.9

Design of Switching Circuits

Two types of problems may be solved by applying binary Boolean algebra to switching circuits. The first is the simplification of a given network—this problem was considered in previous sections of this chapter. The second problem is that of designing a switching circuit with given closure conditions.

In some cases, when the closure conditions are very simple, we may be able to write down the design of the network immediately by using our knowledge of series and parallel forms. For example, suppose we want to design a network containing three switches x, y, and z such that the circuit is closed provided x is closed and either y or z is closed. The switching circuit required is represented by $x \cdot (y + z)$; the network is shown in Fig. 3.25.

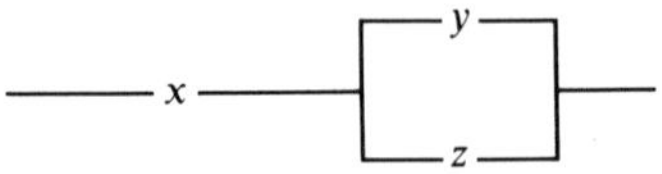

Figure 3.25

Unfortunately, most switching circuit designs cannot be so readily obtained. We shall illustrate a general method of designing a network to satisfy certain given closure conditions by considering those networks with at most three types of switches, x, y, and z, and their negatives.

A binary Boolean function is defined whenever the values of the function are known for all possible values of the variables. For example, $B(x, y)$ is completely specified when the values $B(0, 0)$, $B(0, 1)$, $B(1, 0)$, and $B(1, 1)$ are known. It can be shown that every binary Boolean function $B(x, y)$ can be written in the following *standard* or *canonical form*:

$$B(x, y) = B(1, 1) \cdot x \cdot y + B(1, 0) \cdot x \cdot y' + B(0, 1) \cdot x' \cdot y + B(0, 0) \cdot x' \cdot y'$$

Example 1

Write the binary Boolean function $B(x, y)$ whose values are given in the following table:

x	y	$B(x, y)$
1	1	1
1	0	1
0	1	0
0	0	1

Since $B(1, 1) = 1$, $B(1, 0) = 1$, $B(0, 1) = 0$, and $B(0, 0) = 1$, then

$$B(x, y) = x \cdot y + x \cdot y' + x' \cdot y'$$

This function can be simplified by using the properties of binary Boolean algebra:

$B(x, y) = x \cdot (y + y') + x' \cdot y'$	Distributive property
$= x \cdot 1 + x' \cdot y'$	Example 1 of Sec. 3.7
$= x + x' \cdot y'$	Example 1 of Sec. 3.4
$= x + y'$	Example 2 of Sec. 3.7 and the principle of duality

The switching circuit that actualizes this binary Boolean function is shown in Fig. 3.26.

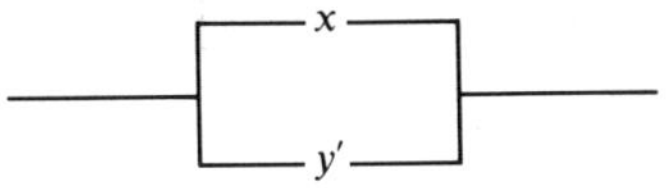

Figure 3.26

Example 2

Design a switching circuit that contains two switches and a light bulb so that the bulb is controlled from either switch.

Let x and y represent the switches. Consider the switches closed (the values of x and y are both 1) and the light on [the value of $B(1, 1)$ is 1]. The first row of the table of values of $B(x, y)$ has been arbitrarily constructed. Now, if the state of a single switch is changed, then the light goes out [the circuit is open and the value of $B(x, y)$ is 0]. This information is indicated in rows two and three of the table. Finally, row four indicates a change of state of a single switch from the positions of rows two or three. Thus, when both switches are open (the values of x and y are both 0), the light is on again [the circuit is closed and the value of $B(0, 0)$ is 1].

x	y	$B(x, y)$
1	1	1
1	0	0
0	1	0
0	0	1

From the table,

$$B(x, y) = x \cdot y + x' \cdot y'$$

The circuit design required is shown in Fig. 3.27.

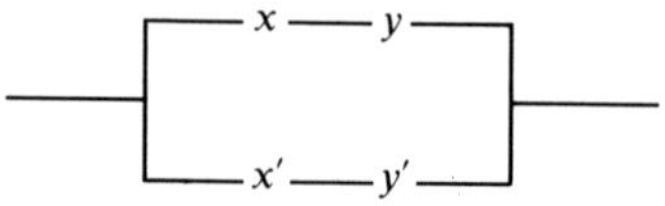

Figure 3.27

In actual practice the circuit is constructed using two single-pole, double-throw switches, as shown in Fig. 3.28.

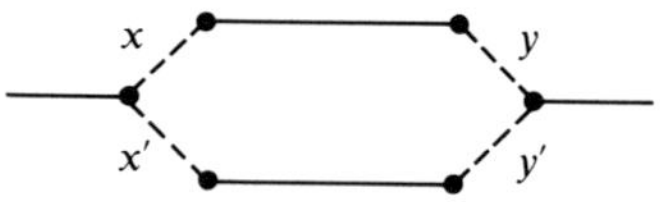

Figure 3.28

It can be shown that every binary Boolean function $B(x, y, z)$ can be written in canonical form, using eight terms:

$$\begin{aligned} B(x, y, z) = {} & B(1, 1, 1) \cdot x \cdot y \cdot z + B(1, 1, 0) \cdot x \cdot y \cdot z' \\ & + B(1, 0, 1) \cdot x \cdot y' \cdot z + B(1, 0, 0) \cdot x \cdot y' \cdot z' \\ & + B(0, 1, 1) \cdot x' \cdot y \cdot z + B(0, 1, 0) \cdot x' \cdot y \cdot z' \\ & + B(0, 0, 1) \cdot x' \cdot y' \cdot z + B(0, 0, 0) \cdot x' \cdot y' \cdot z' \end{aligned}$$

Example 3

Design the following switching circuit using three switches, x, y, and z. If x is closed, the network is closed except when y and z are both open. If x is open, the network will be closed only when y is closed.

Let $B(x, y, z)$ be the binary Boolean function representing the network. The values of $B(x, y, z)$ are given in the following table:

x	y	z	$B(x, y, z)$
1	1	1	1
1	1	0	1
1	0	1	1
1	0	0	0
0	1	1	1
0	1	0	1
0	0	1	0
0	0	0	0

Note that the first four rows of the table indicate the network is closed [$B(x, y, z)$ has the value 1] when x is closed ($x = 1$) except when both y and z are open ($y = z = 0$). The last four rows of the table indicate that the network is closed [$B(x, y, z)$ has the value 1] only when x is open ($x = 0$) and y is closed ($y = 1$).

The function representing the network can now be written in canonical form and simplified:

$$\begin{aligned} B(x, y, z) &= x \cdot y \cdot z + x \cdot y \cdot z' + x \cdot y' \cdot z \\ &\quad + x' \cdot y \cdot z + x' \cdot y \cdot z' \\ &= x \cdot y \cdot (z + z') + x \cdot y' \cdot z \\ &\quad + x' \cdot y \cdot (z + z') \\ &= x \cdot y + x \cdot y' \cdot z + x' \cdot y \\ &= (x + x') \cdot y + x \cdot y' \cdot z \\ &= y + x \cdot y' \cdot z \\ &= y + x \cdot z \end{aligned}$$

The switching circuit represented by this binary Boolean function is shown in Fig. 3.29.

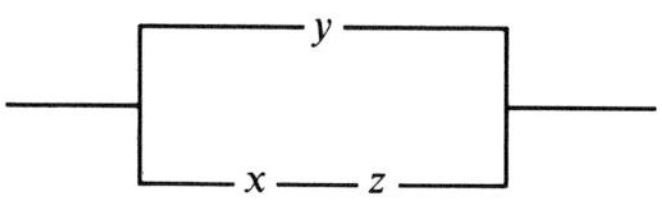

Figure 3.29

Exercises

In Exercises 1 and 2 write the binary Boolean function $B(x, y)$, whose values are given in each table, and simplify the function.

1.

x	y	$B(x, y)$
1	1	1
1	0	1
0	1	1
0	0	0

2.

x	y	$B(x, y)$
1	1	1
1	0	1
0	1	0
0	0	1

In Exercises 3 and 4 write the binary Boolean function $B(x, y, z)$, whose values are given in each table, and simplify the function.

3.

x	y	z	$B(x, y, z)$
1	1	1	1
1	1	0	1
1	0	1	1
1	0	0	0
0	1	1	0
0	1	0	0
0	0	1	0
0	0	0	1

4.

x	y	z	$B(x, y, z)$
1	1	1	0
1	1	0	1
1	0	1	1
1	0	0	0
0	1	1	1
0	1	0	0
0	0	1	0
0	0	0	0

In Exercises 5 through 11 design the indicated switching circuit.

5. The network represented by $B(x, y)$ in Exercise 1.
6. The network represented by $B(x, y)$ in Exercise 2.
7. The network represented by $B(x, y, z)$ in Exercise 3.
8. The network represented by $B(x, y, z)$ in Exercise 4.
9. A network using three switches, x, y, and z. If x is closed, the network is closed except when y or z (but not both) are closed. If x is open, the network is closed except when y and z are both closed or both open.
10. A network using three switches, x, y, and z. The network is closed if at least two of the three switches are closed; otherwise the network is open. (The network represents a simple majority voting machine.)
11. A network using four switches, w, x, y, and z. The network is closed if w is closed and any of the other three switches is closed. The network is open under all other conditions.

In Exercises 12 and 13 write each binary Boolean function $B(x, y)$ in canonical form.

12. $x + y'$
13. $x + x' \cdot y$

In Exercises 14 and 15 write each binary Boolean function $B(x, y, z)$ in canonical form.

14. $x \cdot y' + x \cdot z + x \cdot y$
15. $(x + y) \cdot (x + y') \cdot (x' + z)$
16. Write the 16 possible binary Boolean functions $B(x, y)$.
17. Show that the binary Boolean function $B(x, y)$ has x as a factor if x has the value 1 in every row for which the function is 1.
18. Repeat Exercise 17 for $B(x, y, z)$.
19. Show that the binary Boolean function

$$B(x, y) = (x + y') \cdot (x' + y)$$

also has the desired properties of Example 2. Design the network represented by $B(x, y)$.

20. Design a three-switch circuit for a secret voting machine for a panel consisting of the mayor and two councilmen. A light is turned on if a simple majority vote "yes" by pressing a button. However, the mayor has veto power.
21. Three partners, Tom, Dick, and Harry, hold an insurance policy which pays $100,000 to the partnership if either Tom or Dick dies, provided that either Tom or Harry also dies. Design a switching circuit that represents when the insurance money will be paid.
22. Use the network in Exercise 21 to determine the truth values of statements (*a*) to (*c*):
    (*a*) The insurance money is paid if and only if two of the three partners die.
    (*b*) The insurance money is paid if and only if Tom dies.
    (*c*) The insurance money is paid if Tom dies, or if both Dick and Harry die.
23. Switches x and y each control independently the flow of oil to a furnace. A third switch z stops the flow of oil automatically when the pilot light goes out. Design a network for switches x, y, and z so that either switch x or y will change the state of the flow of oil, except that oil will never flow when the pilot light is out.

3.10

Boolean Algebra

A symbolic formulation of the algebra of statements was developed by George Boole (1815–1864). The abstract mathematical system for which this algebra is but one of many useful models has become known as *Boolean algebra.* The algebra of sets is another model of Boolean algebra. In this chapter we studied binary Boolean algebra and the algebra of switching circuits; both algebras are very special models of Boolean algebra. In 1904, the British mathematician E. V. Huntington set forth the definition and required properties of a Boolean algebra:

Basic Properties of a Boolean Algebra

In the following statements let a, b, and c be any elements of some set β; let $\oplus$ and $\odot$ be two operations defined on β:

Closure properties: $a \oplus b \in \beta$ and $a \odot b \in \beta$

Commutative properties: $a \oplus b = b \oplus a$ and $a \odot b = b \odot a$

Existence of identity elements: There exists an element e such that $a \oplus e = e \oplus a = a$; there exists an element μ such that $a \odot \mu = \mu \odot a = a$.

Distributive properties: $a \oplus (b \odot c) = (a \oplus b) \odot (a \oplus c)$ and $a \odot (b \oplus c) = (a \odot b) \oplus (a \odot c)$

Existence of complements: For each a there exists an element a' such that $a \oplus a' = a' \oplus a = \mu$ and $a \odot a' = a' \odot a = e$.

Whenever we specify the elements of the set β and describe the operations $\oplus$ and $\odot$, we are placing an interpretation on the system. If the above properties are satisfied, then the interpretation is a model of Boolean algebra. For example, let β be the set of all sets; that is, a, b, and c are any three sets. Furthermore, let $\oplus$ represent the union operation and $\odot$ represent the intersection operation. Since the union of two sets is a set, and the intersection of two sets is a set, the closure property is valid. We already showed in Chap. 2 that both the operations of union and intersection are commutative and distributive. Furthermore, the null set plays the role of e; the universal set β plays the role of μ. The complement of each set exists and satisfies the last required property of a Boolean algebra.

Keyed Summary

In this chapter we studied binary Boolean arithmetic and functions and their applications to switching circuits. The concepts and methods presented are part of the body of knowledge called *Boolean algebra.*

3.1 Isomorphisms

An isomorphism is a one-to-one correspondence between sets that preserves the operations defined on them.

3.2 Binary Boolean Arithmetic

Table 3.1 Binary Boolean Addition Table

+	0	1
0	0	1
1	1	1

Table 3.2 Binary Boolean Multiplication Table

·	0	1
0	0	0
1	0	1

If one statement is obtained from another by an interchange of certain terms, then each statement is the dual of the other.

Principle of duality: In binary Boolean arithmetic, the dual of every statement obtained by interchanging the numbers 0 and 1 while also interchanging the operations $+$ and $\cdot$ is true if and only if the original statement is true.

3.3 Binary Boolean Functions

Two binary Boolean functions are equivalent if and only if the values of the functions are equal for all possible replacement values of the variables.

Absorption properties:

$$x \cdot (x + y) = x$$
$$x + (x \cdot y) = x$$

3.4 Properties of Binary Boolean Algebra

Idempotent properties:

$$x + x = x$$
$$x \cdot x = x$$

Commutative properties:

$$x + y = y + x$$
$$x \cdot y = y \cdot x$$

Associative properties:

$$(x + y) + z = x + (y + z)$$
$$(x \cdot y) \cdot z = x \cdot (y \cdot z)$$

Distributive properties:

$$x \cdot (y + z) = (x \cdot y) + (x \cdot z)$$
$$x + (y \cdot z) = (x + y) \cdot (x + z)$$

3.5 Switching Circuits

A mechanical or electrical system (network) consists of (1) a circuit, or path, along which a fluid, gas, or electrical energy flows, and (2) switches that control the flow.

Mechanical or electrical switches are two-state (bistable) devices; that is, such switches are either closed or open.

Two electrical switches, denoted by x and y, connected in parallel form:

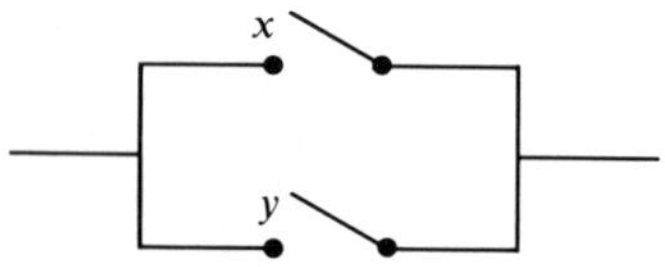

Figure 3.2

Properties of the parallel switches x and y:

Table 3.3

x	y	Parallel circuit
Open	Open	Open
Open	Closed	Closed
Closed	Open	Closed
Closed	Closed	Closed

Two electrical switches, denoted by x and y, connected in series form:

Figure 3.6

Properties of the series switches x and y:

Table 3.4

x	y	Series circuit
Open	Open	Open
Open	Closed	Open
Closed	Open	Open
Closed	Closed	Closed

3.6 The Algebra of Switching Circuits

There is an isomorphism between binary Boolean algebra and the algebra of switching circuits, where

$$\begin{aligned} 0 &\leftrightarrow \text{open} \\ 1 &\leftrightarrow \text{closed} \\ \text{Addition } (+) &\leftrightarrow \text{parallel circuit} \\ \text{Multiplication } (\cdot) &\leftrightarrow \text{series circuit} \end{aligned}$$

A switch or circuit has the value 0 when open and the value 1 when closed.

A parallel circuit is denoted by $x + y$; a series circuit is denoted by $x \cdot y$.

Each binary Boolean function represents a switching circuit; each switching circuit actualizes a function.

Two switching circuits are equivalent if and only if the values of the two circuits are equal for all possible values of the switches.

3.7 Complements in Binary Boolean Algebra

The variable denoted by x' is the complement of x provided the value of x' is 1 when the value of x is 0; the value of x' is 0 when the value of x is 1.

In binary Boolean algebra, the complement of a function can be obtained by interchanging the operations of $+$ and $\cdot$ throughout while also interchanging each variable and its complement.

3.8 More About Switching Circuits

A pair of switches can be designed so that when one switch is open, the second switch is closed; when the first switch is closed, the second switch is open. Each switch is called the negative of the other.

3.9 Design of Switching Circuits

Every binary Boolean function $B(x, y)$ can be written in the following standard or canonical form:

$$B(x, y) = B(1, 1) \cdot x \cdot y + B(1, 0) \cdot x \cdot y' + B(0, 1) \cdot x' \cdot y + B(0, 0) \cdot x' \cdot y'$$

Every binary Boolean function $B(x, y, z)$ can be written in the following standard or canonical form:

$$\begin{aligned} B(x, y, z) = {} & B(1, 1, 1) \cdot x \cdot y \cdot z + B(1, 1, 0) \cdot x \cdot y \cdot z' \\ & + B(1, 0, 1) \cdot x \cdot y' \cdot z + B(1, 0, 0) \cdot x \cdot y' \cdot z' \\ & + B(0, 1, 1) \cdot x' \cdot y \cdot z + B(0, 1, 0) \cdot x' \cdot y \cdot z' \\ & + B(0, 0, 1) \cdot x' \cdot y' \cdot z + B(0, 0, 0) \cdot x' \cdot y' \cdot z' \end{aligned}$$

3.10 Boolean Algebra

The algebra of statements, the algebra of sets, binary Boolean algebra, and the algebra of switching circuits are models of a more general mathematical system called Boolean algebra.

The basic properties of a Boolean algebra include the closure properties, the commutative properties, the existence of identity elements, the distributive properties, and the existence of complements.

Building Your Mathematical Vocabulary

Absorption properties (3.3)
Associative properties (3.4)
Binary Boolean arithmetic (3.2)
Binary Boolean functions (3.3)
Bistable devices (3.5)
Boolean algebra (3.4, 3.10)
Canonical form (3.9)
Commutative properties (3.4)
Complement (3.7)
Distributive properties (3.4)
Dual (3.2)
Equivalent (3.3, 3.6)
Idempotent properties (3.4)
Isomorphism (3.1)
Negative (3.8)
One-to-one correspondence (3.1)
Parallel circuit (3.6)
Parallel form (3.5)
Principle of duality (3.2)
Series circuit (3.6)
Series form (3.5)
Standard form (3.9)
Two-state devices (3.5)
Value (3.6)

4

*Elements of Probability**

4.1

First Fundamental Principle of Enumeration

Many problems that require the counting of ways in which elements of one or more sets may be selected arise in mathematics (as in everyday affairs). For example, how many three-digit positive even integers can be written using the digits 1, 3, 4, and 8 with no digits repeated? In how many ways can a basketball coach form a team if there are three candidates on the squad for each of the five positions? How many different automobile license plates are possible if the plates for a particular state are identified by two letters of the English alphabet followed by a three-digit number greater than 99? To answer these questions we need to develop principles of enumeration or counting that do not involve the listing of all possibilities.

Example 1

It is possible to drive from town X to town Y by means of three routes a, b, and c; it is possible to drive from town Y to town Z by means of two routes d and e. In how many ways can a person travel from town X to town Z by way of town Y?

We first need to select one member from a set of three elements $\{a, b, c\}$, which represents the routes from town X to town Y; we then need to also select one member from a set of two elements $\{d, e\}$, which represents the routes from town Y to town Z.

---

* Portions of this chapter are from "Elements of Probability" in *Elementary Mathematics: Number Systems and Algebra* by Anthony J. Pettofrezzo and Donald W. Hight. Copyright © 1970 by Scott, Foresman and Company.

A convenient method for listing all possible sequences of routes is a *tree diagram*, as in Fig. 4.1. This tree diagram represents two successive selections. The number of the end branches of the tree is equal to the number of possible routes from town X to town Z by way of town Y. The diagram shows clearly that each of three

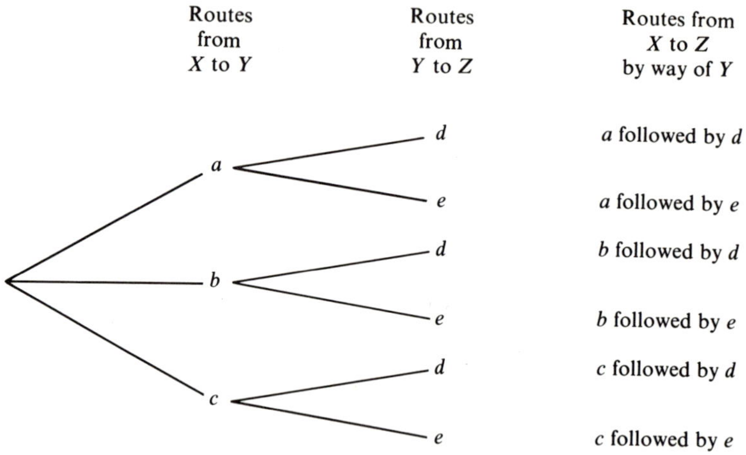

Figure 4.1

routes from X to Y can be followed by either one of two routes from Y to Z. Thus there are 3×2, that is, 6, possible different routes from town X to town Z by way of town Y.

Example 2

Find how many two-digit positive odd integers can be written using the digits 1, 2, 3, 4, and 5 if repetition of the digits is permitted.

The tens digit may be chosen from the digits 1, 2, 3, 4, and 5. For our two-digit number to be odd, the units digit must be 1, 3, or 5. The tree diagram in Fig. 4.2 shows there are 5×3, that is, 15, possible different numbers that satisfy the conditions of the problem.

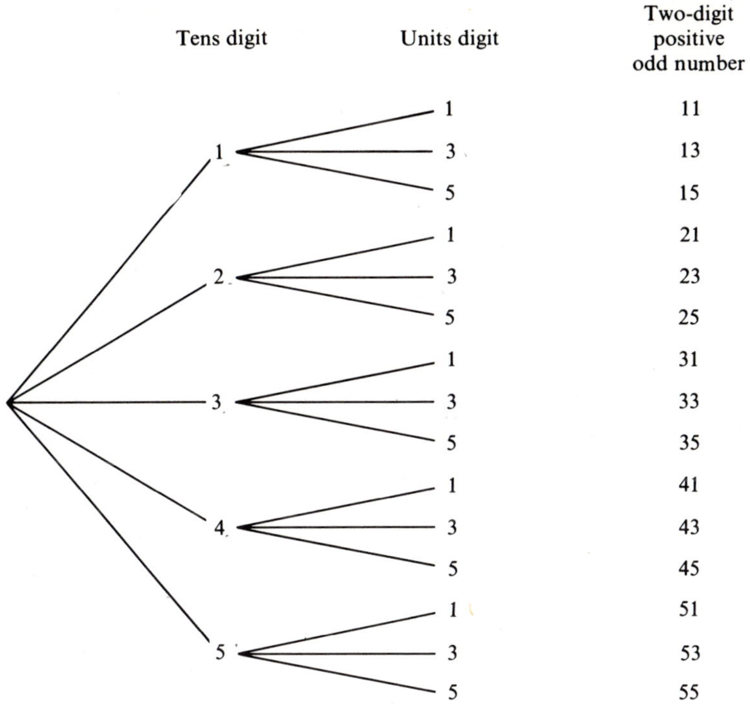

Figure 4.2

Examples 1 and 2 illustrate the first of two fundamental principles of enumeration.

First Fundamental Principle of Enumeration

If a selection can be made in* m *ways, and if another selection can be made in* n *ways, then both selections can be made in* mn *ways.

This principle can be extended for any number of successive selections. Note carefully that an extension of this principle is concerned with a sequence of selections, that is, one selection followed by another, then another, and another, and so forth. Example 3 illustrates an extension of this first principle.

Example 3

A coed selects clothes that can be worn together: three skirts, four blouses, and two cardigan sweaters. How many different complete outfits will she have?

We can use the first fundamental principle of enumeration and determine that there are 3×4 ways of selecting a skirt and a blouse. Then with each of these basic outfits we consider the two cardigans and apply the first principle again:

$$(3 \times 4) \times 2 = 24$$

Thus there are 24 complete outfits that she could wear. This same result could have been obtained from the tree diagram shown in Fig. 4.3.

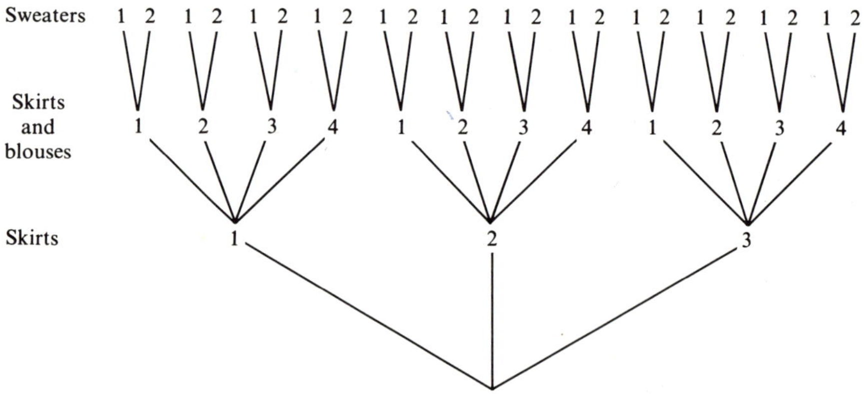

Figure 4.3

Exercises

Let $A = \{a, b, c\}$, $B = \{r, s, t, u, v\}$, and $C = \{x, y\}$. In Exercises 1 through 6 determine the number of selections that can be made by choosing an element from each of the sets as indicated.

1. A followed by B
2. A followed by C
3. B followed by C
4. C followed by B
5. A followed by B, followed by C
6. C followed by B, followed by A

7. An ice-cream parlor offers twelve flavors of ice cream and five kinds of toppings. How many different sundaes can be served?
8. If a girl has four sweaters and six scarfs, how many different sweater-scarf combinations can she wear?
9. Two signal flags are to be arranged in order on a vertical staff. If five different flags are available, how many different two-flag signals can be made?
10. A club has two candidates for president, five for vice-president, and three for secretary-treasurer. Determine how many different executive councils could be elected.
11. Determine how many registration cards can be made by using exactly three of the digits 0, 1, 2, 3, 4, 5, 6, 7, 8, and 9 if each digit may be used more than once.
12. Determine how many fraternity names can be made by using exactly three of the twenty-four letters of the Greek alphabet if each letter can be used more than once.
13. In how many ways can a basketball coach form a team if there are three candidates on the squad for each of the five positions?
14. How many different automobile license plates are possible if the plates for a particular state are identified by two letters of the English alphabet followed by a three-digit number greater than 99?
15. How many three-digit positive odd integers can be written using the digits 1, 2, 3, 4, 5, 6, 7, 8, and 9 with no digits repeated?
16. Determine how many three-digit positive integers are multiples of 5.
17. Some stores use code symbols on their price tags to indicate the cost to the store of their merchandise. If each symbol consists of a letter followed by two digits that may be any digit from 0 through 9 and that may also be the same, how many different symbols are possible?

4.2

Second Fundamental Principle of Enumeration

In Sec. 4.1 we were concerned with counting the number of ways a sequence of selections could be made. Specifically, if a sequence of two selections was to be made, we were concerned with counting the number of ways both selections could be made. However, a different situation occurs when we want to count the number of ways in which either of two selections can be made. The following examples are concerned with the solution of problems of this type; these examples illustrate the use of the *second fundamental principle of enumeration.*

Example 1

A classroom representative is to be chosen from a group of 16 boys and 12 girls. In how many different ways can a representative be selected?

A boy can be chosen in 16 ways; a girl can be chosen in 12 ways. Obviously, either a boy or a girl will be chosen. The number of ways that a boy or a girl can be chosen is $16 + 12$; that is, there are 28 ways of selecting a classroom representative.

Example 2

Find how many different positive integers with at least two digits can be formed using the set of three digits $\{4, 5, 7\}$ if repetition of the digits is not permitted.

We can form either two-digit positive integers or three-digit positive integers. By the first fundamental principle of enumeration we can form 3×2, that is, 6, two-digit positive integers; we can form $3 \times 2 \times 1$, that is, 6, three-digit positive integers. Because we cannot select exactly two digits and exactly three digits at the same time, these selections are called *mutually exclusive.* That is, the selection of only two digits to form a number excludes (does not permit) the selection of three digits to form a number, and similarly, if three digits are selected, we cannot form a two-digit number using all the digits.

Since we want to form two- or three-digit positive integers, we add the number of ways in which each can be formed. Therefore there are $6 + 6$, that is, 12, different positive integers with at least two digits chosen from the set $\{4, 5, 7\}$ without repetition of the digits; these numbers are 45, 47, 54, 57, 74, 75, 457, 475, 547, 574, 745, and 754.

Second Fundamental Principle of Enumeration

If a selection can be made in m ***ways, and if another selection can be made in*** n ***ways, then either of the two selections can be made in*** $m + n$ ***ways.***

As is the case with the first fundamental principle of enumeration, the second fundamental principle of enumeration can be generalized. The second principle and its extension are concerned with the making of a single selection from among two or more possible selections. For instance, in Example 1 we made a selection of 1 of 16 boys or a selection of 1 of 12 girls; in Example 2 we made a selection of one of six two-digit numbers or a selection of one of six three-digit numbers.

The following example illustrates an extension of the second principle.

Example 3

Find how many different signals can be made by arranging in order on a vertical staff at least two flags if four different flags are available.

We can form a signal by using two, three, or four flags. First consider the number of ways a two-flag signal can be made. By the first fundamental principle of enumeration, we can select in order two different flags from the set of four in 4×3, that is, 12, ways. Remember that once a flag is selected for the first position in the signal, we can select the second flag in only one of three ways since only three flags remain. In a similar manner, using the first fundamental principle of enumeration, we can select in order three flags in $4 \times 3 \times 2$, that is, 24, ways; we can select in order four flags in $4 \times 3 \times 2 \times 1$, that is, 24, ways.

Because we cannot make two-, three-, and four-flag signals at the same time, these selections are mutually exclusive. For example, if a two-flag signal is made, then neither a three- nor a four-flag signal can be made. Since we want to make a two- or a three- or a four-flag signal, we add the number of ways in which each can be made. Hence, $12 + 24 + 24$, that is, 60, different signals can be made by arranging in order at least two flags if four different flags are available.

Exercises

1. How many positive integers with at least three digits can be formed using the set of four digits {1, 2, 8, 9} if repetition of the digits is not permitted?
2. How many positive integers with at least three digits can be formed using the set of four digits {1, 2, 8, 9} if repetition of the digits is permitted?

3. How many positive integers with not more than two digits can be formed using the set of five digits {1, 2, 3, 4, 5} if repetition of the digits is not permitted?
4. How many positive integers with not more than two digits can be formed using the set of five digits {1, 2, 3, 4, 5} if repetition of the digits is permitted?
5. How many different signals can be made by arranging in order on a vertical staff one, two, three, or four flags if four different flags are available?
6. How many different signals can be made by arranging in order on a vertical staff not more than two flags if ten different flags are available?
7. Assume that there are four each of four different kinds of flags. How many different signals can be made by arranging in order on a vertical staff one, two, three, or four flags if flags of the same kind can be flown?
8. Determine how many identification cards can be made using exactly three of the nine digits 1, 2, 3, 4, 5, 6, 7, 8, and 9 or exactly two of the twenty-six letters of the English alphabet if repetition of the digits and letters is not permitted.
9. Determine how many identification cards can be made using exactly three of the nine digits 1, 2, 3, 4, 5, 6, 7, 8, and 9 or exactly two of the twenty-six letters of the English alphabet if repetition of the digits and letters is permitted.

4.3

Permutations

Consider the following example, which makes use of the first fundamental principle of enumeration.

Example 1

A track coach needs to select 4 boys from among his squad of 15 to run in the 1-mile relay event. How many different relay teams can he assemble for the event?

In a relay event the order in which the boys run is important. The coach can select the first runner for the relay team in 15 ways. Having chosen the first runner, the coach can select the second runner from among the remaining 14 boys. By applying the first fundamental principle of enumeration the coach can assemble 15×14, that is, 210, different arrangements of the first two runners. To go with each of these arrangements he can select the third runner from among the remaining 13 boys. Hence the coach can assemble 120×13, that is, 2,730, different arrangements of the first three runners for the relay team. To go with each of these arrangements, the coach can select the last runner (the anchor man) from among the remaining 12 boys. Thus the coach can assemble $2{,}730 \times 12$, that is, 32,760, different arrangements of four runners for a relay team.

Notice that an extension of the first fundamental principle of enumeration could have been applied directly to obtain the $15 \times 14 \times 13 \times 12$ different arrangements.

In this example we were interested in finding the number of different arrangements of the elements of a subset of a given set. Consider a second similar problem.

Example 2

How many different arrangements of two books on a shelf can be made if four books are available?

Let a, b, c, and d represent the four books. First consider the number of subsets containing two books that could be selected from the set of four books. The six possible subsets are $\{a, b\}$, $\{a, c\}$, $\{a, d\}$, $\{b, c\}$, $\{b, d\}$, and $\{c, d\}$. Each subset represents a possible combination of two of the four books. Although

$$\{a, b\} = \{b, a\}$$

the arrangement of the books on the shelf in the order "a to the left of b" is different from the order "b to the left of a." Thus, in this case, for each combination there are two different possible arrangements of books. The 12 possible arrangements are:

a, b	a, d	b, d
b, a	d, a	d, b
a, c	b, c	c, d
c, a	c, b	d, c

Consider a set of n different elements. A selection of r elements from this set without regard to order is called a *combination*; that is, a combination is a subset of a set. A selection of r elements from this set with regard to order is called a *linear arrangement* or *linear permutation*. We shall refer to a linear permutation as simply a *permutation*.

In Example 1, we found the number of permutations that can be made from a set of 15 boys taken 4 at a time. In Example 2, we found the number of permutations from a set of four books taken two at a time. The symbol ${}_nP_r$ is used to denote the total number of permutations from a set of n elements taken r at a time. Thus

$${}_{15}P_4 = 15 \times 14 \times 13 \times 12 = 32{,}760$$

and

$${}_4P_2 = 4 \times 3 = 12 \qquad \text{(Why?)}$$

To determine the total number of permutations from a set of n distinguishable elements taken r at a time, we make use of the first fundamental principle of enumeration:

${}_nP_1 = n$ since n elements are available for selection
${}_nP_2 = n(n-1)$ since the first element may be selected in n ways and the second element may be selected in $(n-1)$ ways
${}_nP_3 = n(n-1)(n-2)$ (Why?)
${}_nP_4 = n(n-1)(n-2)(n-3)$ (Why?)
....

and, in general,

$${}_nP_r = n(n-1)(n-2)\cdots(n-r+1)$$

The products of successive positive integers, such as those that occur in the evaluation of ${}_nP_r$, may be conveniently expressed by the use of *factorial notation.* If n is a positive integer, then we define

$$n! = n(n-1)(n-2)(n-3)\cdots(3)(2)(1)$$

where $n!$ is read "n factorial." For example,

$$1! = 1$$
$$2! = 2 \times 1 = 2$$
$$3! = 3 \times 2 \times 1 = 6$$
$$4! = 4 \times 3 \times 2 \times 1 = 24$$
....

Since $(n-r)!/(n-r)! = 1$,

$${}_nP_r = n(n-1)(n-2)\cdots(n-r+1) \times \frac{(n-r)!}{(n-r)!}$$

$$= \frac{n(n-1)(n-2)\cdots(n-r+1)(n-r)\cdots(3)(2)(1)}{(n-r)!}$$

that is,

$${}_nP_r = \frac{n!}{(n-r)!}$$

Suppose we used this formula to evaluate ${}_nP_n$:

$${}_nP_n = \frac{n!}{(n-n)!} = \frac{n!}{0!}$$

However, by definition,

$${}_nP_n = n!$$

So that the formula will be valid for all nonnegative integral values of r less than or equal to n, we define $0! = 1$.

Example 3

Evaluate $8!/(2!3!4!)$.

$$\frac{8!}{2!3!4!} = \frac{8 \times 7 \times 6 \times 5 \times 4 \times 3 \times 2 \times 1}{(2 \times 1) \times (3 \times 2 \times 1) \times (4 \times 3 \times 2 \times 1)}$$

$$= \frac{8 \times 7 \times 6 \times 5}{(2 \times 1) \times (3 \times 2 \times 1)} = \frac{8 \times 7 \times 5}{2 \times 1}$$

$$= 4 \times 7 \times 5 = 140$$

Example 4

How many different signals can be made by arranging in order on a vertical staff three of ten different flags?

The number of permutations of ten elements taken three at a time is ${}_{10}P_3$. Using the formula for ${}_nP_r$, we have

$${}_{10}P_3 = \frac{10!}{(10-3)!} = \frac{10!}{7!} = 10 \times 9 \times 8 = 720$$

Hence 720 different signals can be made by arranging in order on a vertical staff three of ten different flags.

Example 5

In how many different ways can four girls and four boys stand in a row if girls and boys must alternate?

This is not a simple problem of finding ${}_8P_8$ because girls and boys must alternate. For example, even though the arrangement

boy, boy, girl, boy, girl, girl, girl, boy

is included in the ${}_8P_8$ permutations, the arrangement does not satisfy the conditions of the problem.

We must select a girl or a boy to fill the first position. This selection can be made in one of two ways. Once the choice is made, then the sex of the different positions in the row is determined. The girls can be arranged in their positions in one of ${}_4P_4$ ways, that is, 4! ways; the boys can be arranged in their positions in one of ${}_4P_4$ ways, that is, 4! ways. Hence, the total number of different ways four girls and four boys can stand in a row if girls and boys must alternate is

$$2 \times 4! \times 4!$$

that is, 1,152.

Exercises

In Exercises 1 through 12 evaluate each expression.

1. $\dfrac{5!}{4!}$
2. $\dfrac{6!}{1!}$
3. $\dfrac{11!}{8!}$
4. $\dfrac{11!}{9!}$
5. ${}_5P_5$
6. ${}_5P_4$
7. ${}_5P_1$
8. ${}_9P_7$
9. $\dfrac{8!}{5!3!}$
10. $\dfrac{11!}{8!3!}$
11. $\dfrac{7!}{(7-2)!}$
12. $\dfrac{6!}{(6-4)!4!}$
13. Find the number of linear arrangements that can be made of seven distinguishable elements taken four at a time.
14. Find the number of permutations from a set of six distinguishable elements that are taken three at a time.
15. How many different batting orders are possible for a baseball team of nine players?
16. How many different batting orders are possible for a baseball team if the pitcher bats ninth?

17. Of 20 girls competing in a beauty contest, one will be the winner, another will be first runner-up, and another will be second runner-up. How many different selections are possible?

18. How many different sets of student councils could be elected if a class of 30 students elects a president, vice-president, and secretary?

19. In how many different ways can five boys and four girls be arranged in a row if boys and girls must alternate?

20. Six persons enter a room in which four chairs are arranged in a row. In how many different ways could the chairs be filled?

21. Six persons enter a room in which ten chairs are arranged in a row. In how many different ways could the persons sit in the row?

22. Determine whether statements (*a*) to (*d*) are true or false for all positive integers n and r, where $n > r$: (*a*) $(n - r)! = n! - r!$; (*b*) $(n + r)! = n! + r!$; (*c*) $(nr)! = n!\,r!$; (*d*) $(2n)!/n! = (2n/n)!$.

23. Prove that $n!$ is a multiple of 10 if and only if $n \geq 5$.

24. An experimenter requires at least 10,000 permutations of n different objects, taken all at a time. How small can n be?

25. The total number of permutations of n distinguishable elements is $n!$. Show that the total number of permutations of n elements, where k of them are indistinguishable, is $n!/k!$.

26. Use the results of Exercise 25 to determine the total number of arrangements of the letters of the word "book." List the arrangements.

27. Use the results of Exercise 25 to determine the total number of arrangements of the letters of the word "maximum."

28. The total number of permutations of n elements of which k_1 are alike, k_2 others are alike, k_3 others are alike, . . . , and k_r others are alike is expressed by $n!/(k_1!\,k_2!\,k_3! \cdots k_r!)$. How many arrangements of the letters of the word "Mississippi" are possible?

29. In how many ways can ten different objects be divided among A, B, and C if A must receive three objects, B five objects, and C two objects?

4.4

Combinations

We have already defined a combination as a selection of a number of elements without regard to order. For example, consider a committee of 4 people chosen from a group of 30. Unless we specify the various offices that each member of the committee will hold, the order in which the members are selected has no importance. In such a case, we are interested in the possible number of combinations of 30 elements taken 4 at a time.

Example 1

Find how many different three-member committees can be formed from a group of five girls.

Let a, b, c, d, and e denote the five girls, respectively. The subsets of $\{a, b, c, d, e\}$ that contain three elements are:

$$\{a, b, c\},\quad \{a, b, d\},\quad \{a, b, e\},\quad \{a, c, d\},$$
$$\{a, c, e\},\quad \{a, d, e\},\quad \{b, c, d\},\quad \{b, c, e\},$$
$$\{b, d, e\},\quad \{c, d, e\}$$

There are ten subsets. In other words, there are ten combinations of five elements taken three at a time. Therefore, the number of different three-member committees that can be formed from a group of five girls is ten.

In Example 1, suppose that one member of the committee is designated as president and a second member is designated as secretary. How many different committee arrangements are possible? For example, in the committee consisting of girls a, b, and c we might have one of the following arrangements:

President	Secretary	Third member
a	b	c
a	c	b
b	a	c
b	c	a
c	a	b
c	b	a

In each combination of Example 1, such as the set $\{a, b, c\}$, the elements may be arranged in one of ${}_3P_3$, that is, 3!, ways. Hence, there are $10 \times 3!$, that is, 60, different committee arrangements. Note that

$${}_5P_3 = \frac{5!}{(5-3)!} = \frac{5!}{2!} = 5 \times 4 \times 3 = 60$$

If the symbol ${}_nC_r$ is used to denote the total number of combinations from a set of n elements taken r at a time, then

$$ {}_5P_3 = {}_5C_3 \times {}_3P_3 $$

that is,

$$ {}_5P_3 = {}_5C_3 \times 3! $$

and

$$ {}_5C_3 = \frac{{}_5P_3}{3!} $$

The above discussion illustrates a relationship between the number of combinations of n elements taken r at a time and the number of permutations of n elements taken r at a time. In general, every combination of r elements from a set of n elements has associated with it $r!$ permutations of the r elements. Thus, by the first fundamental principle of enumeration,

$$ {}_nP_r = {}_nC_r \times r! $$

Therefore,

$$ {}_nC_r = \frac{{}_nP_r}{r!} $$

and

$$ {}_nC_r = \frac{n!}{r!(n-r)!} $$

Example 2

A student must do five of ten problems on an examination. Find how many different ways the student can select the five problems.

The number of subsets of five elements that can be formed from a set of ten elements is ${}_{10}C_5$; that is,

$$ {}_{10}C_5 = \frac{10!}{5!(10-5)!} = \frac{10!}{5!5!} $$

$$ = \frac{10 \times 9 \times 8 \times 7 \times 6}{5 \times 4 \times 3 \times 2 \times 1} = 252 $$

Hence the student can select the five problems 252 ways.

Example 3

How many different committees consisting of two boys and two girls can be formed from a group of four boys and seven girls?

We can select two boys from the set of four boys in ${}_4C_2$ different ways; we can select two girls from the set of seven girls in ${}_7C_2$ different ways. Using the first fundamental principle of enumeration, we can select two boys and two girls in ${}_4C_2 \times {}_7C_2$ different ways:

$$
\begin{aligned}
{}_4C_2 \times {}_7C_2 &= \frac{4!}{2!(4-2)!} \times \frac{7!}{2!(7-2)!} \\
&= \frac{4!}{2!2!} \times \frac{7!}{2!5!} = \frac{4 \times 3}{2 \times 1} \times \frac{7 \times 6}{2 \times 1} \\
&= 6 \times 21 = 126
\end{aligned}
$$

Hence 126 different committees consisting of two boys and two girls can be formed from a group of four boys and seven girls.

Exercises

In Exercises 1 through 8 evaluate each expression.

1. ${}_5C_1$
2. ${}_5C_4$
3. ${}_8C_2$
4. ${}_8C_6$
5. ${}_7C_7$
6. ${}_7C_0$
7. ${}_nC_n$ for every positive integer n
8. ${}_nC_1$ for every positive integer n
9. In how many different ways can eight club members select a committee of three persons to represent the club?
10. In how many different ways can 2 co-captains be selected from a team of 11 football players? ${}_{11}C_2$

11. Determine the number of different ways 3 finalists can be selected from a group of 20 girls entered in a beauty contest.
12. Determine the number of different ways two vegetables can be selected if six vegetables are available in a cafeteria.
13. On a quiz the students are to do the first five problems and any five of the next seven problems. How many different tests can be completed by a student?
14. How many different salads can be made consisting of lettuce and three other vegetables if a total of six vegetables are available?
15. If a bag contains 12 beads, find the number of different ways 2 beads can be selected.
16. A bag contains five beads. (*a*) In how many different ways can four beads be selected? (*b*) In how many different ways can four beads be selected if one particular bead must be among them?
17. Six candidates run for two councilman offices. If a voter can mark his ballot either for one or for two candidates, in how many ways can he cast his vote?
18. In how many ways can a committee of three be selected from seven people if one person is always included?
19. A bag contains five red beads and four white beads. In how many different ways can four beads be selected in which two beads are red and two beads are white?
20. Determine how many different committees can be formed from a group of 30 students and 4 teachers if each committee is to consist of 3 teachers and 2 students.
21. How many different subsets that contain at least five elements can be selected from a set of eight elements?
22. Find the number of different committees of five members that can be selected from a group of nine democrats and seven Republicans if each committee must consist of not more than two Democrats.
23. Prove that ${}_nC_r = {}_nC_{n-r}$ for every $n \geq r$.
24. Prove that ${}_nC_{r-1} + {}_nC_r = {}_{n+1}C_r$ for every $n \geq r \geq 1$.
25. Consider a *Pascal triangle*, as shown in the accompanying figure. (*a*) Show that the $(r + 1)$th element of the $(n + 1)$th row is ${}_nC_r$. (*b*) Extend the Pascal triangle in the figure by writing down the elements of the next two rows.

1
1 1
1 2 1
1 3 3 1
1 4 6 4 1
1 5 10 10 5 1

26. Show that

(a) $(a + b)^3 = {}_3C_0\, a^3 + {}_3C_1\, a^2b + {}_3C_2\, ab^2 + {}_3C_3\, b^3$

(b) $(a + b)^5 = {}_5C_0\, a^5 + {}_5C_1\, a^4b + {}_5C_2\, a^3b^2 + {}_5C_3\, a^2b^3 + {}_5C_4\, ab^4 + {}_5C_5\, b^5$

(c) $(a + b)^n = {}_nC_0\, a^n + {}_nC_1\, a^{n-1}b + {}_nC_2\, a^{n-2}b^2 + \cdots + {}_nC_{n-1}\, ab^{n-1} + {}_nC_n b^n$

27. Use the results of Exercise 26 to show that

$${}_nC_0 + {}_nC_1 + {}_nC_2 + \cdots + {}_nC_{n-1} + {}_nC_n = 2^n$$

(The total number of possible subsets of a set of n elements is 2^n.)

28. A hospitality shop specializes in gift baskets of fruit. If seven kinds of fruit are available, how many different assortments of one or more kinds of fruit can the shop provide?

4.5

Mathematical Probability

In everyday conversation words such as "probable," "likely," "probability," and "chance" occur frequently. Usually these words are used in a nonnumerical sense. For example, we might hear someone say "It will probably rain today" or "The Dodgers are likely to win the pennant this year." The statements involve judgments that are founded upon experience or, in some cases, wishful thinking on the part of the speaker. The speaker does not attach a numerical measure to such statements. We might reasonably ask, "How probable is it that it will rain today?" or "How likely is it that the Dodgers will win the pennant this year?" What we would like to have then is a numerical measure associated with each of the events "It rains today" and "The Dodgers win the pennant this year" such that by comparing these numbers we could tell at a glance which of the two events has a better chance of occurring.

Consider a situation involving the word "likely" and the idea of mathematical probability. Before every football game the captains of the opposing teams meet with the referee in the center of the field to toss a coin. One captain chooses to call "heads" or "tails." The choice of who first receives the ball depends upon the outcome of the event of tossing the coin. It appears reasonable to assume that the coin is *equally likely* to fall heads as tails. If this were not the case, for example, if tails were more likely to occur, each captain would insist upon being the one to predict the outcome of the toss. Hence (obviously) the captains agree that the *probability* or *chance* that heads occurs is the same as the probability or chance that tails occurs.

Although we may have a vague idea of what is meant by the probability that a certain event occurs, it is the purpose of this section to supply a workable definition of the concept. One of the first elementary definitions of probability was stated by the famous French mathematician Poincaré and will be used here:

The* mathematical probability *of an event is the ratio of the number of outcomes favorable to that event to the total number of possible outcomes, provided that the possible outcomes can be assumed to be equally likely.

Example 1

A box contains three red and two white beads of the same size and shape; that is, they are indistinguishable to the touch. Consider an experiment in which one bead is chosen at *random* (by chance). What is the probability that the bead chosen is white?

It seems reasonable to assume that each bead is equally likely to be chosen. The event "a white bead is chosen" has two outcomes favorable to it because there are two white beads in the box. There are a total of five possible outcomes, both favorable and unfavorable to the event. Hence the probability that the bead chosen is white is $\frac{2}{5}$.

Example 2

An ordinary game die, that is, a cube whose six faces are marked with 1, 2, 3, 4, 5, and 6 dots, respectively, is thrown. What is the probability that an odd number appears on the top face?

We agree to consider the outcomes of this experiment to be the number of dots on the top face of the die when the die comes to rest. We further agree that there are only six possible outcomes: 1, 2, 3, 4, 5, and 6, and that these outcomes are equally likely. Three of the six outcomes are favorable to the event "an odd number appears": 1, 3, and 5. Thus the probability that an odd number appears on the top face is $\frac{3}{6}$, that is, $\frac{1}{2}$.

Example 3

If two coins are tossed, what is the probability that two heads appear?

Consider the two coins as being of different denominations: a nickel and a dime. (This is really not necessary but will be easier for us to discuss in terms of the outcomes of the experiment.) The possible outcomes for this experiment may be summarized by the following table:

Nickel	Dime	Outcome
Heads	Heads	2 heads
Heads	Tails	1 head, 1 tail
Tails	Heads	1 head, 1 tail
Tails	Tails	2 tails

Note that the toss of a nickel may result in a head or a tail. Regardless of the result of tossing a nickel, the toss of a dime may also result in a head or a tail. The result of one toss does not affect the other. Each of the four outcomes appear to be equally likely to occur. Since only one of the four outcomes produces two heads, the probability that two heads appear is $\frac{1}{4}$.

Example 4

A bag contains five red, four white, and three blue beads. Two beads are selected at random at one time.
(*a*) What is the probability that they are both red?
(*b*) What is the probability that they are not both red?

(*a*) There is a total of $5 + 4 + 3$, that is, 12, different beads in the bag. We can select 2 beads from among the 12 in one of ${}_{12}C_2$ ways, that is, in 66 ways since

$$
\begin{aligned}
{}_{12}C_2 &= \frac{12!}{2!\,(12-2)!} \\
&= \frac{12!}{2!\,10!} = \frac{12 \times 11}{2 \times 1} = 66
\end{aligned}
$$

It is reasonable to assume that these 66 possible outcomes are equally likely because the selection of beads is made at random. We can select two red beads from among the five red beads in one of ${}_5C_2$ ways; that is, in ten ways since

$$ {}_5C_2 = \frac{5!}{2!(5-2)!} $$

$$ = \frac{5!}{2!\,3!} = \frac{5 \times 4}{2 \times 1} = 10 $$

Since there are 10 ways of selecting two red beads and 66 ways of selecting any two beads, the probability that both beads selected are red is $\frac{10}{66}$, that is, $\frac{5}{33}$.

(*b*) Because there are 66 ways of selecting any two beads and 10 ways of selecting two red beads, there are 66 − 10, that is, 56, ways of selecting two beads which are not both red. Note carefully that among these 56 are selections including one red bead and one bead of another color. Hence the probability that the beads selected are not both red is $\frac{56}{66}$, that is, $\frac{28}{33}$.

Exercises

1. An ordinary die is thrown. Determine the probability that the number which appears on the top face is (*a*) 3; (*b*) 6; (*c*) 0; (*d*) less than 3; (*e*) less than 6; (*f*) greater than 0.
2. A game device like that shown in the accompanying figure has five regions in which the spinner is equally likely to stop. Consider the outcome of a single spin to be the number of the region in which the spinner stops. Determine the probability that the number which occurs is (*a*) 3; (*b*) 0; (*c*) odd; (*d*) even; (*e*) less than or equal to 3; (*f*) greater than 4.

3. Consider two successive outcomes of the game device described in Exercise 2.
(*a*) In how many ways can two numbers occur?
(*b*) What is the probability that each outcome is 5?
(*c*) What is the probability that the sum of the outcomes is 9?
(*d*) What is the probability that the outcomes are equal?

4. A box contains five red, four white, and three blue beads. Three beads are selected at random at one time.
   (*a*) In how many ways can three beads be selected?
   (*b*) What is the probability that the three beads selected are blue?
   (*c*) What is the probability that the three beads selected are white?
   (*d*) What is the probability that the three beads selected are the same color?
5. Determine the probability that a student selected at random from a class of 17 boys and 18 girls is a boy.
6. If two coins are tossed, what is the probability that a head and a tail appear?
7. If two coins are tossed, what is the probability that neither is a tail?
8. For every 2,000 cash receipts, a supermarket prints stars on three receipts that entitle the recipients to a prize. What is the probability that a shopper finds a star on his or her receipt?
9. Consider four beads selected at random at one time from a box containing five red, four white, and three blue beads.
   (*a*) Find the probability that the four beads selected are white.
   (*b*) Find the probability that the four beads selected are blue.
   (*c*) Find the probability that two of the beads selected are blue and two of the beads are white.
   (*d*) Find the probability that exactly three of the four beads selected are red.
10. What is wrong with reasoning that the probability of a person being born in 1 of the 50 states of the United States is $\frac{1}{50}$?
11. If 4 out of 15 persons that were in a plane are known to have survived a crash, what is the probability that two particular persons survived?
12. There are 52 whole weeks and 2 days in a leap year. If a leap year is chosen at random, what is the chance that it contains 53 Sundays?

4.6

Sample Spaces

To determine the probability that a certain event occurs, we must first identify the possible outcomes. It will be convenient to label the set of all possible equally likely outcomes of an experiment as a *sample space S*. (A sample space for a particular problem plays the role of the universal set in that problem.) Each outcome of an experiment is called a *sample point* or, more briefly, a *point*. The total number of points in a sample space S is denoted by $n(S)$. Any subset of a sample space is called an *event*. The number of elements in an event A is denoted by $n(A)$. The mathematical probability of an event A may be represented by $P(A)$ and expressed as

$$P(A) = \frac{n(A)}{n(S)}$$

A very important property to remember in selecting a sample space S for an experiment is that the points of S must be equally likely to occur. For instance, in Example 1 of Sec. 4.5 we might have reasoned incorrectly that the sample space S could be expressed as

$$S = \{\text{red, white}\}$$

and that the event A could be expressed as

$$A = \{\text{red}\}$$

Then we would have concluded that the probability of event A was

$$P(A) = \frac{n(A)}{n(S)} = \frac{1}{2}$$

The events "red" and "white," that is, "a red bead is chosen" and "a white bead is chosen," respectively, are not equally likely—there are more red beads than white beads. A correct sample space S for that example would be

$$S = \{\text{red}_1, \text{red}_2, \text{red}_3, \text{white}_1, \text{white}_2\}$$

where names of the colors of the beads contain subscripts to make them distinguishable.

Example 1

Find a sample space for the experiment of tossing three coins. Determine the probability that all three coins are heads.

Each of the three coins can fall one of two ways: heads or tails. Thus, by the first fundamental principle of enumeration, there are $2 \times 2 \times 2$, that is, eight, ways the three coins can fall. A tree diagram for the sequence of possible outcomes when three coins are tossed is shown in Fig. 4.4. A sample space S is the set of eight possible outcomes denoted by the ordered triples of h's (heads) and t's (tails):

$$S = \{(h, h, h), (h, h, t), (h, t, h), (h, t, t), (t, h, h), (t, h, t), (t, t, h), (t, t, t)\}$$

The event A of tossing three heads consists of the single outcome (h, h, h). Thus, the probability of tossing three heads is $\frac{1}{8}$ since

$$P(A) = \frac{n(A)}{n(S)} = \frac{1}{8}$$

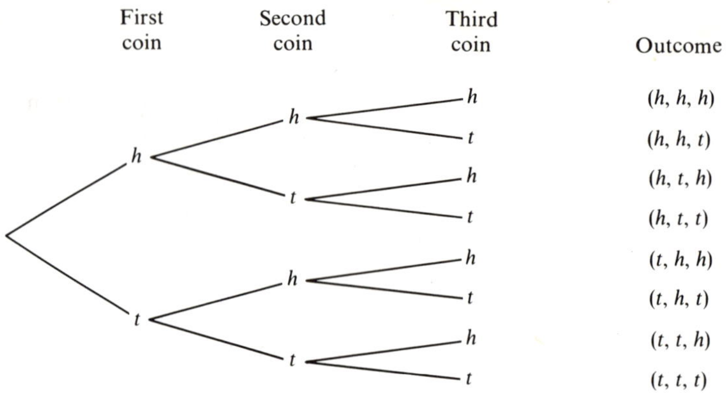

Figure 4.4

Example 2

Consider a sample space for the experiment of tossing three coins, and determine the probability of tossing at least two heads.

A sample space S was determined in Example 1:

$$S = \{(h, h, h), (h, h, t), (h, t, h), (h, t, t), (t, h, h), (t, h, t), (t, t, h), (t, t, t)\}$$

Let A be the event that at least two heads occur. Then

$$A = \{(h, h, h), (h, h, t), (h, t, h), (t, h, h)\}$$

Therefore, the probability of tossing at least two heads when three coins are tossed is $\frac{1}{2}$ since

$$P(A) = \frac{n(A)}{n(S)} = \frac{4}{8} = \frac{1}{2}$$

Example 3

Determine a sample space for the experiment of throwing a pair of dice. Determine the probability that the sum of the face values is 5.

A sample space S is the Cartesian set

$$\{1, 2, 3, 4, 5, 6\} \times \{1, 2, 3, 4, 5, 6\}$$

The points of S can be represented as an array:

$$\begin{array}{cccccc} (1, 1), & (1, 2), & (1, 3), & (1, 4), & (1, 5), & (1, 6) \\ (2, 1), & (2, 2), & (2, 3), & (2, 4), & (2, 5), & (2, 6) \\ (3, 1), & (3, 2), & (3, 3), & (3, 4), & (3, 5), & (3, 6) \\ (4, 1), & (4, 2), & (4, 3), & (4, 4), & (4, 5), & (4, 6) \\ (5, 1), & (5, 2), & (5, 3), & (5, 4), & (5, 5), & (5, 6) \\ (6, 1), & (6, 2), & (6, 3), & (6, 4), & (6, 5), & (6, 6) \end{array}$$

The first member of each ordered pair represents the outcome of one die; the second member represents the outcome of the other die.

The event A that the sum of the face values is 5 is represented by

$$\{(1, 4), (2, 3), (3, 2), (4, 1)\}$$

Hence the probability that the sum of the face values is 5 is $\frac{1}{9}$ since

$$P(A) = \frac{n(A)}{n(S)} = \frac{4}{36} = \frac{1}{9}$$

Example 4

Six beads are numbered 1 through 6 and placed in a box. If two beads are selected at random from the box, what is the probability that their sum is 6?

Let x represent the number on the first bead selected; let y represent the number on the second bead selected. Then the ordered pairs of numbers (x, y) make up a sample space S. Since x represents one of six numbers and y represents one of five numbers, the number of elements of S is 30. The event A that the sum of the numbers is 6 is represented by

$$\{(1, 5), (2, 4), (4, 2), (5, 1)\}$$

Hence the probability that the sum of the numbers on the beads is 6 is $\frac{2}{15}$ since

$$P(A) = \frac{n(A)}{n(S)} = \frac{4}{30} = \frac{2}{15}$$

The elements of S are represented geometrically by the points in Fig. 4.5; the elements of A are represented by the heavy dots.

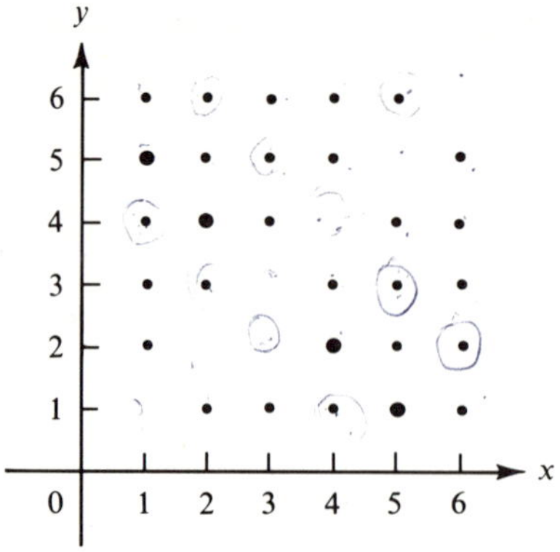

Figure 4.5

Exercises

In Exercises 1 through 4 consider a box that contains five otherwise indistinguishable beads numbered 1 through 5.

1. (*a*) Find a sample space for the experiment of selecting a bead at random.
   (*b*) Determine the probability that the bead selected is numbered 3.
   (*c*) Determine the probability that the bead selected has an odd number.
2. (*a*) Find a sample space for the experiment of selecting two beads at random at one time.
   (*b*) Determine the probability that the two beads selected are numbered 2 and 3, respectively.
   (*c*) Determine the probability that the sum of the numbers of the two beads selected is 6.
3. (*a*) Find a sample space for the experiment of selecting two beads at random if the first bead is replaced before the second bead is selected.
   (*b*) Determine the probability that the first bead selected is numbered 1 and the second bead selected is numbered 5.
   (*c*) Determine the probability of selecting the same bead twice in a row.
4. (*a*) Find a sample space for the experiment of selecting two beads at random but not at the same time.
   (*b*) Determine the probability that the first bead selected is numbered 1 and the second bead selected is numbered 5.
   (*c*) Determine the probability that the sum of the numbers of the two beads selected is 6.
5. If three coins are tossed, find the probability that one head and two tails appear.

6. (*a*) Find a sample space for the experiment of tossing four coins.
   (*b*) Determine the probability that all four coins are heads.
   (*c*) Determine the probability that exactly two heads appear.
   (*d*) Determine the probability that at least two heads appear.
7. If a pair of dice are thrown, determine the probability that the sum of the face values is (*a*) 7; (*b*) 3; (*c*) 8; (*d*) 11.
8. If a pair of dice are thrown, find the probability that the sum of the face values is (*a*) at least 5; (*b*) not more than 8; (*c*) an odd number; (*d*) an even number.
9. If a pair of dice are thrown, find the probability that the face values of the two dice are equal.
10. Describe four different sample spaces for the experiment in which a card is chosen at random from an ordinary deck of bridge cards. (*Hint*: Consider suit, color, and denomination.)
11. What is the probability of choosing a king if a card is chosen at random from an ordinary deck of bridge cards?
12. What is the probability of choosing a king if a card is chosen at random from the set of spades in an ordinary deck of bridge cards?
13. Find the probability of choosing a face card if a card is chosen at random from an ordinary deck of bridge cards.
14. If a card is selected at random from an ordinary deck of bridge cards, determine the probability that the card is a (*a*) red card; (*b*) 7; (*c*) red card numbered 7.
15. Find the probability of being dealt a five-card hand containing three aces and two kings.
16. Find the probability of being dealt a 13-card bridge hand containing eight spades and five hearts.
17. Four married couples are to play bridge. If a man must select a woman at random to be his partner, what is the probability that no man has his wife as a partner?
18. A medical research team wants to compare the effectiveness of five drugs A, B, C, D, and E. Their facilities allow them to test only three drugs at a time in an experiment.
   (*a*) How many different experiments can be performed using three drugs?
   (*b*) If one experiment is chosen at random, what is the probability that drug A is considered in the experiment?
   (*c*) If one experiment is chosen at random, what is the probability that both drugs A and D are considered in the experiment?
   (*d*) If one experiment is chosen at random, what is the probability that either drug A or drug D is considered in the experiment?
19. Assume that a clock stops at a random time. What is the probability that the minute hand comes to rest after the hour mark 2 and before the hour mark 4?

4.7

The Algebra of Probability

By definition the mathematical probability of an event A is the ratio of the number of elements in A to the number of elements in a sample space S. Since A is a subset of S, the number of elements in A ranges from zero to the number of elements in S; that is,

$$0 \leq n(A) \leq n(S)$$

Therefore, dividing by the number of elements in S, we have

$$0 \leq \frac{n(A)}{n(S)} \leq 1$$

Hence

$$0 \leq P(A) \leq 1$$

the probability of an event is a numerical value between 0 and 1.

If $A = S$, then the event A is certain to occur, and its probability is 1:

$$\frac{n(S)}{n(S)} = 1$$

If $A = \varnothing$, the empty set, then the occurrence of event A is impossible, and its probability is 0:

$$\frac{n(\varnothing)}{n(S)} = 0$$

The occurrence of an event is often called a *success*; the nonoccurrence of an event is called a *failure*. Recall that the set of elements of a universal set which are not elements of A is called the complement of A and is often denoted by $\bar{A}$, and recall that $A \cup \bar{A}$ is the universal set. For any event A, we may think of the set S as the universal set. Then

$$A \cup \bar{A} = S$$

Now, the sets A and $\bar{A}$ are disjoint sets since no element belongs to both sets. Therefore,

$$n(A) + n(\bar{A}) = n(S)$$

and

$$\frac{n(A)}{n(S)} + \frac{n(\bar{A})}{n(S)} = 1$$

$$P(A) + P(\bar{A}) = 1$$

That is, for any event A the probability of success plus the probability of failure is 1.

Notice that in Example 4 of Sec. 4.5 we could have considered obtaining two red beads a success and not obtaining two red beads a failure. Part (b) could have been completed by using the above principle and subtracting $\frac{5}{33}$ from 1.

Example 1

If three coins are tossed, what is the probability that at least one head appears?

A sample space S consists of the eight points described in Example 1 of Sec. 4.6. Let A be the event that at least one head appears. Then $\bar{A}$ is the event that all three coins are tails. Since the number of elements in $\bar{A}$ is one, the outcome (t, t, t), then $P(\bar{A}) = \frac{1}{8}$. Since

$$P(A) + P(\bar{A}) = 1$$

then

$$\begin{aligned} P(A) &= 1 - P(\bar{A}) \\ &= 1 - \tfrac{1}{8} = \tfrac{7}{8}. \end{aligned}$$

Hence the probability is $\frac{7}{8}$ that at least one head appears if three coins are tossed.

Example 2

In the darkness, Paul reached into a drawer that contained eight white socks and two black socks, and he selected three socks at random. Determine the probability
(a) that he selected at least two socks of the same color;
(b) that he selected at least two white socks.

(a) Since three socks are selected, at least two of them must be of the same color. Hence, the event of selecting at least two socks of the same color is certain to occur; the probability of this event is 1. (b) Since there are ten socks in the drawer, there are ${}_{10}C_3$, that is, 120, ways of selecting three socks. Thus the sample space S consists of 120 sets of socks. If we let A represent the event that at least two white socks are selected, then $\bar{A}$ represents the event that one white sock and two black socks are selected. There are eight ways of selecting one white sock and two black socks. (Why?) Thus

$$P(\bar{A}) = \frac{n(\bar{A})}{n(S)} = \frac{8}{120} = \frac{1}{15}$$

Since

$$P(A) + P(\bar{A}) = 1$$

then

$$\begin{aligned} P(A) &= 1 - P(\bar{A}) \\ &= 1 - \tfrac{1}{15} = \tfrac{14}{15} \end{aligned}$$

Hence the probability is $\frac{14}{15}$ that Paul selected at least two white socks.

Exercises

1. If a pair of dice are thrown, what is the probability that the sum of the face values is (*a*) at least 2; (*b*) less than 2?
2. If two coins are tossed, what is the probability of obtaining (*a*) at least one tail; (*b*) no tails?
3. If the probability is 0.4 that a radio receiver on a moon surveyor will operate for more than a week, what is the probability that it will operate for not more than a week?
4. If the probability of winning a game is 0.3, what is the probability of losing or tying the game?
5. If the probability is $\frac{1}{4}$ that two heads appear when two coins are tossed, what event has probability $\frac{3}{4}$?
6. If the probability is $\frac{33}{36}$ that the sum of the face values is less than 11 when a pair of dice are thrown, what event has probability $\frac{3}{36}$?
7. Find the probability of not obtaining three heads when three coins are tossed?
8. A pair of dice are thrown. Determine the probability that the two face values are not equal.
9. Determine the probability that at least one face value is odd if a pair of dice are thrown.
10. A card is chosen at random from an ordinary deck of bridge cards. Determine the probability that the card chosen is neither a seven of hearts nor a nine of spades.

11. Two socks are selected at random from a drawer containing eight white socks and four black socks. Determine (*a*) the probability that at least one white sock is selected; (*b*) the probability that at least one black sock is selected.
12. Consider two disjoint sets A and B with m elements and n elements, respectively. If an element is selected at random from the set $A \cup B$, express probabilities (*a*) to (*d*) in terms of m and n: (*a*) $P(A)$; (*b*) $P(B)$; (*c*) $P(A) + P(B)$; (*d*) $P(\bar{A})$.
13. A work-study committee of three people is to be chosen at random from a group of six members of management and four members of a labor union. What is the probability that at least one labor union member is on the committee?

4.8

Mathematical Odds

The *mathematical odds in favor* of the occurrence of an event A is defined to be the ratio of the probability of A to the probability of not A, that is,

$$\frac{P(A)}{P(\bar{A})}$$

The *mathematical odds against* the occurrence of an event A is defined to be the reciprocal of this ratio:

$$\frac{P(\bar{A})}{P(A)}$$

Example 1

Find the mathematical odds in favor of obtaining two heads in two tosses of a coin.

Let A be the event of obtaining two heads in two tosses of a coin. Then $P(A) = \frac{1}{4}$ and $P(\bar{A}) = 1 - \frac{1}{4} = \frac{3}{4}$. Therefore, the mathematical odds in favor of A are

$$\frac{P(A)}{P(\bar{A})} = \frac{1/4}{3/4} = \frac{1}{3}$$

that is, 1 to 3.

Example 2

A pair of dice are thrown. Find the mathematical odds against 7 being the sum of the face values.

Let A be the event that the sum of the face values is 7. Using the sample space of 36 points illustrated in Example 3 of Sec. 4.6,

$$A = \{(1, 6), (2, 5), (3, 4), (4, 3), (5, 2), (6, 1)\}$$

Thus the number of elements in A is 6, and

$$P(A) = \tfrac{6}{36} = \tfrac{1}{6}$$

Then

$$P(\bar{A}) = 1 - \tfrac{1}{6} = \tfrac{5}{6}$$

Hence

$$\frac{P(\bar{A})}{P(A)} = \frac{5/6}{1/6} = \frac{5}{1}$$

That is, the mathematical odds against the sum of the face values being 7 are 5 to 1.

Exercises

1. A box contains five red, four white, and three blue beads of the same size and shape. If one bead is chosen at random, find:
   (*a*) the mathematical odds in favor of the bead being red
   (*b*) the mathematical odds against the bead being red
   (*c*) the mathematical odds in favor of the bead being white
   (*d*) the mathematical odds in favor of the bead being blue
2. A box contains six red, three white, and three blue beads. If two beads are chosen at random at one time, find:
   (*a*) the mathematical odds in favor of both beads being red
   (*b*) the mathematical odds against both beads being red
   (*c*) the mathematical odds in favor of both beads being white
   (*d*) the mathematical odds in favor of both beads being blue
3. If three coins are tossed, find:
   (*a*) the mathematical odds in favor of obtaining at least one head
   (*b*) the mathematical odds in favor of obtaining three heads or three tails
4. If a pair of dice are thrown, find:
   (*a*) the mathematical odds in favor of the sum of the face values being 6
   (*b*) the mathematical odds against the sum of the face values being 11

5. Two defective light bulbs were accidentally placed in a box containing eight nondefective light bulbs.
   (*a*) What are the mathematical odds in favor of selecting at random a nondefective light bulb?
   (*b*) If two light bulbs are selected at random at one time, what are the mathematical odds in favor of at least one light bulb being nondefective?
   (*c*) If a defective light bulb is selected and not replaced in the box, what are the mathematical odds in favor of the next bulb selected at random being a nondefective light bulb?
   (*d*) What is the fewest number of bulbs that should be selected to be certain of obtaining a nondefective bulb?
6. A card is chosen at random from an ordinary deck of bridge cards. Determine:
   (*a*) the mathematical odds in favor of the card being a face card
   (*b*) the mathematical odds in favor of the card being a spade
   (*c*) the mathematical odds against the card being the seven of hearts
   (*d*) the mathematical odds against the card being an ace
7. A combination lock contains three dials with 40 settings each. What are the mathematical odds against accidentally setting the dials on the combination?
8. If the mathematical odds in favor of the occurrence of event A are r/s, what is the probability that event A will occur?

4.9

The Addition Law of Probability

Recall that two or more events are called mutually exclusive if they have no points in common, that is, if the sets of points representing the events are disjoint sets. The events A and $\bar{A}$ are mutually exclusive events—both events cannot occur at the same time. As another example, in throwing a die the six possible outcomes in terms of their face values represent six mutually exclusive events. In determining a sample space for an experiment we always consider a set of mutually exclusive events.

There are problems in which we are interested in determining the probability that one of two mutually exclusive events occurs in an experiment.

Example 1

If a pair of dice are thrown, what is the probability that the sum of the face values is 5 or 8?

By the second fundamental principle of enumeration, the number of ways of obtaining 5 or 8 is equal to the number of ways of obtaining 5 plus the number of ways of obtaining 8. By an examination of the sample space in Example 3 of Sec. 4.6 we can determine the points associated with event A, "the sum of the face values is 5," or event B, "the sum of the face values is 8." We find that

$$A \cup B = \{(1, 4), (2, 3), (3, 2), (4, 1), (2, 6), (3, 5), (4, 4), (5, 3), (6, 2)\}$$

The probability of obtaining 5 or 8 is expressed as

$$P(A \cup B) = \frac{n(A \cup B)}{n(S)} = \frac{9}{36} = \frac{1}{4}$$

Note that if events A and B are mutually exclusive events, then

$$n(A \cup B) = n(A) + n(B)$$

Therefore,

$$\frac{n(A \cup B)}{n(S)} = \frac{n(A)}{n(S)} + \frac{n(B)}{n(S)}$$

$$P(A \cup B) = P(A) + P(B)$$

That is, *if events A and B are mutually exclusive, then the probability of A or B is equal to the probability of A plus the probability of B.* This statement is a special form of a general rule known as the *addition law of probability.*

Example 2

If A and B are mutually exclusive events with the probability of event A equal to $\frac{1}{4}$ and the probability of event B equal to $\frac{1}{3}$, find the probability of A or B.

By the special form of the addition law of probability,

$$P(A \cup B) = P(A) + P(B)$$
$$= \tfrac{1}{4} + \tfrac{1}{3} = \tfrac{7}{12}$$

We shall now consider the more general form of the addition law of probability. If events A and B are *not* mutually exclusive, then $P(A \cup B) \neq P(A) + P(B)$. For example, consider the event A, "the sum is a multiple of 4," and the event B, "the sum is greater than 9," when a pair of dice are thrown. Then $n(A \cup B) = 14$, $n(A) = 9$, $n(B) = 6$, and

$$P(A \cup B) \neq P(A) + P(B)$$

since

$$\tfrac{14}{36} \neq \tfrac{9}{36} + \tfrac{6}{36}$$

Note that the set $A \cap B$ contains a point, "the sum is 12"; that is, 12 is a multiple of 4 and also greater than 9. Since the point in $A \cap B$ is counted among the points of A and among the points of B,

$$n(A \cup B) = n(A) + n(B) - n(A \cap B)$$

and

$$P(A \cup B) = P(A) + P(B) - P(A \cap B)$$
$$\tfrac{14}{36} = \tfrac{9}{36} + \tfrac{6}{36} - \tfrac{1}{36}$$

Recall from Exercise 5 of Sec. 2.9 that, for any two sets A and B,

$$n(A \cup B) = n(A) + n(B) - n(A \cap B)$$

Therefore,

$$\frac{n(A \cup B)}{n(S)} = \frac{n(A)}{n(S)} + \frac{n(B)}{n(S)} - \frac{n(A \cap B)}{n(S)}$$

$$P(A \cup B) = P(A) + P(B) - P(A \cap B)$$

This statement is called the *addition law of probability*.

Example 3

A card is chosen from an ordinary deck of bridge cards. What is the probability that either a spade or a face card is chosen?

Consider the suit and denomination of each of the 52 cards as points of our sample space. Let A be the event that a spade is chosen; let B be the event that a face card is chosen. Note that A and B are not mutually exclusive events. The set $A \cap B$ contains the jack of spades, the queen of spades, and the king of spades. Now, $n(S) = 52$, $n(A) = 13$, $n(B) = 12$, and $n(A \cap B) = 3$. Therefore, $P(A) = \frac{13}{52}$, $P(B) = \frac{12}{52}$, and $P(A \cap B) = \frac{3}{52}$. By the addition law of probability,

$$P(A \cup B) = P(A) + P(B) - P(A \cap B)$$
$$= \tfrac{13}{52} + \tfrac{12}{52} - \tfrac{3}{52} = \tfrac{11}{26}$$

The probability is $\frac{11}{26}$ that a spade or face card is chosen.

Exercises

In Exercises 1 through 6 identify the events as being mutually exclusive or not mutually exclusive.

1. Obtaining an odd number and obtaining an even number when a die is thrown.
2. Obtaining an odd number and obtaining a 5 when a die is thrown.
3. Selecting a face card and selecting a heart when a card is selected from an ordinary deck of bridge cards.
4. Selecting a face card and selecting a number card when a card is selected from an ordinary deck of bridge cards.
5. Obtaining exactly one head and obtaining exactly one tail when three coins are tossed.
6. Obtaining at least two heads and obtaining at least one tail when three coins are tossed.
7. If A and B are mutually exclusive events with $P(A) = \frac{1}{2}$ and $P(B) = \frac{1}{3}$, find $P(A \cup B)$.
8. Given $P(A) = \frac{2}{3}$, $P(B) = \frac{1}{4}$, and $P(A \cap B) = \frac{1}{12}$, find $P(A \cup B)$.
9. Show that if $P(A) = \frac{2}{5}$ and $P(B) = \frac{3}{4}$, then A and B cannot be mutually exclusive events.
10. Consider three mutually exclusive events A, B, and C that exhaust the sample space. If $P(\bar{A}) = \frac{5}{6}$ and $P(B) = \frac{1}{3}$, find $P(C)$.
11. If a pair of dice are thrown, what is the probability that the sum of the face values is 4 or 9?
12. If a pair of dice are thrown, what are the mathematical odds in favor of the sum of the face values being 7 or 11?
13. What is the probability of obtaining four heads or four tails when four coins are tossed?
14. What are the mathematical odds against obtaining four heads or four tails when four coins are tossed?
15. Determine the probability of selecting a red face card or a spade number card from a deck of bridge cards.
16. Determine the probability of selecting a queen of spades or an ace of any suit from a deck of bridge cards.
17. One card is drawn from each of two ordinary decks of bridge cards. Find the probability that at least one of the two cards is a spade.
18. Determine the probability of selecting a red face card or heart face card from a deck of bridge cards.

19. A bag contains 7 limes and 11 lemons. If a person picks two pieces of fruit from the bag, what is the probability that they are both (*a*) lemons; (*b*) the same kind of fruit?

20. Three members of a committee are chosen at random from a group of four teachers, three lawyers, and six doctors. What is the probability that there will be a majority of one profession on the committee?

21. Consider a box that contains five otherwise indistinguishable beads numbered 1, 2, 3, 4, and 5, respectively, from which two beads are selected at random at one time.
    (*a*) Determine the probability of selecting two beads, the sum of whose assigned numbers is 5 or 6.
    (*b*) What are the mathematical odds in favor of selecting two beads, the sum of whose assigned numbers is 3 or 4?
    (*c*) What are the mathematical odds in favor of selecting two beads that are both odd or both even?

4.10

Conditional Probability

In Sec. 2.9 we considered the problem of 47 students in a certain high school studying mathematics, biology, and French. The Venn diagram in Fig. 4.6 indicates the number of these students studying each subject.

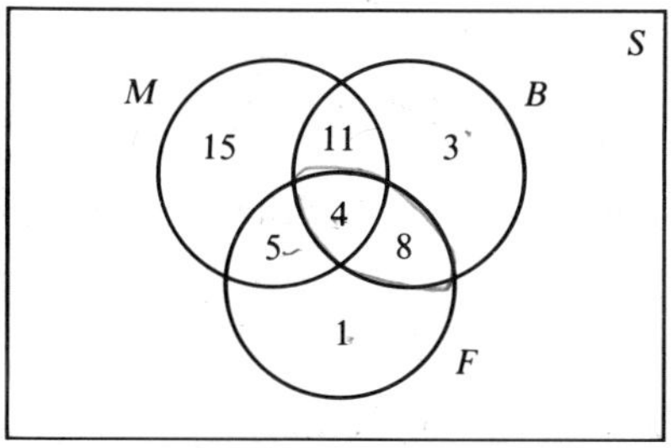

Figure 4.6

Let M be the event that a student studies mathematics, B the event that he studies biology, and F the event he studies French. If a student is selected at random, what is the probability that he studies mathematics? Our sample space consists of the

47 students; thus, $n(S) = 47$. There are $15 + 5 + 11 + 4$, that is, 35, students studying mathematics. Hence,

$$P(M) = \frac{n(M)}{n(S)} = \frac{35}{47}$$

Imagine that we select a student at random and identify that student as one who studies biology. What is the probability that he also studies mathematics? The situation has changed; our sample space is no longer the set of 47 students in the high school. Our sample space has been restricted to the set of 26 students who study biology, as indicated by the shaded region of the Venn diagram in Fig. 4.7.

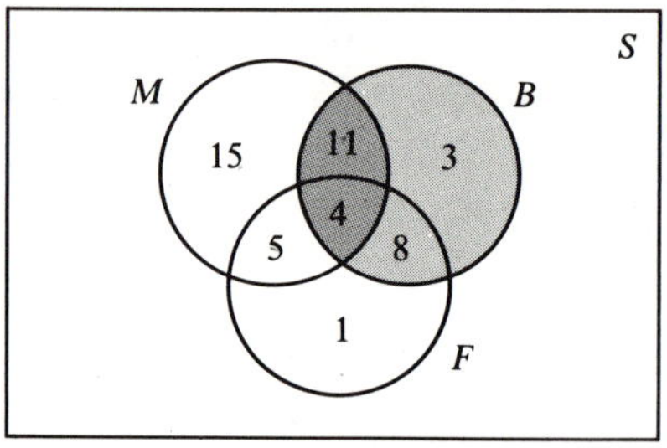

Figure 4.7

Among these 26 students, those who also study mathematics are contained in the heavily shaded region. There are $11 + 4$, that is, 15, students who study mathematics among the 26 students who study biology. Hence, the probability that a student studies mathematics given that he studies biology is $\frac{15}{26}$. The probability $\frac{15}{26}$ just calculated is called the *conditional probability* of M given B and is denoted as $P(M|B)$. It is determined by considering the set B as a restricted sample space; the points favorable to the event "M given B" are contained in $M \cap B$. Thus,

$$P(M|B) = \frac{n(M \cap B)}{n(B)}$$

In general, given two events A and B of the same sample space,

$$P(A|B) = \frac{n(A \cap B)}{n(B)}$$

provided $n(B) \neq 0$. Since A and B are subsets of the same sample space,

$$\frac{n(A \cap B)}{n(B)} = \frac{n(A \cap B)/n(S)}{n(B)/n(S)} = \frac{P(A \cap B)}{P(B)}$$

$$P(A|B) = \frac{P(A \cap B)}{P(B)} \qquad \text{where } P(B) \neq 0$$

This statement defines the *conditional probability* of A given B, which is denoted by $P(A \mid B)$.

The idea of conditional probability is not really a new one. All probabilities are determined only after some conditions that restrict the possibilities (sample space) to be considered have been stated. Furthermore, note that the conditional probability $P(A \mid B)$ is defined completely by the number of elements in the two subsets B and $A \cap B$ of the original sample space.

Example 1

Assume that each child born is equally likely to be either a boy or a girl. If a family has three children, find the probability that all are boys if at least two of the children are boys.

Let b represent a boy and g represent a girl. The points of an appropriate sample space S may be represented by each branch of the tree diagram in Fig. 4.8. The first letter of each ordered

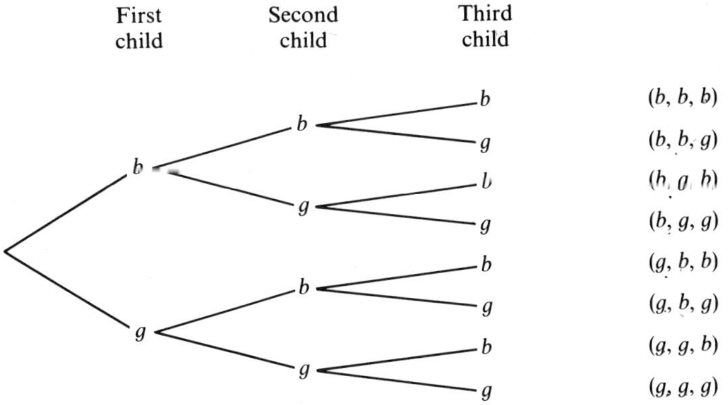

Figure 4.8

triple represents the oldest child, the second letter represents the middle child, and the third letter represents the youngest child. Let A represent the event that all the children are boys; let B represent the event that at least two of the children are boys. Then $A = \{(b, b, b)\}$, $B = \{(b, b, b), (b, b, g), (b, g, b), (g, b, b)\}$, $A \cap B =$

$\{(b, b, b)\}$. Furthermore, $n(B) = 4$, $n(A \cap B) = 1$, and $n(S) = 8$. Therefore,

$$P(A \mid B) = \frac{P(A \cap B)}{P(B)} = \frac{1/8}{4/8} = \frac{1}{4}$$

That is, the probability that all the children are boys given the fact that at least two are boys is $\frac{1}{4}$.

Example 2

A pair of dice are thrown. Find the probability that the sum of the face values is greater than 8 given that at least one of the faces is a 6.

An original sample space S for the experiment of throwing a pair of dice consists of the 36 points of ordered pairs in the array shown in Fig. 4.9.

(1, 1),	(1, 2),	(1, 3),	(1, 4),	(1, 5),	(1, 6),
(2, 1),	(2, 2),	(2, 3),	(2, 4),	(2, 5),	(2, 6),
(3, 1),	(3, 2),	(3, 3),	(3, 4),	(3, 5),	(3, 6),
(4, 1),	(4, 2),	(4, 3),	(4, 4),	(4, 5),	(4, 6),
(5, 1),	(5, 2)	(5, 3),	(5, 4),	(5, 5),	(5, 6),
(6, 1),	(6, 2),	(6, 3),	(6, 4),	(6, 5),	(6, 6).

Figure 4.9

Let A be the event that the sum of the face values is greater than 8; let B be the event that at least one of the faces is a 6. The points associated with event A are contained in the triangular region of the array; the points associated with event B are contained in the shaded region of the array. The points (3, 6), (4, 6), (5, 6), (6, 3), (6, 4), (6, 5), and (6, 6) are associated with the event $A \cap B$. Since $n(A \cap B) = 7$ and $n(B) = 11$, then

$$P(A \mid B) = \frac{n(A \cap B)}{n(B)} = \frac{7}{11}$$

That is, the probability that the sum of the face values is greater than 8 given that at least one of the face values is a 6 is $\frac{7}{11}$.

Note that $P(A|B) \neq P(A)$ in both Examples 1 and 2. In Example 1,

$$P(A|B) = \tfrac{1}{4} \qquad \text{while} \qquad P(A) = \tfrac{1}{8}$$

In Example 2,

$$P(A|B) = \tfrac{7}{11} \qquad \text{while} \qquad P(A) = \tfrac{5}{18}$$

When $P(A|B) = P(A)$, then A and B are called *independent events*. An event A is independent of event B if the occurrence of B does not affect the occurrence of A.

Example 3

Show that if a card is selected at random from an ordinary deck of bridge cards, the event of obtaining a five is independent of the event of obtaining a club.

Consider the sample space S to be the set of 52 cards identified by suit and denomination. Then $n(S) = 52$. Let A represent the event of obtaining a five; let B represent the event of obtaining a club. Then $n(A) = 4$, $P(A) = \frac{4}{52} = \frac{1}{13}$, $n(B) = 13$, and $P(B) = \frac{13}{52} = \frac{1}{4}$. Now $A \cap B$ represents the event of obtaining the five of clubs; $n(A \cap B) = 1$ and $P(A \cap B) = \frac{1}{52}$. By the definition of conditional probability,

$$P(A|B) = \frac{P(A \cap B)}{P(B)} = \frac{1/52}{1/4} = \frac{1}{13}$$

Since $P(A)$ also equals $\frac{1}{13}$,

$$P(A|B) = P(A)$$

Thus A and B are independent events.

Our common knowledge of the composition of an ordinary deck of bridge cards should have suggested to us that events A and B of Example 3 were independent events. Suit and denomination are independent properties of a card. A deck of cards contains one card of each denomination for each suit. We should be familiar with other independent events. For example, the outcome of each toss of a coin is an event independent of the outcome of previous tosses; the outcome of each throw of a die is an event independent of the outcome of throwing another die.

Exercises

In Exercises 1 through 6 consider the problem presented in this section of 47 students in a certain high school who are studying mathematics, biology, and French.

1. What is the probability that a student studies French given that he studies biology?
2. What is the probability that a student studies biology given that he studies mathematics?
3. What is the probability that a student studies mathematics given that he studies both biology and French?
4. What is the probability that a student studies mathematics given that he studies biology or French?
5. What is the probability that a student studies biology or French given that he studies mathematics?
6. What is the probability that a student studies biology and mathematics given that he studies French?
7. If a family has three children, find the probability that the oldest child is a boy if at least two of the children are boys.
8. A pair of dice are thrown. Find the probability that the sum of the face values is less than 6 given that one and only one face value is a 4.
9. A pair of dice are thrown. Find the probability that the sum of the face values is 7 given that at least one of the face values is a 4.
10. A bridge player has been dealt nine spades and four hearts. What is the probability that his partner has the remaining four spades?
11. Consider a set of 25 disks such that on each one we have marked one of the pairs of letters AC, AD, BC, or BD. The frequencies are shown below at the right and correspond to the table at the left:

	C	D
A	AC	AD
B	BC	BD

	C	D
A	7	3
B	5	10

Find (*a*) $P(C)$; (*b*) $P(D|A)$; (*c*) $P(B \cup D)$; (*d*) $P(A|B \cup D)$; (*e*) $P(C|B)$; (*f*) $P(C|A \cup B)$.

12. A group of students consists of 300 boys and 200 girls. One-third of the boys are freshmen and the remainder are sophomores. One-fourth of the girls are freshmen and the remainder are sophomores. If a student is chosen at random, find the probability that (*a*) the student is a freshman girl; (*b*) the student is either a freshman or a girl; (*c*) the student is a freshman if we know she is a girl; (*d*) the student is a girl if we know she is a freshman.

4.11

The Multiplication Law of Probability

The definition of conditional probability can be stated in an alternate useful form. Let A and B be any two events for which $P(B) \neq 0$. By the definition of conditional probability,

$$P(A \mid B) = \frac{P(A \cap B)}{P(B)}$$

Therefore,

$$P(A \cap B) = P(B) \cdot P(A \mid B)$$

This statement is called the *multiplication law of probability*, which simply states that to determine the fraction of successes that are in both A and B, we first determine the fraction of successes in B and multiply this by the fraction of successes which, being in B, are also in S.

If $P(A) \neq 0$, then $P(B \mid A) = \dfrac{P(B \cap A)}{P(A)} = \dfrac{P(A \cap B)}{P(A)}$

and

$$P(A \cap B) = P(A) \cdot P(B \mid A) = P(B) \cdot P(A \mid B)$$

Example 1

A box contains nine red and six white beads. Determine the probability of selecting a red bead and then a white bead if the bead drawn first is not replaced.

Let R represent the event of selecting a red bead on the first draw; let W represent the event of selecting a white bead on the second draw. The $R \cap W$ represents the event of selecting a red bead followed by a white bead. Because there are 9 red beads among the 15 beads, $P(R) = \frac{9}{15} = \frac{3}{5}$. Assuming that a red bead was selected on the first draw, there would be 6 white beads among the remaining 14 beads; hence $P(W|R) = \frac{6}{14} = \frac{3}{7}$. By the multiplication law of probability,

$$P(R \cap W) = P(R) \cdot P(W|R) = \frac{3}{5} \cdot \frac{3}{7} = \frac{9}{35}$$

Again, we may interpret the above product in the following manner. When we select two beads without replacement from a box of nine red beads and six white beads, $\frac{3}{5}$ of the time we shall select a red bead on the first draw, and, when we do, $\frac{3}{7}$ of those times we shall also select a white bead on the second draw.

Example 2

Consider the box of beads in Example 1. Find the probability of selecting (*a*) two red beads; (*b*) two white beads; (*c*) a white and then a red bead. (*d*) Show that the sum of the probabilities obtained in (*a*) to (*c*) and in Example 1 is equal to 1.

$$(a)\ P(R \cap \overline{W}) = P(R) \cdot P(\overline{W}|R) = \frac{3}{5} \cdot \frac{4}{7} = \frac{12}{35}$$

$$(b)\ P(\overline{R} \cap W) = P(\overline{R}) \cdot P(W|\overline{R}) = \frac{2}{5} \cdot \frac{5}{14} = \frac{1}{7}$$

$$(c)\ P(\overline{R} \cap \overline{W}) = P(\overline{R}) \cdot P(\overline{W}|\overline{R}) = \frac{2}{5} \cdot \frac{9}{14} = \frac{9}{35}$$

$$(d)\ P(R \cap W) + P(R \cap \overline{W}) + P(\overline{R} \cap W) + P(\overline{R} \cap \overline{W}) = \frac{9}{35} + \frac{12}{35} + \frac{1}{7} + \frac{9}{35} = 1$$

If A and B are independent events, then $P(A|B) = P(A)$ and, similarly, $P(B|A) = P(B)$. In this case the multiplication law of probability may be stated simply as

$$P(A \cap B) = P(A) \cdot P(B)$$

In general, if $A_1, A_2, A_3, \ldots, A_n$ are n independent events, then

$$P(A_1 \cap A_2 \cap A_3 \cap \cdots \cap A_n) = P(A_1) \cdot P(A_2) \cdot P(A_3) \cdots P(A_n)$$

Example 3

For a certain type of seed the probability of a single seed germinating is 0.6. Find the smallest number of seeds that should be planted such that the probability of at least one seed germinating is 0.95.

It is assumed that whether or not a particular seed germinates is an event independent of what other seeds do. The probability that one seed does not germinate is 0.4; the probability that two seeds do not germinate is $(0.4)^2$, that is, $(0.4)(0.4)$; the probability that three seeds do not germinate is $(0.4)^3$, that is, $(0.4)(0.4)(0.4)$; $\cdots$; in general, the probability that n seeds do not germinate is $(0.4)^n$. Therefore, the probability that at least one seed among n seeds germinates is given by the expression $1 - (0.4)^n$. For the probability of at least one seed germinating to be 0.95 or greater,

$$1 - (0.4)^n \geq 0.95$$
$$0.05 \geq (0.4)^n$$

(The solution to this inequality may be found by trial and error or by the use of logarithms.) The smallest counting number that satisfies this inequality is 4. Thus, at least four seeds should be planted to be 95 percent certain that at least one seed will germinate.

The multiplication law is most useful when two or more events occur in succession, regardless of whether or not the events are independent. When the probabilities of the possible outcomes of each event can be determined, based upon the conditions at that time in the sequence, the sequence of events is a random process called a *stochastic process*. We have been studying stochastic processes whenever two or more events have been considered in sequential order. For example, selecting several beads from a box of colored beads, regardless of whether or not each bead drawn is replaced prior to the next selection, is a stochastic process. The successive tosses of a coin, the successive throws of a pair of dice, and the successive selections of a card from a deck of cards are stochastic processes. Many questions concerning stochastic processes can best be answered by use of a tree diagram, as illustrated in Fig. 4.10 in Example 4.

Example 4

One urn contains three white, five red, and two black balls. A second urn contains eight white, three red, and four black balls. An urn is selected by chance, and then a ball is selected at random. Find the probability of selecting a red ball.

Because each urn is equally likely to be chosen, and there is a total of 8 red balls among the 25 balls, one might incorrectly assume that the probability of selecting a red ball is $\frac{8}{25}$.

Consider a sequence of two experiments in which we first choose an urn and then select a ball, as indicated in the tree diagram of Fig. 4.10.

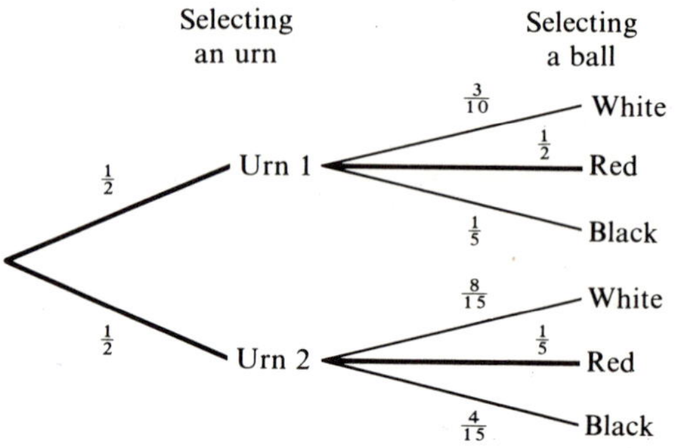

Figure 4.10

The paths of each selection are labeled with the associated probabilities. The paths that are heavily shaded result in the final selection of a red ball. The probability of selecting urn 1 and then a red ball is given by the product $\frac{1}{2} \cdot \frac{1}{2}$, that is, $\frac{1}{4}$; the probability of selecting urn 2 and then a red ball is given by the product $\frac{1}{2} \cdot \frac{1}{5}$, that is, $\frac{1}{10}$. Since these selections are mutually exclusive, the probability of selecting a red ball by the random process described is given by $\frac{1}{4} + \frac{1}{10}$, that is, $\frac{7}{20}$.

Note that the probability of selecting a white ball is $\frac{1}{2} \cdot \frac{3}{10} + \frac{1}{2} \cdot \frac{8}{15}$, that is, $\frac{5}{12}$; the probability of selecting a black ball is $\frac{1}{2} \cdot \frac{1}{5} + \frac{1}{2} \cdot \frac{4}{15}$, that is, $\frac{7}{30}$. Since either a red, white, or black ball must be selected, we may check our results by noting that

$$P(\text{red}) + P(\text{white}) + P(\text{black}) = \frac{7}{20} + \frac{5}{12} + \frac{7}{30} = 1$$

Exercises

1. If a die is thrown two times, what is the probability of first throwing a 3 and then a 4?
2. If a single card is selected from each of three separate decks of bridge cards, what is the probability that all will be hearts?
3. What is the probability of drawing a king and then a queen from an ordinary deck of bridge cards without replacement?

4. A box contains six white and four black balls. If three balls are drawn in succession and none are replaced, what is the probability that the first two will be white and the third will be black?

5. Steven and Paul draw in turn from an urn containing nine black and three white balls. If the balls are not replaced after being drawn and the winner is the player who first draws a white ball, what is Steve's chance of winning if he has the first draw?

6. If Donna and Robby each throw three coins, what is the chance that they throw the same number of heads?

7. What is the chance that the ace of hearts lies next to the ace of spades in a well-shuffled deck?

8. If the probability of a person winning a prize is $\frac{1}{4}$, what is the probability that the person does not win on two successive attempts?

9. If the probability that a husband lives 20 years more is 0.8 and the probability that his wife lives 20 years more is 0.6, what is the probability that 20 years from now the husband or wife or both will be living?

10. A die weighted so that the chance of throwing a number is proportional to that number and an unbiased die are thrown together. What is the chance of obtaining a 9 on a single throw of the pair of dice?

11. Each of three urns contains nine slips of paper numbered 1 to 9. If one slip is drawn from each urn, what is the chance that the sum of the numbers on the slips is odd?

12. If, on the average, two out of every three persons have had athlete's foot and three out of every eight have had an impacted wisdom tooth, what is the probability that at least one of four persons selected has had both?

13. A bag contains only three coins; one is coined with two heads, and the other two have both a head and a tail. A coin is drawn at random from the bag. If the coin is tossed four times in succession and each time a head turns up, what is the probability that this coin is the coin with two heads?

14. A population consists of equal numbers of males and females. Among the males, $\frac{2}{3}$ are color blind; among the females, $\frac{4}{9}$ are color blind. If a person selected at random turns out to be color blind, what is the probability that the person is male?

15. How many dice have to be thrown to render the probability of no ace less than $\frac{1}{3}$?

16. If the probability of a marksman hitting the bullseye is $\frac{2}{5}$, how many shots must be fired to be 90 percent certain of at least one hit?

17. Find an expression for the probability that at least 2 people in a group of 25 people have the same birthday with regard to month and day. Assume a year of 365 days.

4.12

Mathematical Expectation

Quite often a set of numbers is described by a single number that represents the set in some way. For example, the average of the set of numbers {3, 7, 10, 16} is 9. The average of a set of numbers is the quotient of the sum of the numbers and the number of elements in the set. Thus,

$$\frac{3 + 7 + 10 + 16}{4} = 9$$

The topic of this section, *mathematical expectation*, is a type of average. Before defining what we mean by mathematical expectation, let us consider an example that contains the idea.

Example 1

A light bulb manufacturer can make a profit of \$0.10 on each nondefective bulb. However, if a light bulb is defective, he loses \$0.50. What is the average profit per light bulb that the manufacturer can expect if 95 percent of the light bulbs are nondefective?

If 95 percent of the light bulbs are nondefective, the manufacturer will make a profit of \$0.10 × 95, that is, \$9.50, for every 100 bulbs produced. If 5 percent of the light bulbs are defective, the manufacturer will lose \$0.50 × 5, that is, \$2.50, for every 100 bulbs produced. Thus his net profit will be \$9.50 – \$2.50, that is, \$7.00, for every 100 bulbs produced. Hence the manufacturer can expect an average profit of \$0.07 per light bulb.

The calculations are summarized by means of the following table:

Light bulb	Probability	Profit	Probability × Profit
Nondefective	0.95	\$0.10	\$0.0950
Defective	0.05	–\$0.50	–\$0.0250
Sum (expected profit per light bulb):			\$0.07

Consider a set of mutually exclusive events $A_1, A_2, A_3, \ldots, A_n$ that form a sample space S for an experiment or game. Let $p_1, p_2, p_3, \ldots, p_n$ be the probabilities of these events, respectively. If numerical values $v_1, v_2, v_3, \ldots, v_n$ may be attached to each event, respectively, then the *mathematical expectation* E of the experiment or game is defined as

$$E = p_1 v_1 + p_2 v_2 + p_3 v_3 + \cdots + p_n v_n$$

Example 2

Determine the mathematical expectation E for the experiment of throwing a single die.

The mutually exclusive events are the possible face values 1, 2, 3, 4, 5, and 6. The probability of each event is $\frac{1}{6}$. Therefore, by the definition of mathematical expectation,

$$\begin{aligned} E &= \tfrac{1}{6} \times 1 + \tfrac{1}{6} \times 2 + \tfrac{1}{6} \times 3 + \tfrac{1}{6} \times 4 + \tfrac{1}{6} \times 5 + \tfrac{1}{6} \times 6 \\ &= \tfrac{1}{6}(1 + 2 + 3 + 4 + 5 + 6) = \tfrac{1}{6}(21) = 3.5 \end{aligned}$$

Hence the mathematical expectation for the experiment of throwing a single die is that the face value is 3.5.

Example 3

The Fibonacci Construction Company knows they will spend $100 preparing a bid on a street repair job that should yield a $1,000 profit. They are one of five companies that have an equal chance of receiving the contract for the job. What is their mathematical expectation in submitting a bid?

The probability of event A, "the Fibonacci Company will receive the contract for the job," is $\frac{1}{5}$; the value associated with receiving the contract is $900 ($1,000 minus the $100 expenses). The probability of event $\bar{A}$, "they will not receive the contract for the job," is $\frac{4}{5}$; the value associated with not receiving the contract is −$100 (the cost of preparing a bid). Thus

$$\begin{aligned} E &= \tfrac{1}{5} \times 900 + \tfrac{4}{5} \times (-100) \\ &= 180 - 80 = 100 \end{aligned}$$

Therefore, the mathematical expectation of the Fibonacci Company in submitting a bid is $100.

Example 4

Steven and Paul play a game in which a pair of coins are tossed. Steven collects $0.09 if two heads appear; Paul collects $0.03 if two heads do not appear. What is Steven's expected average gain per toss of the coins?

The calculations are summarized by means of the following table:

Event	Probability of the event	Steven's gain	Probability × Gain
Two heads	$\frac{1}{4}$	$0.09	$0.0225
Not two heads	$\frac{3}{4}$	– $0.03	– $0.0225
Sum (Steven's expected gain per toss):			0

In Example 4, Steven's expected gain per toss is 0. If we compute Paul's expected gain per toss, we would find that it is also 0. The game in Example 4 is an example of a fair game. A game is called a *fair game* if each player has a mathematical expectation of zero.

Exercises

1. Amy has 1 of 1,000 tickets distributed in a contest that has a $100 prize.
   (*a*) If the ticket was free, what is her mathematical expectation?
   (*b*) If Amy paid $1.00 for the ticket, what is her mathematical expectation?
   (*c*) If Amy paid $0.10 for the ticket, what is her mathematical expectation?
   (*d*) If tickets cost $1.00 and Amy bought two of them (to increase her chances, she thought), what is her mathematical expectation?
2. A nickel, two dimes, one quarter, and one half-dollar are wrapped in five identical packages.
   (*a*) What is the mathematical expectation when one package is selected at random?
   (*b*) What would be a fair price to pay for the privilege of selecting a package?
   (*c*) What would be the mathematical expectation when one package is selected at random if three empty packages were added?
   (*d*) If the two dimes were in the same package, what would be the mathematical expectation when one package is selected at random?

3. Steven and Paul play a game called Tong, in which they simultaneously show their right hands with either one, two, or three fingers extended. Paul scores two points if both boys show an equal number of fingers, and Steven scores one point if they do not. What is Paul's mathematical expectation?
4. A glasscutter can sell a particular pane of glass that cost him \$1.20 for \$3.00. If the probability of breaking the glass is 0.10, what is his mathematical expectation?
5. If the mathematical odds are 2:1 against John winning a \$600 scholarship, what is his mathematical expectation?
6. If the mathematical odds are 1:5 in favor of John winning a \$10 contest, what is his mathematical expectation if he paid \$1 to enter the contest?
7. Determine the expected sum of the face values when a pair of dice are thrown.
8. A bead is selected from a box of five indistinguishable beads numbered 1, 2, 3, 4, and 5, respectively. Determine the mathematical expectation for the experiment of selecting a bead at random.
9. A box contains five otherwise indistinguishable beads numbered 1, 2, 3, 4, and 5, respectively.
   (*a*) If two beads are selected at random at one time, what is the expected sum of the numbers on the beads?
   (*b*) If two beads are selected at random, but the first bead is replaced before the second bead is selected, what is the expected sum of the numbers on the beads?
10. Determine the expected number of heads when three coins are tossed.
11. Suppose the manufacturer in Example 1 could reduce production costs and thereby make a profit of \$0.12 on each nondefective light bulb. If his loss is still \$0.50 on a defective light bulb, what is the average profit per light bulb that the manufacturer can expect?
12. A manufacturer of small motors can make a profit of \$3.50 on each nondefective motor. However, if a motor is defective, he loses \$15.75. What is the average profit per motor that the manufacturer can expect if 92 percent of the motors are nondefective?
13. Consider a 100-question multiple-choice test. Each question has four suggested answers, only one of which is correct. Robby took the test but randomly checked answers without reading the questions.
    (*a*) If one point is credited for each correct answer, what is Robby's expected score on the test?
    (*b*) If one point is credited for each correct answer and $-\frac{1}{3}$ of a point is credited for each incorrect answer, what is Robby's expected score?
14. Scott and Kim play a game in which three coins are tossed. Scott agrees that Kim scores 3 points if exactly two heads appear; Kim agrees that Scott scores 5 points if exactly two heads do not appear. What is Scott's expected average gain per toss of the coins?

15. Betty and Tony play a game in which each person tosses either one or two dimes. Betty wins all of Tony's coins that have the same face value, heads or tails, as her own; Tony wins all of Betty's coins that do not match his in face value.
    (*a*) What is Betty's mathematical expectation if each person tosses one dime?
    (*b*) What is Betty's mathematical expectation if she tosses two dimes and Tony tosses one dime?
16. A football coach must decide whether to try for a touchdown (6 points) with a probability of success of 0.35 or try for a field goal (3 points) with a probability of success of 0.75. Which decision has the highest expected gain?
17. Two players each put \$1.00 in a pot. They then decide to throw a pair of dice alternately, the first one to throw a 7 winning the pot. How much should the player who starts add to the pot to make this a fair game?
18. Bert is a gambler betting against the house on the roll of a pair of dice. If the house odds are 1:1 for less than 7, 1:1 for more than 7, and 4:1 for 7, and Bert is betting \$5.00 on each possibility, what is the house's expectation?
19. Peter and Paul make alternate tosses of a (possibly biased) coin. Peter begins. The winner is the one who first gets heads. The probability of heads is p.
    (*a*) What is the probability that Peter wins?
    (*b*) What is the probability that Paul wins?
    (*c*) If $p = \frac{1}{2}$ and Peter's stake is \$1, what should Paul's stake be in order that the game be a fair game?
    (*d*) What is the probability that at least n tosses will be made before the game ends?

4.13

Empirical Probability

In some experiments it may not be possible to determine a set of equally likely outcomes. For example, it is not possible to list a set of equally likely outcomes to apply our definition of mathematical probability to weather forecasting, athletic events, or the study of life insurance. However, a typical weather forecast might read "There is a 30 percent chance of rain today." The meteorologist has beeen able to attach a numerical measure to his prediction of rain by determining what proportion of the times it has rained in the past when the temperature, barometric reading, wind, and other factors were almost identical to those of the day on which he is making the prediction.

Consider a simple experiment in which a thumbtack is dropped on a smooth, hard surface. The tack can come to rest in one of two positions, as illustrated in Fig. 4.11, which we call "up" and "down." (We assume that the tack cannot land on its point since the surface on which it is dropped is smooth and hard.) Generally the events up and down will not be equally likely to occur. However,

$$\text{Probability of up} + \text{probability of down} = 1$$

It seems reasonable to assume that the probability of up is some fixed ratio for any particular tack but different for different-size tacks. Indeed, the length of the shaft of the tack is one important factor that determines the probability of up. For instance, if the length of the shaft is zero, the tack approximates a coin, and we would expect the probability of up to be approximately $\frac{1}{2}$.

Figure 4.11

How can we determine the probability of up for a particular tack? We can answer this question empirically, that is, by experimentation and observation. Consider a fixed number of trials of dropping the tack, and then record the number of successes, that is, the number of times the tack came to rest up. The ratio of successes to trials can be used to estimate the probability of up. Of course, we would expect our estimate to improve with an increase in the number of trials. For example, we would place more confidence in an estimate obtained from 500 trials than in one obtained from 100 trials.

If we perform an experiment and find that a particular tack lands up 59 times and down 41 times, the empirical probability of up according to our experiment is $59/(59 + 41)$, that is, $\frac{59}{100}$. In general, if an event occurs m times in an experiment that involves n trials, the *empirical probability* of the event is defined as equal to m/n.

We intuitively interpret a baseball player's batting average, a basketball player's field goal or free-throw average, and a quarterback's passing average as empirical probabilities. We expect more from the player with a higher average, and a professional player's salary for the current year is usually based on his record over the past years.

In addition, life insurance rates are computed from empirical probabilities based on data recorded in mortality tables. A mortality table contains a numerical record of a particular set of persons alive at each of several different ages. Thus, according to these tables, the empirical probability of living from age x to a certain age y is

$$\frac{\text{The number of persons alive at age } y}{\text{The number of persons alive at age } x}$$

Table 4.1 contains some selected data from a mortality table that was constructed from records of a particular set of 100,000 persons (such information is available from the U.S. Department of Health, Education and Welfare, National Center for Health Statistics.)

Table 4.1

Age	Number alive
0	100,000
10	96,936
15	96,733
20	96,280
40	93,096
60	79,152
70	60,910
85	19,861

Example

Use Table 4.1 to determine the probability that a person of age 15 will be alive at age 60.

In the experience recorded in Table 4.1, there were 96,733 persons alive at age 15 and 79,152 persons alive at age 60. Thus the probability that a person of age 15 will be alive at age 60 is 79,152/96,733, that is, approximately 0.82.

Exercises

1. A carpet tack is tossed 200 times. If the tack comes to rest "up" 32 times, what is the empirical probability that it will come "up" on any particular toss?
2. A upholstery tack lands up 82 times and down 200 times. What is the empirical probability that it will land up on any particular toss?
3. A weather bureau has been correct in 84 of their last 120 predictions. Find the empirical probability that their prediction for tomorrow will be correct.

4. In 1,000 cases, when certain weather conditions were the same as today, it rained on the following day 203 times. On this basis, what is the probability it will rain tomorrow?
5. A girl has earned an A in 18 of the 25 courses she has taken. On this basis, what is the probability she will get an A in any particular course?
6. A teacher who has taught algebra to 1,500 students has given A's to 200. Of the 90 students taking algebra this year, how many students might be expected to receive an A?
7. Of 200 firecrackers that were lighted, 16 failed to explode. How many of the next 50 should one expect to explode?
8. Use Table 4.1 to determine the probability that a person of age 15 will be alive at age 85.
9. Use Table 4.1 to determine the probability that a person of age 70 will be alive at age 85.
10. Under a \$1,000 one-year term life insurance policy, the insurance company guarantees to pay to a person's beneficiary \$1,000 if the person dies within the year. If the probability that a man of age 25 will live at least one more year is 0.99, what is the insurance company's mathematical expectation if they sell him a \$1,000 one-year term life insurance policy for \$20?

4.14

Probability Functions

The mathematical probability $P(A)$ of an event A is a *probability function.* The domain of $P(A)$ consists of the possible sets A and S; the range of $P(A)$ is the set of real numbers between 0 and 1.

Throughout our discussion of probability we required that the sample space considered consist of a set of equally likely points. In many problems it may not be desirable or practical to describe such a sample space. Also, it is desirable to redefine our concept of a probability function $P(A)$ for the domain to be a set of numbers. Whenever a one-to-one correspondence between a set of mutually exclusive events that exhaust a sample space and an appropriate set of real numbers can be established, we can change the domain of the probability function from a set of events to a set of numbers. The variable x that represents any one of the elements of this set of real numbers for an experiment is called a *random variable.*

Let x be a random variable whose values $x_1, x_2, x_3, \ldots, x_n$ correspond to n mutually exclusive and exhaustive events $A_1, A_2, A_3, \ldots, A_n$, respectively. The function $f(x)$ for which

$$f(x_i) = P(A_i)$$

is called a *probability function.* A probability function defined over the subsets of a sample space has certain properties:

1. $f(x_i) \geq 0$ for $i = 1, 2, 3, \ldots, n$.
2. $f(x_1) + f(x_2) + \cdots + f(x_n) = 1$.
3. If A and B are events corresponding to x_i and x_j, respectively, then $P(A \cup B) = f(x_i) + f(x_j)$.

Property (1) states that the value of the probability function must be nonnegative for each value of the random variable. Property (2) states that the sum of the values of the probability function for all possible values of the random variable x must be equal to 1. Property (3) states that the events corresponding to the values of the random variable must be mutually exclusive.

Example 1

Describe a probability function for the experiment of tossing three coins.

A sample space for this experiment is

$$S = \{(h, h, h), (h, h, t), (h, t, h), (h, t, t), (t, h, h), (t, h, t), (t, t, h), (t, t, t)\}$$

Let x be a random variable whose value is equal to the number of heads obtained when three coins are tossed. Then we may consider a new sample space S', where

$$S' = \{0, 1, 2, 3\}$$

Each value of the random variable x is an element of S'. Furthermore, associated with each value of x is a set of mutually exclusive events corresponding to a subset of the sample space S. The event (t, t, t) is associated with the value $x = 0$; the events (h, t, t), (t, h, t), and (t, t, h) are associated with the value $x = 1$; the events (h, h, t), (h, t, h), and (t, h, h) are associated with the value $x = 2$; the event (h, h, h) is associated with the value $x = 3$. A probability function $f(x)$ for this experiment is such that

$$f(0) = \tfrac{1}{8} \qquad f(1) = \tfrac{3}{8} \qquad f(2) = \tfrac{3}{8} \qquad f(3) = \tfrac{1}{8}$$

Note that $f(x)$ is a possible probability function since $f(x) \geq 0$ for $x = 0, 1, 2, 3$ and $f(0) + f(1) + f(2) + f(3) = 1$.

Example 2

Six beads are numbered 1 through 5 and placed in a box. Two beads are then selected at random. Describe a probability function for the sum of the values obtained.

A sample space for this experiment is the following set of pairs of numbers:

$$S = \{(1, 2), (1, 3), (1, 4), (1, 5), (2, 3), (2, 4), (2, 5), (3, 4), (3, 5), (4, 5)\}$$

However, the present experiment is concerned with the set

$$S' = \{3, 4, 5, 6, 7, 8, 9\}$$

of possible sums of the values of the beads. Associated with each sum is a set of events corresponding to a subset of the sample space S:

$$\begin{aligned}
&\{(1, 2)\} && \leftrightarrow 3\\
&\{(1, 3)\} && \leftrightarrow 4\\
&\{(1, 4), (2, 3)\} && \leftrightarrow 5\\
&\{(1, 5), (2, 4)\} && \leftrightarrow 6\\
&\{(2, 5), (3, 4)\} && \leftrightarrow 7\\
&\{(3, 5)\} && \leftrightarrow 8\\
&\{(4, 5)\} && \leftrightarrow 9
\end{aligned}$$

These events are mutually exclusive and exhaustive (include all possibilities); therefore, we may use the set S' as a domain for the random variable x for this experiment.

The following table of values defines the probability function $f(x)$ for this experiment:

x	3	4	5	6	7	8	9
$f(x)$	$\frac{1}{10}$	$\frac{1}{10}$	$\frac{1}{5}$	$\frac{1}{5}$	$\frac{1}{5}$	$\frac{1}{10}$	$\frac{1}{10}$

A probability function may also be described by means of a graph. For example, consider the probability function of Example 1. Rectangles whose heights are equal to $f(x_i)$ may be constructed on intervals of one unit centered about x_i on a horizontal axis as shown in Fig. 4.12. Each rectangle constructed has an area numerically equal to $f(x_i)$, and the numerical value of the total area of the graph is equal to 1. This type of graph is called a *probability histogram*.

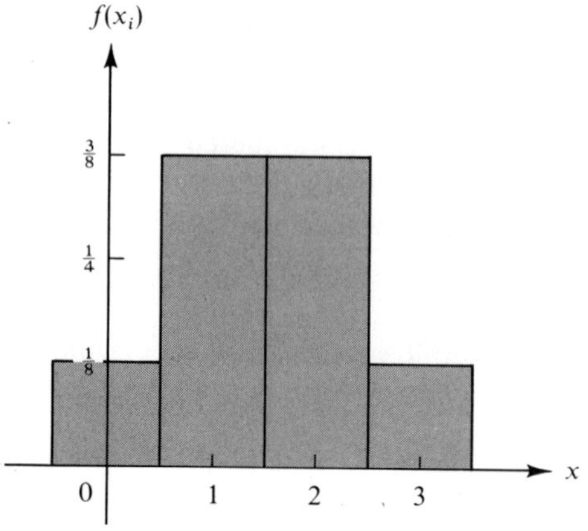

Figure 4.12

Example 3

Determine a probability function for the random variable representing the sum of the face values that may be obtained when a pair of dice are thrown. Construct a probability histogram for the function.

An appropriate sample space S of 36 equally likely outcomes when a pair of dice are thrown was described in Example 3 of Sec. 4.6. Let x be a random variable representing the sum of the face values of the dice. Then

$$x \in \{2, 3, 4, 5, 6, 7, 8, 9, 10, 11, 12\}$$

Recall that there is only one way of obtaining the sum 2, two ways of obtaining the sum 3, three ways of obtaining the sum 4, four ways of obtaining the sum 5, five ways of obtaining the sum 6, six ways of obtaining the sum 7, five ways of obtaining the sum 8, four ways of obtaining the sum 9, three ways of obtaining the sum 10, two ways of obtaining the sum 11, and one way of obtaining the sum 12.

The following table of values defines the probability function $f(x)$ for this experiment:

x	2	3	4	5	6	7	8	9	10	11	12
$f(x)$	$\frac{1}{36}$	$\frac{1}{18}$	$\frac{1}{12}$	$\frac{1}{9}$	$\frac{5}{36}$	$\frac{1}{6}$	$\frac{5}{36}$	$\frac{1}{9}$	$\frac{1}{12}$	$\frac{1}{18}$	$\frac{1}{36}$

Finally, the probability histogram for this function is shown in Fig. 4.13.

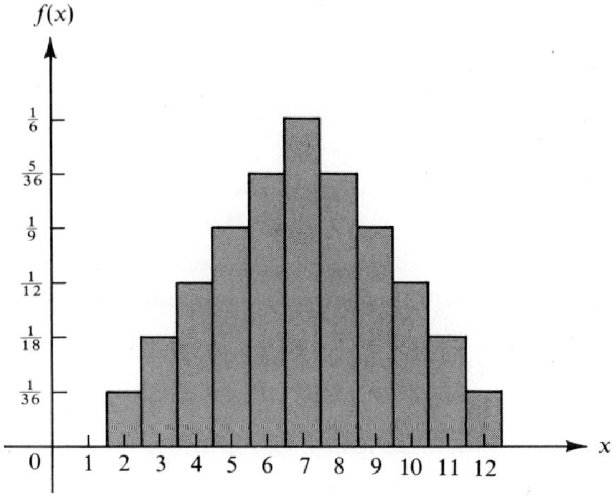

Figure 4.13

Exercises

1. Describe a domain for a random variable for the number of heads when (*a*) five coins are tossed; (*b*) *n* coins are tossed.
2. Find a domain for a random variable representing the sum of the face values that may be obtained when (*a*) three dice are thrown; (*b*) *n* dice are thrown.
3. Construct a table of values for the probability function for the experiment of tossing five coins.
4. Construct a table of values for the probability function for the experiment of throwing three dice.

In Exercises 5 through 9 determine which of the given functions may be considered probability functions of the random variable *x*.

5.

x	1	2	3
$f(x)$	$\frac{5}{12}$	$\frac{1}{3}$	$\frac{1}{4}$

6.

x	1	3	5	7
$f(x)$	1	0	0	1

7. $f(x) = \frac{1}{4}$ for $x = 1, 2, 3, 4$
8. $f(x) = {}_3C_x (\frac{2}{3})^x (\frac{1}{3})^{3-x}$ for $x = 0, 1, 2, 3$
9. $f(x) = \dfrac{{}_6C_x \cdot {}_3C_{2-x}}{{}_9C_2}$
10. Find k such that $f(x)$ may be considered a probability function if $f(x) = x/k$ for $x = 1, 2, 3, 4, 5$. Construct a probability histogram for $f(x)$.
11. A die is weighted so that the probability of a particular face value is proportional to that number. Describe by means of a table of values a probability function for the face values of the die.
12. A box contains three red and two white beads. Let a random variable x represent the number of selections without replacement needed to obtain a red bead. Find the probability function $f(x)$. Construct a probability histogram for $f(x)$.
13. A box contains five toys. A child randomly takes out a toy and then replaces it; he does this three times. Let x be the number of different toys he takes out. Find a set of values for the probability function of x.
14. The *mathematical expectation of a random variable* x, denoted by $E(x)$, is defined as

$$E(x) = x_1 f(x_1) + x_2 f(x_2) + \cdots + x_n f(x_n)$$

$E(x)$ is often called the *mean* of x. Find the mean number of heads for the experiment of tossing three coins.
15. Find the mathematical expectation of the random variable in Example 3 (see Exercise 14).
16. A box contains five red and three white beads. If three beads are selected simultaneously and at random, find the expected number of white beads.
17. If a probability function is constant for all values of the random variable x, then $f(x)$ may be expressed in the form

$$f(x) = \frac{1}{n} \qquad \text{for } x = 1, 2, 3, \ldots, n$$

and is called a *uniform probability function.* (*a*) Show that the probability function for x, where x is the face value obtained when a die is thrown, is a uniform probability function. (*b*) Find the mean of x (see Exercise 14).

4.15

Binomial Probability Functions

In this section we shall describe two important and related types of probability functions. Consider an experiment consisting of two mutually exclusive events A and $\bar{A}$: if $P(A) = p$, then $P(\bar{A}) = 1 - p$. A random variable for such an experiment may have only two values. As a matter of convenience, let the value 1 correspond to the event A; let the value 0 correspond to the event $\bar{A}$. Then

$$f(1) = p \qquad \text{and} \qquad f(0) = 1 - p$$

The probability function $f(x)$ may be expressed as

$$f(x) = p^x(1 - p)^{1-x} \qquad \text{for } x = 0, 1$$

A random variable x whose value is either 0 or 1 is called a *Bernoulli random variable*; its probability function (noted above) is called a *Bernoulli probability function*. An experiment with a Bernoulli random variable is called a *Bernoulli experiment*.

Example 1

A die has four faces colored red and two faces colored green. Determine a probability function for the outcome of a single throw of the die.

Let A be the event that a red face is obtained when the die is thrown. Then $\bar{A}$ is the event that a green face is obtained. Let the value 1 of the Bernoulli random variable x correspond to the event A; let the value 0 correspond to the event $\bar{A}$. Then $f(1) = \frac{2}{3}$ and $f(0) = \frac{1}{3}$. The Bernoulli probability function $f(x)$ may be expressed as

$$f(x) = (\tfrac{2}{3})^x(\tfrac{1}{3})^{1-x} \qquad \text{for } x = 0, 1$$

This experiment of a single throw of the die is a Bernoulli experiment.

Example 2

An experiment consists of a sequence of five successive throws of the colored die in Example 1. Find a probability function for the number of times a red face is obtained.

This experiment consists of a sequence of five independent trials of a Bernoulli experiment. On each Bernoulli trial consider the event that a red face is obtained to be a success. Let x represent the number of successes in the five trials. The probability of x successes followed by $5 - x$ failures is given by $(\frac{2}{3})^x(\frac{1}{3})^{5-x}$. However, a total of x successes may result from several different sequences of successes and failures. Because there are ${}_5C_x$ such sequences of successes and failures that total x successes and because these sequences are mutually exclusive events, the probability of exactly x successes is given by ${}_5C_x(\frac{2}{3})^x(\frac{1}{3})^{5-x}$. Hence the probability function for the number of times a red face is obtained when the colored die is thrown five times is

$$f(x) = {}_5C_x(\tfrac{2}{3})^x(\tfrac{1}{3})^{5-x} \qquad \text{for } x = 0, 1, 2, 3, 4, 5$$

The following table of values shows this probability function:

x	0	1	2	3	4	5
$f(x)$	$\frac{1}{243}$	$\frac{10}{243}$	$\frac{40}{243}$	$\frac{80}{243}$	$\frac{80}{243}$	$\frac{32}{243}$

The probability function determined in Example 2 is an example of a second type of probability function that we now discuss in general. Consider an experiment that consists of a sequence of n trials of a Bernoulli experiment, with success on each trial equal to p. A random variable that represents the number of successes in such an experiment is called a *binomial random variable*. The possible values of a binomial random variable are 0, 1, 2, . . . , n. The probability function for such a random variable is called a *binomial probability function* and is defined by

$$f(x) = {}_nC_x p^x(1-p)^{n-x} \qquad \text{for } x = 0, 1, 2, \ldots, n$$

The term binomial probability function derives from the fact that each value of $f(x)$ is a term of the expansion of the binomial expression $(a+b)^n$, where $a = p$ and $b = 1 - p$.

Example 3

If 70 percent of the people of a state favor the incumbent for reelection as governor, find the probability that a random sample of eight voters favor him.

Each voter sampled is a Bernoulli trial with probability of success ("favor the incumbent for reelection") equal to $\frac{7}{10}$. Polling a random sample of eight voters may be considered a sequence of eight such Bernoulli trials. Let x represent the number of successes; then the binomial probability function to be considered is given by

$$f(x) = {}_8C_x(\tfrac{7}{10})^x(\tfrac{3}{10})^{8-x}$$

The sample will favor the governor for reelection if five, six, seven, or eight successes are obtained. Hence, the probability that the random sample of eight voters favor him is given by

$$_8C_5\left(\tfrac{7}{10}\right)^5\left(\tfrac{3}{10}\right)^3 + {_8C_6}\left(\tfrac{7}{10}\right)^6\left(\tfrac{3}{10}\right)^2 + {_8C_7}\left(\tfrac{7}{10}\right)^7\left(\tfrac{3}{10}\right)^1 + {_8C_8}\left(\tfrac{7}{10}\right)^8\left(\tfrac{3}{10}\right)^0$$

An evaluation of each term of this expression gives us the approximate sum $0.254 + 0.296 + 0.198 + 0.058$, that is, 0.806. Therefore, the probability that the random sample of eight voters will favor him is approximately 0.806.

Exercises

1. What is the probability that if seven coins are tossed, exactly five will fall heads and two will fall tails?
2. Construct a probability histogram for the binomial probability function in Example 2.
3. If one tosses five coins together, what is the probability that he will obtain at least four heads?
4. Find a probability function for the number of heads obtained when six coins are tossed. Construct a probability histogram for this function.
5. If the Cardinals win, on the average, two of every three games they play with the Giants, what is the probability that the Cardinals win at least three games in a five-game series with the Giants?
6. If the probability that a marksman will hit a certain target is 0.1, find the value of his expectation if for each hit in three trials he will receive $20.
7. Show that the mathematical expectation of a binomial random variable is np (see Exercise 14 of Sec. 4.14).
8. A box contains three red, two white, and five blue beads. Let a random variable x represent the number of blue beads selected if three beads are selected simultaneously and at random. (*a*) Find the probability function $f(x)$. (*b*) Determine the expected number of blue beads selected.
9. A thumbtack that can fall up with a probability of $\frac{2}{3}$ is tossed four times. Find the probability that the events "up" and "down" occur equally often.
10. At State University the ratio of men to women is 3:1. In a random group of ten students, find the "most likely" number of men. (That is, what number of men has the greatest probability of occurring?)
11. A submarine captain has the option of firing two, three, or four torpedoes at an enemy ship during a 1-minute interval. If he fires only two, each torpedo has a probability of 0.5 of hitting the enemy ship. The probability of each

torpedo hitting the enemy ship is reduced to 0.4 if he fires three and 0.3 if he fires four torpedoes in the same time interval. If two hits are necessary to sink the enemy ship, how many torpedoes should the captain fire?

12. A machine manufactures 50 items per hour. On the average, four of these items will be defective. If a sample of three items is selected at random, what is the probability that no more than one is defective?

13. Let x be a random variable representing the number of trials necessary for the first success in a sequence of Bernoulli trials. Show that the probability function of x is given by

$$f(x) = (1 - p)^{x-1}p \qquad \text{for } x \geq 1$$

where p is the probability of success on a single Bernoulli trial. This probability function is called a *geometric probability function*.

14. If a pair of dice are tossed until a seven appears, find the probability that a seven appears on the xth toss, where $x = 1, 2, 3, \ldots, n$ (see Exercise 13).

Keyed Summary

In this chapter we studied the fundamental principles of enumeration, permutations, combinations, mathematical probability, and probability functions. The concepts and methods presented are part of the body of knowledge called the *algebra of probability*.

4.1 First Fundamental Principle of Enumeration

If a selection can be made in m ways, and if another selection can be made in n ways, then both selections can be made in mn ways.

4.2 Second Fundamental Principle of Enumeration

If a selection can be made in m ways, and if another selection can be made in n ways, then either of the two selections can be made in $m + n$ ways.

4.3 Permutations

A selection without regard to order is a combination; that is, a combination is a subset of a set.

A selection with regard to order is a linear arrangement or linear permutation. A linear permutation is called simply a permutation.

If n is a positive integer, then

$$n! = n(n-1)(n-2) \cdots (3)(2)(1)$$

The symbol $n!$ is read "n factorial." By definition, $0! = 1$.

The total number of permutations from a set of n distinguishable elements taken r at a time is denoted by ${}_nP_r$, and

$${}_nP_r = n(n-1)(n-2) \cdots (n-r+1) = \frac{n!}{(n-r)!}$$

The total number of permutations of n elements, where k of them are indistinguishable, is $n!/k!$.

The total number of permutations of n elements of which k_1 are alike, k_2 others are alike, k_3 others are alike, ..., and k_r others are alike is expressed by

$$\frac{n!}{k_1!\,k_2!\,k_3! \cdots k_r!}$$

4.4 Combinations

The total number of combinations from a set of n elements taken r at a time is denoted by ${}_nC_r$, and

$${}_nC_r = \frac{{}_nP_r}{r!} = \frac{n!}{r!\,(n-r)!}$$

4.5 Mathematical Probability

The mathematical probability of an event is the ratio of the number of outcomes favorable to that event to the total number of possible outcomes, provided that the possible outcomes can be assumed to be equally likely.

4.6 Sample Spaces

The set of all possible equally likely outcomes of an experiment is a sample space S.

Each outcome of an experiment is called a sample point or, more briefly, a point. The total number of points in a sample space S is denoted by $n(s)$.

Any subset of a sample space is called an event. The number of elements in an event A is denoted by $n(A)$.

The mathematical probability of an event A may be represented by $P(A)$ and expressed as

$$P(A) = \frac{n(A)}{n(S)}$$

4.7 *The Algebra of Probability*

The probability of an event A is a numerical value between 0 and 1; that is,

$$0 \leq P(A) \leq 1$$

If $A = S$, then $P(A) = 1$; if $A = \varnothing$, then $P(A) = 0$.

The occurrence of an event is often called a success; the nonoccurrence of an event is called a failure.

For any event A the probability of success plus the probability of failure is 1; that is,

$$P(A) + P(\bar{A}) = 1$$

4.8 *Mathematical Odds*

The mathematical odds in favor of the occurrence of an event A is defined as the ratio of the probability of A to the probability of not A; that is, $P(A)/P(\bar{A})$.

The mathematical odds against the occurrence of an event A is defined $P(\bar{A})/P(A)$.

4.9 *The Addition Law of Probability*

The addition law of probability states that

$$P(A \cup B) = P(A) + P(B) - P(A \cap B)$$

If events A and B are mutually exclusive, then

$$P(A \cup B) = P(A) + P(B)$$

4.10 *Conditional Probability*

The conditional probability of A given B is denoted by $P(A | B)$ and is defined by

$$P(A | B) = \frac{P(A \cap B)}{P(B)} \qquad \text{where } P(B) \neq 0$$

When $P(A | B) = P(A)$, then A and B are called independent events.

4.11 The Multiplication Law of Probability

The multiplication law of probability states that

$$P(A \cap B) = P(B) \cdot P(A|B) = P(A) \cdot P(B|A)$$

If A and B are independent events, then

$$P(A \cap B) = P(A) \cdot P(B)$$

In general, if $A_1, A_2, A_3, \ldots, A_n$ are independent events, then

$$P(A_1 \cap A_2 \cap A_3 \cap \cdots \cap A_n) = P(A_1) \cdot P(A_2) \cdot P(A_3) \cdots P(A_n)$$

When the probabilities of the possible outcomes of each event in a sequence of events can be determined, based upon conditions at that time in the sequence, the sequence of events is a random process called a stochastic process.

4.12 Mathematical Expectation

Let $A_1, A_2, A_3, \ldots, A_n$ be a set of mutually exclusive events that form a sample space S for an experiment or game. Let $p_1, p_2, p_3, \ldots, p_n$ be the probabilities of these events, respectively. If numerical values $v_1, v_2, v_3, \ldots, v_n$ may be attached to each event, respectively, then the mathematical expectation E of the experiment or game is defined as

$$E = p_1 v_1 + p_2 v_2 + p_3 v_3 + \cdots + p_n v_n$$

A game is called a fair game if each player has a mathematical expectation of zero.

4.13 Empirical Probability

The empirical probability of an event is the ratio of the number of successes to the number of trials, which is determined by experimentation and observation.

4.14 Probability Functions

The mathematical probability $P(A)$ of an event A is a probability function. The domain of $P(A)$ consists of the possible sets A and S; the range of $P(A)$ is the set of real numbers between 0 and 1.

Let x be a random variable whose values $x_1, x_2, x_3, \ldots, x_n$ correspond to n mutually exclusive and exhaustive events $A_1, A_2, A_3, \ldots, A_n$, respectively. The function $f(x)$, for which

$$f(x_i) = P(A_i)$$

is called a probability function.

A probability function defined over the subsets of a sample space has three certain properties:

1. $f(x_i) \geq 0$ for $i = 1, 2, 3, \ldots, n$.
2. $f(x_1) + f(x_2) + \cdots + f(x_n) = 1$.
3. If A and B are events corresponding to x_i and x_j, respectively, then $P(A \cup B) = f(x_i) + f(x_j)$.

A probability function may be described by means of a graph called a probability histogram.

4.15 Binomial Probability Functions

An experiment consisting of two mutually exclusive events A and $\bar{A}$ is called a Bernoulli experiment. A random variable x with value 1 corresponding to event A and with value 0 corresponding to event $\bar{A}$ is called a Bernoulli random variable. The probability function

$$f(x) = p^x(1-p)^{1-x} \qquad \text{for } x = 0, 1$$

where p is the probability of event A, is called a Bernoulli probability function.

A random variable x that represents the number of successes in an experiment consisting of a sequence of n Bernoulli trials is called a binomial random variable. The probability function

$$f(x) = {}_nC_x p^x(1-p)^{n-x} \qquad \text{for } x = 0, 1, 2, \ldots, n$$

where p is the probability of success on each Bernoulli trial, is called a binomial probability function.

Building Your Mathematical Vocabulary

Addition law of probability (4.9)
Bernoulli experiment (4.15)
Bernoulli probability function (4.15)
Bernoulli random variable (4.15)
Binomial probability function (4.15)
Binomial random variable (4.15)
Combination (4.3, 4.4)
Conditional probability (4.10)
Empirical probability (4.13)
Event (4.6)
Factorial notation (4.3)
Failure (4.7)
Fair game (4.12)
First fundamental principle of enumeration (4.1)
Geometric probability function (4.15)
Independent events (4.10)
Linear arrangement (4.3)
Linear permutation (4.3)
Mathematical expectation (4.12, 4.14)
Mathematical odds (4.8)
Mathematical probability (4.5)

Mean of a random variable (4.14)
Multiplication law of probability (4.11)
Mutually exclusive (4.2)
Pascal triangle (4.4)
Permutation (4.3)
Point (4.6)
Probability function (4.14)
Probability histogram (4.14)
Random (4.5)
Random variable (4.14)
Sample point (4.6)
Sample space (4.6)
Second fundamental principle of enumeration (4.2)
Stochastic process (4.11)
Success (4.7)
Tree diagram (4.1)
Uniform probability function (4.14)

5

Algebra of Matrices

5.1

Matrices

In many branches of mathematics it is necessary to consider a set of numbers in a *rectangular array*; in fact, in many everyday activities sets of numbers arranged in rows and columns are used for comparison purposes and the keeping of records. For example, consider the following performance record of a Pony League coach's three starting pitchers during the first half of the season:

	Innings	Runs	Hits	Strikeouts	Walks
Steve Tepp	35	10	26	63	15
Paul Fressi	22	11	24	30	8
Robby Pett	9	3	8	7	5

If the pattern in which this information is to be recorded is clearly defined in advance, the rectangular array of numbers may be presented simply as

$$\begin{pmatrix} 35 & 10 & 26 & 63 & 15 \\ 22 & 11 & 24 & 30 & 8 \\ 9 & 3 & 8 & 7 & 5 \end{pmatrix}$$

Rectangular arrays of elements such as

$$\begin{pmatrix} 3 & 0 \\ 1 & 2 \end{pmatrix} \qquad \begin{pmatrix} 4 & -1 & 5 \\ 3 & 7 & 2 \end{pmatrix} \qquad \begin{pmatrix} a & b \\ c & d \\ e & f \end{pmatrix}$$

are called *matrices* (singular: *matrix*). A matrix of m rows and n columns is called a matrix of *order* m by n. For example, the matrix $\begin{pmatrix} 4 & -1 & 5 \\ 3 & 7 & 2 \end{pmatrix}$ is a matrix of order 2 by 3: the elements of the first row are 4, -1, and 5; the elements of the second row are 3, 7, and 2; the elements of the first column are 4 and 3; the elements of the second column are -1 and 7; and the elements of the third column are 5 and 2. If the number of rows is equal to the number of columns, then the matrix is called a *square matrix*. The order of a square matrix of n rows and n columns is simply said to be n. Thus, the matrix $\begin{pmatrix} 3 & 0 \\ 1 & 2 \end{pmatrix}$ is a square matrix of order 2.

Example 1

Determine the order of matrices (a) to (d):

(a) $\begin{pmatrix} 5 & 1 & 5 \\ 2 & 3 & -4 \end{pmatrix}$ (b) $\begin{pmatrix} 3 & 6 \\ 4 & -2 \end{pmatrix}$

(c) $\begin{pmatrix} 3 & 1 \\ -5 & 1 \\ 0 & 2 \end{pmatrix}$ (d) $\begin{pmatrix} 0 & 0 & 0 \\ 1 & 0 & 0 \\ 2 & 4 & 6 \end{pmatrix}$

(a) 2 by 3; (b) 2; (c) 3 by 2; (d) 3.

If the elements of a matrix are real numbers or real variables, then the matrix is called a *real matrix*. Hereafter, unless stated otherwise, by "matrix" we shall mean a real matrix.

It will often be convenient to denote matrices by capital letters such as A, B, C, ... (The context of a discussion will indicate the order of the matrices.) Generally, if A is a real matrix of order m by n, we may write

$$A = \begin{pmatrix} a_{11} & a_{12} & \cdots & a_{1n} \\ a_{21} & a_{22} & \cdots & a_{2n} \\ \cdots & \cdots & \cdots & \cdots \\ a_{m1} & a_{m2} & \cdots & a_{mn} \end{pmatrix}$$

The element of the ith row and jth column is designated by a_{ij}. Thus, the elements of the ith row are $a_{i1}, a_{i2}, \ldots, a_{in}$; the elements of the jth column are $a_{1j}, a_{2j}, \ldots, a_{mj}$. It is sometimes convenient to name matrix A by the symbol $((a_{ij}))$.

Example 2

Construct a matrix $((a_{ij}))$ of order 3 by 2, where $a_{ij} = 5i - j^2$.

Since $a_{ij} = 5i - j^2$, then

$$\begin{aligned} a_{11} &= 5(1) - (1)^2 = 4 & a_{12} &= 5(1) - (2)^2 = 1 \\ a_{21} &= 5(2) - (1)^2 = 9 & a_{22} &= 5(2) - (2)^2 = 6 \\ a_{31} &= 5(3) - (1)^2 = 14 & a_{32} &= 5(3) - (2)^2 = 11 \end{aligned}$$

Therefore,

$$((a_{ij})) = \begin{pmatrix} 4 & 1 \\ 9 & 6 \\ 14 & 11 \end{pmatrix}$$

Equality of Matrices

Two matrices A and B are* equal *if and only if they are of the same order and corresponding elements are equal.

In other words, $((a_{ij})) = ((b_{ij}))$ if and only if matrices $((a_{ij}))$ and $((b_{ij}))$ are of the same order and $a_{ij} = b_{ij}$ for all pairs of values i and j. For example, consider the matrices

$$A = \begin{pmatrix} a & b & c \\ d & e & f \end{pmatrix} \quad \text{and} \quad B = \begin{pmatrix} 3 & 0 & 5 \\ 1 & -2 & 4 \end{pmatrix}$$

Both matrices are of the same order. Then $A = B$ if and only if $a = 3, b = 0, c = 5, d = 1, e = -2$, and $f = 4$.

Example 3

Determine whether or not the matrices of the following pairs are equal:

(*a*) $\begin{pmatrix} 3 & 1 & 2 \\ 4 & -2 & 5 \end{pmatrix}$ and $\begin{pmatrix} 3 & 1 \\ 4 & -2 \end{pmatrix}$

(*b*) $\begin{pmatrix} 1 & 3 \\ 5 & 7 \end{pmatrix}$ and $\begin{pmatrix} 1 & 5 \\ 3 & 7 \end{pmatrix}$

(*c*) $\begin{pmatrix} 2 & 4 \\ 4 & 1 \end{pmatrix}$ and $\begin{pmatrix} 2 & 4 \\ 4 & 1 \end{pmatrix}$

(*d*) $\begin{pmatrix} 5 & 2 \\ 4 & 1 \\ 0 & 0 \end{pmatrix}$ and $\begin{pmatrix} 5 & 2 \\ 4 & c \\ a & b \end{pmatrix}$

(*a*) These matrices cannot be equal because they are not of the same order. (*b*) Although these matrices are of the same order, they are not equal because corresponding elements are not equal. (*c*) These matrices are equal. (*d*) These matrices are equal if and only if $a = b = 0$ and $c = 1$.

Some matrices are particularly interesting. Any matrix, all of whose elements are zero, is called a *null* or *zero matrix*. There exists a null matrix of order m by n

for every possible combination of integral values of m and n. For example, $\begin{pmatrix} 0 & 0 \\ 0 & 0 \end{pmatrix}$ is the null matrix of order 2; $\begin{pmatrix} 0 & 0 & 0 \\ 0 & 0 & 0 \end{pmatrix}$ is the null matrix of order 2 by 3. Every null matrix is denoted by 0.

Exercises

1. Determine the order of matrices (a) to (d):

   (a) $\begin{pmatrix} 5 & -1 & 2 \\ 0 & 1 & 4 \end{pmatrix}$ $\quad$ (b) $\begin{pmatrix} 7 & 3 \\ 5 & 11 \end{pmatrix}$

   (c) $\begin{pmatrix} 1 & 0 & 0 \\ 0 & 3 & 0 \\ 0 & 0 & 1 \end{pmatrix}$ $\quad$ (d) $\begin{pmatrix} 6 & 2 \\ 3 & 1 \\ 7 & 7 \end{pmatrix}$

2. How many elements does a matrix of order m by n have?
3. A matrix has 24 elements. What are the possible orders of A?
4. Let $((a_{ij})) = \begin{pmatrix} 9 & 6 \\ 3 & -2 \\ 1 & 4 \end{pmatrix}$. Find (a) a_{21} ; (b) a_{32}.
5. Write the matrix $((a_{ij}))$ of order 2 for which $a_{11} = 3$, $a_{12} = 7$, $a_{21} = -1$, and $a_{22} = 0$.
6. Construct a square matrix $((a_{ij}))$ of order 3, where $a_{ij} = i^2 + j$.
7. Construct a matrix $((a_{ij}))$ of order 2 by 3, where $a_{ij} = 3i - j + 1$.
8. Construct a matrix $((a_{ij}))$ of order 4 by 3, where $a_{ij} = 2i^2 + 3j$.
9. In the matrix $((a_{ij}))$, describe the position of the elements for which (a) $i = 3$; (b) $j = 2$; (c) $i = j$; (d) $i > j$.
10. Determine whether or not the matrices of the following pairs are equal:

    (a) $\begin{pmatrix} 6+1 & 2 \\ 0 & 5-1 \end{pmatrix}$ and $\begin{pmatrix} 7 & \sqrt{4} \\ 0 & 4 \end{pmatrix}$

    (b) $\begin{pmatrix} 1 & 8 & 0 \\ 2 & 3 & 0 \end{pmatrix}$ and $\begin{pmatrix} 1 & 8 \\ 2 & 3 \end{pmatrix}$

11. Find, if possible, the values of x and y that make the following statements true:

    (a) $\begin{pmatrix} 3 & x \\ 1 & 4 \end{pmatrix} = \begin{pmatrix} 3 & 5 \\ 1 & 4 \end{pmatrix}$ $\quad$ (b) $\begin{pmatrix} x & y \\ 1 & 0 \end{pmatrix} = \begin{pmatrix} 2 & 8 \\ 1 & 0 \end{pmatrix}$

    (c) $\begin{pmatrix} x & 2 & 0 \\ 3 & 1 & 0 \end{pmatrix} = \begin{pmatrix} 5 & 2 \\ 3 & 1 \end{pmatrix}$ $\quad$ (d) $\begin{pmatrix} x & y \\ 0 & x \\ 0 & 3 \end{pmatrix} = \begin{pmatrix} 1 & 2 \\ 0 & 5 \\ 0 & 3 \end{pmatrix}$

12. Find a, b, and c if $A = B$, where

$$A = \begin{pmatrix} a^2 & bc \\ 3a - c & a + 1 \end{pmatrix} \quad \text{and} \quad B = \begin{pmatrix} 1 & 0 \\ 1 & 2 \end{pmatrix}$$

13. Write the null matrix of order 3.

5.2

Addition and Subtraction of Matrices

Consider again the matrix in Sec. 5.1, which represented the pitching performances of three Pony League pitchers during the first half of the season. Suppose that the matrix representing the pitching performances of these players during the second half of the season is

$$\begin{pmatrix} 43 & 8 & 28 & 65 & 12 \\ 33 & 11 & 27 & 51 & 6 \\ 6 & 1 & 2 & 6 & 1 \end{pmatrix}$$

A matrix representing the combined statistics on each of the three pitchers may be obtained by adding the corresponding elements of the two matrices:

$$\begin{pmatrix} 35 & 10 & 26 & 63 & 15 \\ \mathbf{22} & \mathbf{11} & \mathbf{24} & \mathbf{30} & \mathbf{8} \\ 9 & 3 & 8 & 7 & 5 \end{pmatrix} + \begin{pmatrix} 43 & 8 & 28 & 65 & 12 \\ \mathbf{33} & \mathbf{11} & \mathbf{27} & \mathbf{51} & \mathbf{6} \\ 6 & 1 & 2 & 6 & 1 \end{pmatrix} = \begin{pmatrix} 78 & 18 & 54 & 128 & 27 \\ \mathbf{55} & \mathbf{22} & \mathbf{51} & \mathbf{81} & \mathbf{14} \\ 15 & 4 & 10 & 13 & 6 \end{pmatrix}$$

Thus, for example, Paul Fressi pitched 55 innings and yielded 22 runs and 51 hits while striking out 81 batters and walking 14.

The definition of matrix addition is motivated by the consideration of the way in which such rectangular arrays are combined.

Addition of Matrices

If A and B are matrices of the order m by n, then the* sum *A + B is defined as a third matrix C of the same order such that each element of C is the sum of the corresponding elements of A and B.

In other words, if $((a_{ij}))$ and $((b_{ij}))$ are matrices of the same order, then $((a_{ij})) + ((b_{ij})) = ((c_{ij}))$, where $c_{ij} = a_{ij} + b_{ij}$ for all pairs of values i and j. For example, if

$$A = \begin{pmatrix} a & b & c \\ d & e & f \end{pmatrix} \quad \text{and} \quad B = \begin{pmatrix} g & h & i \\ j & k & l \end{pmatrix}$$

then

$$A + B = \begin{pmatrix} a + g & b + h & c + i \\ d + j & e + k & f + l \end{pmatrix}$$

Example 1

Find the sum $A + B$, where

(*a*) $A = \begin{pmatrix} 5 & -2 \\ 6 & 1 \end{pmatrix}$ and $B = \begin{pmatrix} 0 & 3 \\ 2 & 4 \end{pmatrix}$

(*b*) $A = \begin{pmatrix} 2 & 1 & 3 \\ 5 & 4 & -1 \end{pmatrix}$ and $B = \begin{pmatrix} 0 & 0 & 5 \\ 1 & 2 & 4 \end{pmatrix}$

$$(a)\ A + B = \begin{pmatrix} 5 & -2 \\ 6 & 1 \end{pmatrix} + \begin{pmatrix} 0 & 3 \\ 2 & 4 \end{pmatrix}$$

$$= \begin{pmatrix} 5+0 & -2+3 \\ 6+2 & 1+4 \end{pmatrix} = \begin{pmatrix} 5 & 1 \\ 8 & 5 \end{pmatrix}$$

$$(b)\ A + B = \begin{pmatrix} 2 & 1 & 3 \\ 5 & 4 & -1 \end{pmatrix} + \begin{pmatrix} 0 & 0 & 5 \\ 1 & 2 & 4 \end{pmatrix}$$

$$= \begin{pmatrix} 2+0 & 1+0 & 3+5 \\ 5+1 & 4+2 & -1+4 \end{pmatrix} = \begin{pmatrix} 2 & 1 & 8 \\ 6 & 6 & 3 \end{pmatrix}$$

Because each element in the sum of two matrices is the algebraic sum of corresponding elements, it follows that several fundamental properties of the addition of real numbers are also valid for the addition of matrices:

1. *Matrix addition is commutative:* $A + B = B + A$.
2. *Matrix addition is associative:* $(A + B) + C = A + (B + C)$.

Example 2

Verify the associative property of matrix addition, where

$$A = \begin{pmatrix} 2 & 1 \\ 5 & 0 \\ -2 & 3 \end{pmatrix} \qquad B = \begin{pmatrix} 6 & -1 \\ -2 & 3 \\ 0 & 4 \end{pmatrix} \qquad C = \begin{pmatrix} 0 & 2 \\ 2 & 4 \\ 5 & 1 \end{pmatrix}$$

$$(A + B) + C = \left[\begin{pmatrix} 2 & 1 \\ 5 & 0 \\ -2 & 3 \end{pmatrix} + \begin{pmatrix} 6 & -1 \\ -2 & 3 \\ 0 & 4 \end{pmatrix}\right] + \begin{pmatrix} 0 & 2 \\ 2 & 4 \\ 5 & 1 \end{pmatrix}$$

$$= \begin{pmatrix} 8 & 0 \\ 3 & 3 \\ -2 & 7 \end{pmatrix} + \begin{pmatrix} 0 & 2 \\ 2 & 4 \\ 5 & 1 \end{pmatrix} = \begin{pmatrix} 8 & 2 \\ 5 & 7 \\ 3 & 8 \end{pmatrix}$$

$$A + (B + C) = \begin{pmatrix} 2 & 1 \\ 5 & 0 \\ -2 & 3 \end{pmatrix} + \left[\begin{pmatrix} 6 & -1 \\ -2 & 3 \\ 0 & 4 \end{pmatrix} + \begin{pmatrix} 0 & 2 \\ 2 & 4 \\ 5 & 1 \end{pmatrix}\right]$$

$$= \begin{pmatrix} 2 & 1 \\ 5 & 0 \\ -2 & 3 \end{pmatrix} + \begin{pmatrix} 6 & 1 \\ 0 & 7 \\ 5 & 5 \end{pmatrix} = \begin{pmatrix} 8 & 2 \\ 5 & 7 \\ 3 & 8 \end{pmatrix}$$

Hence $(A + B) + C = A + (B + C)$ since

$$\begin{pmatrix} 8 & 2 \\ 5 & 7 \\ 3 & 8 \end{pmatrix} = \begin{pmatrix} 8 & 2 \\ 5 & 7 \\ 3 & 8 \end{pmatrix}$$

The null matrix 0 of order m by n is such that for every matrix A of order m by n,

$$A + 0 = 0 + A = A$$

Thus, the null matrix of order m by n is the *additive identity element* for the set of all matrices of order m by n.

The subtraction of matrices is defined in a manner similar to that used for the addition of matrices.

Subtraction of Matrices

If A and B are matrices of the order m by n, then the* difference *$A - B$ is defined as a third matrix C of the same order such that each element of C is the difference of the corresponding elements of A and B.

For example, if

$$A = \begin{pmatrix} a & b & c \\ d & e & f \end{pmatrix} \quad \text{and} \quad B = \begin{pmatrix} g & h & i \\ j & k & l \end{pmatrix}$$

then

$$A - B = \begin{pmatrix} a - g & b - h & c - i \\ d - j & e - k & f - l \end{pmatrix}$$

Example 3

Find the difference $A - B$, where

(*a*) $A = \begin{pmatrix} 2 & 5 \\ 7 & 4 \end{pmatrix}$ and $B = \begin{pmatrix} 1 & 0 \\ 6 & -3 \end{pmatrix}$

(*b*) $A = \begin{pmatrix} 4 & 5 & 0 \\ 1 & 6 & 4 \end{pmatrix}$ and $B = \begin{pmatrix} 5 & 4 \\ 8 & 2 \end{pmatrix}$

(*a*) $$A - B = \begin{pmatrix} 2 & 5 \\ 7 & 4 \end{pmatrix} - \begin{pmatrix} 1 & 0 \\ 6 & -3 \end{pmatrix}$$
$$= \begin{pmatrix} 2-1 & 5-0 \\ 7-6 & 4-(-3) \end{pmatrix} = \begin{pmatrix} 1 & 5 \\ 1 & 7 \end{pmatrix}$$

(*b*) The difference $A - B$ does not exist because the matrices A and B are of different orders.

Exercises

1. Find the sum of matrices A and B, where
$$A = \begin{pmatrix} 3 & 10 \\ 1 & 1 \\ 5 & 2 \end{pmatrix} \quad \text{and} \quad B = \begin{pmatrix} 2 & -2 \\ 5 & 6 \\ 3 & 0 \end{pmatrix}$$

2. Find $A + B$, where
$$A = \begin{pmatrix} x & y \\ 2x & -y \end{pmatrix} \quad \text{and} \quad B = \begin{pmatrix} 3x & -y \\ -x & 2y \end{pmatrix}$$

3. If the sum of two matrices exists, the matrices are said to be *conformable for addition*.
   (*a*) Under what conditions are A and B conformable for addition if matrix A is of order m by 3 and matrix B is of order 2 by n?
   (*b*) Under what conditions are A and B conformable for addition if matrix A is of order r by 2 and matrix B is of order s by 3?

4. Verify the associative property of matrix addition, where
$$A = \begin{pmatrix} 8 & 7 & 1 \\ 2 & 3 & 0 \end{pmatrix} \quad B = \begin{pmatrix} -1 & 2 & 1 \\ 3 & 0 & 5 \end{pmatrix} \quad C = \begin{pmatrix} 6 & 0 & -4 \\ 2 & 1 & 3 \end{pmatrix}$$

5. Find the difference $A - B$, where
$$A = \begin{pmatrix} 5 & 0 & 6 \\ 2 & 0 & 3 \\ 1 & 5 & 8 \end{pmatrix} \quad \text{and} \quad B = \begin{pmatrix} 3 & -1 & 2 \\ 4 & 0 & 0 \\ 5 & 3 & 1 \end{pmatrix}$$

6. Show that $A - B \neq B - A$ for matrices A and B of Exercise 5.

7. Given the two matrices
$$A = \begin{pmatrix} 6 & 3 & 1 \\ 5 & 1 & -4 \end{pmatrix} \quad \text{and} \quad B = \begin{pmatrix} 2 & 1 & 3 \\ -1 & 2 & 1 \end{pmatrix}$$
find (*a*) $A + B$; (*b*) $A - B$.

8. If $A = \begin{pmatrix} 1 & 2 \\ 3 & 4 \\ 5 & 6 \end{pmatrix}$ and $B = \begin{pmatrix} -3 & -2 \\ 1 & -5 \\ 4 & 3 \end{pmatrix}$, find $D = \begin{pmatrix} p & q \\ r & s \\ t & u \end{pmatrix}$ such that $A + B - D = 0$.

9. Solve the matrix equation

$$\begin{pmatrix} a_{11} & a_{12} & a_{13} \\ a_{21} & a_{22} & a_{23} \end{pmatrix} + \begin{pmatrix} 2 & 1 & 0 \\ 3 & -1 & 4 \end{pmatrix} = \begin{pmatrix} 5 & -4 & 2 \\ 3 & 6 & 1 \end{pmatrix}$$

10. Find the additive inverse of matrices (*a*) and (*b*):

(*a*) $\begin{pmatrix} 6 & 0 & 2 \\ -3 & 1 & -9 \end{pmatrix}$ (*b*) $\begin{pmatrix} a & b \\ c & d \end{pmatrix}$

11. Determine A and B if

$$A + B = \begin{pmatrix} 8 & 1 \\ 5 & 1 \end{pmatrix} \quad \text{and} \quad A - B = \begin{pmatrix} -2 & 3 \\ -3 & 1 \end{pmatrix}$$

12. Consider the following table, which gives price information about four commodities on a trading exchange:

Commodity	June 1 prices	
	High	Low
Sugar	8.25	7.96
Cotton	39.85	39.28
Potatoes	3.31	3.18
Silver	156.00	153.60

(*a*) Express the high prices of the commodities as a *column matrix.*
(*b*) Express the low prices of the commodities as a *column matrix.*
(*c*) Find the difference of the two matrices in (*a*) and (*b*). What does this matrix represent?

13. The following table shows the male-female, undergraduate-graduate enrollments in 1970 for three universities:

University	Undergraduate male	Undergraduate female	Graduate male	Graduate female
Yarvard	3,520	650	1,300	350
Darton	2,600	2,250	650	475
Penale	8,770	5,580	2,320	1,310

(*a*) Write a 3 by 2 matrix M that represents the male undergraduate and graduate enrollments in the three universities.

(*b*) Write a 3 by 2 matrix F that represents the female undergraduate and graduate enrollments in the three universities.
(*c*) Find $M + F$. What does the matrix represent?
(*d*) Find $M - F$. What does the matrix represent?

5.3

Multiplication of a Matrix by a Real Number

Consider the sum of three equal matrices such as

$$\begin{pmatrix} 2 & 1 \\ 5 & 0 \end{pmatrix} + \begin{pmatrix} 2 & 1 \\ 5 & 0 \end{pmatrix} + \begin{pmatrix} 2 & 1 \\ 5 & 0 \end{pmatrix}$$

It is convenient to consider this sum as a product of the real number 3 and the matrix $\begin{pmatrix} 2 & 1 \\ 5 & 0 \end{pmatrix}$; that is,

$$3\begin{pmatrix} 2 & 1 \\ 5 & 0 \end{pmatrix} = \begin{pmatrix} 2 & 1 \\ 5 & 0 \end{pmatrix} + \begin{pmatrix} 2 & 1 \\ 5 & 0 \end{pmatrix} + \begin{pmatrix} 2 & 1 \\ 5 & 0 \end{pmatrix}$$

Since $\begin{pmatrix} 2 & 1 \\ 5 & 0 \end{pmatrix} + \begin{pmatrix} 2 & 1 \\ 5 & 0 \end{pmatrix} + \begin{pmatrix} 2 & 1 \\ 5 & 0 \end{pmatrix} = \begin{pmatrix} 6 & 3 \\ 15 & 0 \end{pmatrix}$, we can define

$$3\begin{pmatrix} 2 & 1 \\ 5 & 0 \end{pmatrix} = \begin{pmatrix} 6 & 3 \\ 15 & 0 \end{pmatrix}$$

Note that

$$3 \cdot 2 = 6 \qquad 3 \cdot 1 = 3$$
$$3 \cdot 5 = 15 \qquad 3 \cdot 0 = 0$$

that is,

$$3\begin{pmatrix} 2 & 1 \\ 5 & 0 \end{pmatrix} = \begin{pmatrix} 3 \cdot 2 & 3 \cdot 1 \\ 3 \cdot 5 & 3 \cdot 0 \end{pmatrix} = \begin{pmatrix} 6 & 3 \\ 15 & 0 \end{pmatrix}$$

The definition of the multiplication of a matrix by a real number is a logical consequence of a consideration of the way in which several addends $A + A + \cdots + A$ may be represented.

Multiplication of a Matrix by a Real Number

Given any matrix A and a real number t, then the* product *tA is a matrix whose elements are the products of t and the corresponding elements of A.

In other words, if t is any real number, then $t((a_{ij})) = ((ta_{ij}))$ for all pairs of values i and j. For example, if $A = \begin{pmatrix} a & b \\ c & d \\ e & f \end{pmatrix}$, then

$$tA = \begin{pmatrix} ta & tb \\ tc & td \\ te & tf \end{pmatrix}$$

By definition, if $A = 0$, then $tA = 0$; if $t = 0$, then $tA = 0$.

Note that

$$A - B = A + (-1)B$$

for any two matrices A and B of the same order. In particular,

$$A - A = A + (-1)(A) = 0$$

Thus, $(-1)A$ is the additive inverse of A.

Example 1

Find the product $4A$, where

$$A = \begin{pmatrix} -3 & 2 & 1 \\ 0 & 6 & 4 \end{pmatrix}$$

$$4A = \begin{pmatrix} 4\cdot(-3) & 4\cdot 2 & 4\cdot 1 \\ 4\cdot 0 & 4\cdot 6 & 4\cdot 4 \end{pmatrix} = \begin{pmatrix} -12 & 8 & 4 \\ 0 & 24 & 16 \end{pmatrix}$$

Example 2

Let $A = \begin{pmatrix} 1 & 2 \\ 0 & 3 \end{pmatrix}$ and $B = \begin{pmatrix} 2 & -1 \\ 4 & 6 \end{pmatrix}$. Show that

(*a*) $(3 + 2)A = 3A + 2A$
(*b*) $2(A + B) = 2A + 2B$
(*c*) $3(2A) = (3 \cdot 2)A$

(*a*) $(3 + 2)A = 5A = 5\begin{pmatrix} 1 & 2 \\ 0 & 3 \end{pmatrix} = \begin{pmatrix} 5 & 10 \\ 0 & 15 \end{pmatrix}$;

$$3A + 2A = 3\begin{pmatrix} 1 & 2 \\ 0 & 3 \end{pmatrix} + 2\begin{pmatrix} 1 & 2 \\ 0 & 3 \end{pmatrix}$$

$$= \begin{pmatrix} 3 & 6 \\ 0 & 9 \end{pmatrix} + \begin{pmatrix} 2 & 4 \\ 0 & 6 \end{pmatrix} = \begin{pmatrix} 5 & 10 \\ 0 & 15 \end{pmatrix};$$

hence, $(3 + 2)A = 3A + 2A$.

$$(b)\ 2(A + B) = 2\left[\begin{pmatrix}1 & 2\\0 & 3\end{pmatrix} + \begin{pmatrix}2 & -1\\4 & 6\end{pmatrix}\right]$$

$$= 2\begin{pmatrix}3 & 1\\4 & 9\end{pmatrix} = \begin{pmatrix}6 & 2\\8 & 18\end{pmatrix};$$

$$2A + 2B = 2\begin{pmatrix}1 & 2\\0 & 3\end{pmatrix} + 2\begin{pmatrix}2 & -1\\4 & 6\end{pmatrix}$$

$$= \begin{pmatrix}2 & 4\\0 & 6\end{pmatrix} + \begin{pmatrix}4 & -2\\8 & 12\end{pmatrix} = \begin{pmatrix}6 & 2\\8 & 18\end{pmatrix};$$

hence, $2(A + B) = 2A + 2B$.

$$(c)\ 3(2A) = 3\left[2\begin{pmatrix}1 & 2\\0 & 3\end{pmatrix}\right] = 3\begin{pmatrix}2 & 4\\0 & 6\end{pmatrix} = \begin{pmatrix}6 & 12\\0 & 18\end{pmatrix};$$

$$(3\cdot 2)A = 6A = 6\begin{pmatrix}1 & 2\\0 & 3\end{pmatrix} = \begin{pmatrix}6 & 12\\0 & 18\end{pmatrix};$$

hence, $3(2A) = (3\cdot 2)A$.

Several fundamental properties of multiplication and addition of real numbers are also valid for the multiplication of a matrix by a real number, as illustrated in Example 2. Let r and s represent any two real numbers, and let A and B represent any two matrices of the same order:

1. $(r + s)A = rA + sA$
2. $r(A + B) = rA + rB$
3. $r(sA) = (rs)A$

Properties (1) and (2) are distributive properties of the multiplication of a matrix by a real number. Property (3) is an associative property of the multiplication of a matrix by a real number.

Exercises

In Exercises 1 through 4 let

$$A = \begin{pmatrix}2 & -7 & 0\\3 & 5 & -4\end{pmatrix} \quad \text{and} \quad B = \begin{pmatrix}3 & 8 & 4\\7 & 2 & 2\end{pmatrix}$$

1. Find $2A$.
2. Find $A - 3B$.
3. Find $2A + B$.

4. Find $3A - 2B$.

5. Verify that $(r + s)A = rA + sA$, where $r = 3$, $s = -1$, and $A = \begin{pmatrix} 4 & 1 & 3 \\ 2 & -1 & 0 \end{pmatrix}$.

6. Verify that $r(A + B) = rA + rB$, where $r = 2$, $A = \begin{pmatrix} 11 & -3 \\ 5 & 4 \end{pmatrix}$, and $B = \begin{pmatrix} 7 & 2 \\ 1 & -3 \end{pmatrix}$.

7. Verify that $r(sA) = (rs)A$, where $r = 2$, $s = 3$, and $A = \begin{pmatrix} 6 & 2 \\ 1 & 0 \\ 4 & 1 \end{pmatrix}$.

8. Two matrices can be added only when they are of the same order. Is there any restriction on the order of a matrix under multiplication by a real number?

9. If $rA = sA$, does $r = s$?

10. Given $A = \begin{pmatrix} -1 & 5 \\ 2 & 3 \end{pmatrix}$ and $B = \begin{pmatrix} 3 & 0 \\ 5 & 6 \end{pmatrix}$. Find C such that $A - 3C = 2B$.

11. The following table shows the number of seats in England's House of Commons held by members of each country in 1955:

Country	Number of seats held in the House of Commons
England	511
Wales	36
Scotland	71
Northern Ireland	12

(*a*) Write the number of seats held by members of each country as a *row matrix*.

(*b*) There are a total of 630 seats. Multiply the row matrix in (*a*) by an appropriate real number to obtain a row matrix indicating the fraction of seats held by members of each country.

5.4

Matrix Multiplication

The way in which the product of two matrices is defined is quite arbitrary. The following definition has proved very useful in terms of the applications to mathematics and the sciences:

Multiplication of Matrices

Let A be a matrix of order m by n; let B be a matrix of order n by r. The* product *AB is defined as a matrix C of order m by r such that the element in the ith row and jth column of C is equal to the sum of the products of the elements in the ith row of A and the corresponding elements in the jth column of B.

In other words, if $((a_{ij}))$ is a matrix of order m by n and $((b_{ij}))$ is a matrix of order n by r, then $((a_{ij}))((b_{ij})) = ((c_{ij}))$, where

$$c_{ij} = a_{i1}b_{1j} + a_{i2}b_{2j} + \cdots + a_{in}b_{nj}$$

for all pairs of values i and j. For example, consider the matrices

$$A = \begin{pmatrix} 3 & 1 & 4 \\ 2 & 5 & 1 \end{pmatrix} \qquad \text{and} \qquad B = \begin{pmatrix} 2 & 4 \\ 0 & 5 \\ 1 & 3 \end{pmatrix}$$

A is of order 2 by **3** and B is of order **3** by 2, so the product AB exists and is of order 2. The element in the first row and first column of AB is formed by summing the products indicated:

$$3 \cdot 2 + 1 \cdot 0 + 4 \cdot 1 = 10$$

that is,

$$AB = \begin{pmatrix} \mathbf{3} & \mathbf{1} & \mathbf{4} \\ 2 & 5 & 1 \end{pmatrix} \begin{pmatrix} \mathbf{2} & 4 \\ \mathbf{0} & 5 \\ \mathbf{1} & 3 \end{pmatrix} = \begin{pmatrix} \mathbf{10} & \cdots \\ \cdots & \cdots \end{pmatrix}$$

The element in the first row and second column of AB is formed by summing the products indicated:

$$3 \cdot 4 + 1 \cdot 5 + 4 \cdot 3 = 29$$

that is,

$$AB = \begin{pmatrix} \mathbf{3} & \mathbf{1} & \mathbf{4} \\ 2 & 5 & 1 \end{pmatrix} \begin{pmatrix} 2 & \mathbf{4} \\ 0 & \mathbf{5} \\ 1 & \mathbf{3} \end{pmatrix} = \begin{pmatrix} 10 & \mathbf{29} \\ \cdots & \cdots \end{pmatrix}$$

The element in the second row and first column of AB is formed by summing the products indicated:

$$2 \cdot 2 + 5 \cdot 0 + 1 \cdot 1 = 5$$

that is,

$$AB = \begin{pmatrix} 3 & 1 & 4 \\ \mathbf{2} & \mathbf{5} & \mathbf{1} \end{pmatrix} \begin{pmatrix} \mathbf{2} & 4 \\ \mathbf{0} & 5 \\ \mathbf{1} & 3 \end{pmatrix} = \begin{pmatrix} 10 & 29 \\ \mathbf{5} & \cdots \end{pmatrix}$$

Finally, the element in the second row and second column of AB is formed by summing the products indicated:

$$2 \cdot 4 + 5 \cdot 5 + 1 \cdot 3 = 36$$

that is,

$$AB = \begin{pmatrix} 3 & 1 & 4 \\ \mathbf{2} & \mathbf{5} & \mathbf{1} \end{pmatrix} \begin{pmatrix} 2 & \mathbf{4} \\ 0 & \mathbf{5} \\ 1 & \mathbf{3} \end{pmatrix} = \begin{pmatrix} 10 & 29 \\ 5 & \mathbf{36} \end{pmatrix}$$

Note (from the definition) that the product AB exists only when the number of columns of matrix A is equal to the number of rows of matrix B. When AB exists, the matrices A and B are said to be *conformable for multiplication.*

Example 1

Find the products AB and BA, if they exist, where $A = \begin{pmatrix} 5 & 2 \\ 3 & 0 \end{pmatrix}$ and $B = \begin{pmatrix} 1 & -2 \\ 0 & 4 \end{pmatrix}$.

The products AB and BA both exist because the matrices are conformable when taken in either order.

$$AB = \begin{pmatrix} 5 & 2 \\ 3 & 0 \end{pmatrix} \begin{pmatrix} 1 & -2 \\ 0 & 4 \end{pmatrix} = \begin{pmatrix} 5\cdot 1 + 2\cdot 0 & 5\cdot(-2) + 2\cdot 4 \\ 3\cdot 1 + 0\cdot 0 & 3\cdot(-2) + 0\cdot 4 \end{pmatrix}$$

$$= \begin{pmatrix} 5 & -2 \\ 3 & -6 \end{pmatrix}$$

$$BA = \begin{pmatrix} 1 & -2 \\ 0 & 4 \end{pmatrix} \begin{pmatrix} 5 & 2 \\ 3 & 0 \end{pmatrix} = \begin{pmatrix} 1\cdot 5 + (-2)\cdot 3 & 1\cdot 2 + (-2)\cdot 0 \\ 0\cdot 5 + 4\cdot 3 & 0\cdot 2 + 4\cdot 0 \end{pmatrix}$$

$$= \begin{pmatrix} -1 & 2 \\ 12 & 0 \end{pmatrix}$$

Hence, $AB \neq BA$ since

$$\begin{pmatrix} 5 & -2 \\ 3 & -6 \end{pmatrix} \neq \begin{pmatrix} -1 & 2 \\ 12 & 0 \end{pmatrix}$$

From the results of Example 1 it is clear that AB is not always equal to BA; that is, matrix multiplication is *not commutative.*

Although matrix multiplication is generally not commutative, two other fundamental properties of the multiplication and addition of real numbers are also valid for matrix multiplication:

1. *Matrix multiplication is associative:* $(AB)C = A(BC)$.
2. *Matrix multiplication is distributive with respect to addition:* $A(B + C) = AB + AC$ *and* $(A + B)C = AC + BC$.

Example 2

Verify the associative property of matrix multiplication, where

$$A = \begin{pmatrix} 2 & 0 \\ 1 & 1 \end{pmatrix} \quad B = \begin{pmatrix} 4 & 0 & 3 \\ -1 & 2 & -2 \end{pmatrix} \quad C = \begin{pmatrix} 1 & 2 \\ 2 & -3 \\ 0 & 1 \end{pmatrix}$$

$$(AB)C = \left[\begin{pmatrix} 2 & 0 \\ 1 & 1 \end{pmatrix}\begin{pmatrix} 4 & 0 & 3 \\ -1 & 2 & -2 \end{pmatrix}\right]\begin{pmatrix} 1 & 2 \\ 2 & -3 \\ 0 & 1 \end{pmatrix}$$

$$= \begin{pmatrix} 8 & 0 & 6 \\ 3 & 2 & 1 \end{pmatrix}\begin{pmatrix} 1 & 2 \\ 2 & -3 \\ 0 & 1 \end{pmatrix} = \begin{pmatrix} 8 & 22 \\ 7 & 1 \end{pmatrix}$$

$$A(BC) = \begin{pmatrix} 2 & 0 \\ 1 & 1 \end{pmatrix}\left[\begin{pmatrix} 4 & 0 & 3 \\ -1 & 2 & -2 \end{pmatrix}\begin{pmatrix} 1 & 2 \\ 2 & -3 \\ 0 & 1 \end{pmatrix}\right]$$

$$= \begin{pmatrix} 2 & 0 \\ 1 & 1 \end{pmatrix}\begin{pmatrix} 4 & 11 \\ 3 & -10 \end{pmatrix} = \begin{pmatrix} 8 & 22 \\ 7 & 1 \end{pmatrix}$$

Hence, $(AB)C = A(BC)$ since

$$\begin{pmatrix} 8 & 22 \\ 7 & 1 \end{pmatrix} = \begin{pmatrix} 8 & 22 \\ 7 & 1 \end{pmatrix}$$

Example 3

Verify the distributive property of matrix multiplication, where

$$A = \begin{pmatrix} 4 & -1 \\ 0 & 2 \\ 1 & 0 \end{pmatrix} \quad B = \begin{pmatrix} 3 & -1 \\ 1 & 2 \end{pmatrix} \quad C = \begin{pmatrix} 0 & 3 \\ 1 & -2 \end{pmatrix}$$

$$A(B + C) = \begin{pmatrix} 4 & -1 \\ 0 & 2 \\ 1 & 0 \end{pmatrix}\left[\begin{pmatrix} 3 & -1 \\ 1 & 2 \end{pmatrix} + \begin{pmatrix} 0 & 3 \\ 1 & -2 \end{pmatrix}\right]$$

$$= \begin{pmatrix} 4 & -1 \\ 0 & 2 \\ 1 & 0 \end{pmatrix}\begin{pmatrix} 3 & 2 \\ 2 & 0 \end{pmatrix} = \begin{pmatrix} 10 & 8 \\ 4 & 0 \\ 3 & 2 \end{pmatrix}$$

$$AB + AC = \begin{pmatrix} 4 & -1 \\ 0 & 2 \\ 1 & 0 \end{pmatrix}\begin{pmatrix} 3 & -1 \\ 1 & 2 \end{pmatrix} + \begin{pmatrix} 4 & -1 \\ 0 & 2 \\ 1 & 0 \end{pmatrix}\begin{pmatrix} 0 & 3 \\ 1 & -2 \end{pmatrix}$$

$$= \begin{pmatrix} 11 & -6 \\ 2 & 4 \\ 3 & -1 \end{pmatrix} + \begin{pmatrix} -1 & 14 \\ 2 & -4 \\ 0 & 3 \end{pmatrix} = \begin{pmatrix} 10 & 8 \\ 4 & 0 \\ 3 & 2 \end{pmatrix}$$

Hence, $A(B + C) = AB + AC$ since

$$\begin{pmatrix} 10 & 8 \\ 4 & 0 \\ 3 & 2 \end{pmatrix} = \begin{pmatrix} 10 & 8 \\ 4 & 0 \\ 3 & 2 \end{pmatrix}$$

Consider the matrices

$$A = \begin{pmatrix} 1 & 0 \\ 1 & 0 \end{pmatrix} \quad \text{and} \quad B = \begin{pmatrix} 0 & 0 \\ 1 & 1 \end{pmatrix}$$

Note that $AB = 0$, but neither A nor B is the null matrix of order 2. Recall that in the algebra of real numbers, if $ab = 0$, then $a = 0$ or $b = 0$. Thus, the cancellation property of multiplication is not valid for matrix multiplication.

Exercises

1. Find AB, where

$$A = \begin{pmatrix} 7 & 0 \\ 1 & 3 \end{pmatrix} \quad \text{and} \quad B = \begin{pmatrix} 5 & 1 & 4 \\ 2 & -1 & 0 \end{pmatrix}$$

2. Find AB, where

$$A = \begin{pmatrix} 1 & 2 & 3 \\ 0 & 1 & 2 \\ 0 & 0 & 1 \end{pmatrix} \quad \text{and} \quad B = \begin{pmatrix} 1 & 2 & 3 \\ 0 & 1 & 2 \\ 0 & 0 & 1 \end{pmatrix}$$

3. Find the products AB and BA, if they exist, where

$$A = \begin{pmatrix} 2 & 0 \\ -1 & 3 \end{pmatrix} \quad \text{and} \quad B = \begin{pmatrix} 3 & 5 \\ 4 & 0 \\ -1 & 2 \end{pmatrix}$$

4. Verify that the multiplication of square matrices is generally not commutative, using

$$A = \begin{pmatrix} 2 & 1 \\ -1 & 3 \end{pmatrix} \quad \text{and} \quad B = \begin{pmatrix} 1 & 3 \\ 4 & 0 \end{pmatrix}$$

5. Show that $AB = BA$, where

$$A = \begin{pmatrix} 3 & 0 & -2 \\ 3 & 1 & -2 \\ -1 & 0 & 1 \end{pmatrix} \quad \text{and} \quad B = \begin{pmatrix} 1 & 0 & 2 \\ -1 & 1 & 0 \\ 1 & 0 & 3 \end{pmatrix}$$

6. Find AB and BA, where

$$A = (a \quad b \quad c) \quad \text{and} \quad B = \begin{pmatrix} a \\ b \\ c \end{pmatrix}$$

7. Given $A = \begin{pmatrix} 2 & 1 \\ 3 & 0 \end{pmatrix}$ and $B = \begin{pmatrix} 1 & 5 \\ 0 & 1 \end{pmatrix}$, show that

$$(A + B)(A + B) \neq A^2 + 2AB + B^2$$

8. Determine under what conditions

$$(A + B)(A - B) = A^2 - B^2.$$

9. Verify the associative property of matrix multiplication, where

$$A = \begin{pmatrix} 2 & 0 & 1 \\ -1 & 3 & 4 \end{pmatrix} \quad B = \begin{pmatrix} 1 & 2 & 0 \\ -1 & 3 & 4 \\ 0 & 2 & -1 \end{pmatrix} \quad C = \begin{pmatrix} 0 & 2 \\ 1 & 2 \\ 0 & 5 \end{pmatrix}$$

10. Verify the distributive property of matrix multiplication with respect to addition, where

$$A = \begin{pmatrix} 2 & 1 \\ -3 & 4 \end{pmatrix} \quad B = \begin{pmatrix} 1 & -1 \\ 2 & 0 \end{pmatrix} \quad C = \begin{pmatrix} 3 & 1 \\ -1 & 2 \end{pmatrix}$$

11. If the orders of matrices A, B, C, and D are 3 by 2, 5 by 2, 4 by 3, and 4 by 5, respectively, state all possible products involving exactly two of these matrices.

12. Let A be of order m by n and B of order r by s. (*a*) Under what conditions does AB exist? (*b*) What is the order of AB, if it exists? (*c*) Under what conditions does BA exist? (*d*) What is the order of BA, if it exists?

13. Can the product of two nonsquare matrices be a square matrix? If so, when? If not, why not?

14. Find two matrices A and B of order 2 for which $AB = 0$, but neither A nor B is the null matrix of that order.

15. Find all matrices A such that

(*a*) $\begin{pmatrix} 3 & 0 \\ 1 & 0 \end{pmatrix} A = \begin{pmatrix} 0 & 0 \\ 0 & 0 \end{pmatrix}$

(*b*) $\begin{pmatrix} 4 & 2 \\ 1 & 1 \end{pmatrix} A = A \begin{pmatrix} 4 & 2 \\ 1 & 1 \end{pmatrix}$

16. Solve the matrix equation

$$\begin{pmatrix} 3 & 1 \\ 4 & 2 \end{pmatrix}\begin{pmatrix} a & b \\ c & d \end{pmatrix} = \begin{pmatrix} 1 & 0 \\ 0 & 1 \end{pmatrix}$$

17. Find A^3, where $A = \begin{pmatrix} 1 & 3 \\ 0 & 2 \end{pmatrix}$.

18. Find $A^2, A^3, A^4, \ldots, A^n$, where $A = \begin{pmatrix} 1 & 1 \\ 1 & 0 \end{pmatrix}$.

    (*Note:* The sequence 1, 1, 2, 3, 5, 8, 13, 21, 34, ..., where each term is the sum of the two preceding terms, is called a *Fibonacci sequence*. Fibonacci sequences occur in certain biological studies.)

19. Show that $A(B + C)D = ABD + ACD$.

20. Show that if $AB = BA$, then $A^2B = BA^2$.

21. Construct a multiplication table for the set of *Pauli matrices*, where $i^2 = -1$:

$$A = \begin{pmatrix} 1 & 0 \\ 0 & 1 \end{pmatrix} \quad B = \begin{pmatrix} 0 & 1 \\ -1 & 0 \end{pmatrix} \quad C = \begin{pmatrix} 0 & -1 \\ 1 & 0 \end{pmatrix} \quad D = \begin{pmatrix} -1 & 0 \\ 0 & -1 \end{pmatrix}$$

$$E = \begin{pmatrix} i & 0 \\ 0 & -i \end{pmatrix} \quad F = \begin{pmatrix} -i & 0 \\ 0 & i \end{pmatrix} \quad G = \begin{pmatrix} 0 & -i \\ -i & 0 \end{pmatrix} \quad H = \begin{pmatrix} 0 & i \\ i & 0 \end{pmatrix}$$

    (*Note:* The Pauli matrices are used in the study of electron spin in atomic physics.)

22. If

$$\begin{pmatrix} x_1 \\ x_2 \end{pmatrix} = \begin{pmatrix} a & b \\ c & d \end{pmatrix}\begin{pmatrix} y_1 \\ y_2 \end{pmatrix} \quad \text{and} \quad \begin{pmatrix} y_1 \\ y_2 \end{pmatrix} = \begin{pmatrix} e & f \\ g & h \end{pmatrix}\begin{pmatrix} z_1 \\ z_2 \end{pmatrix}$$

    find x_1 and x_2 in terms of z_1 and z_2.

23. The following table represents the number of different typewriters produced by companies C_1 and C_2 daily:

Typewriter	Company C_1	Company C_2
Electric	12	10
Standard	22	30
Portable	18	5

    Let $A = \begin{pmatrix} 12 & 10 \\ 22 & 30 \\ 18 & 5 \end{pmatrix}$, and let $D = \begin{pmatrix} 6 \\ 5 \end{pmatrix}$ represent the number of days the companies operate each week; that is, company C_1 operates 6 days/week and company C_2 operates 5 days/week. Find AD, and determine what it represents.

24. A steel mill produces two types of steel: high-grade and low-grade. The units of raw materials needed to make one unit of each type is given in the following table:

Grade	Iron ore	Coke	Limestone	Manganese	Tungsten
High	12	3	6	3	1
Low	10	3	5	1	0

Let $M = \begin{pmatrix} 12 & 3 & 6 & 3 & 1 \\ 10 & 3 & 5 & 1 & 0 \end{pmatrix}$. The dollar cost of one unit of each of the five raw materials is known and can be expressed by the matrix

$$C = \begin{pmatrix} 10 \\ 6 \\ 2 \\ 3 \\ 3 \end{pmatrix} \begin{matrix} \text{iron ore} \\ \text{coke} \\ \text{limestone} \\ \text{manganese} \\ \text{tungsten} \end{matrix}$$

Find MC and interpret the results.

5.5

Special Matrices

In this section we shall study some special matrices and their properties with respect to the operation of matrix multiplication.

The elements of a square matrix D that lie on the *main diagonal* (from the upper left to the lower right corner) are called the *diagonal elements* of D. A square matrix whose nondiagonal elements are all equal to zero is called a *diagonal matrix.* For example,

$$\begin{pmatrix} 2 & 0 \\ 0 & 1 \end{pmatrix} \quad \begin{pmatrix} 3 & 0 \\ 0 & 0 \end{pmatrix} \quad \begin{pmatrix} 4 & 0 & 0 \\ 0 & -1 & 0 \\ 0 & 0 & 5 \end{pmatrix}$$

are diagonal matrices, as is any null matrix of order n.

Example 1

Let $A = \begin{pmatrix} a & b \\ c & d \end{pmatrix}$ be any square matrix of order 2; let $D = \begin{pmatrix} e & 0 \\ 0 & f \end{pmatrix}$ be any diagonal matrix of order 2. Find (*a*) DA; (*b*) AD.

(*a*) $DA = \begin{pmatrix} e & 0 \\ 0 & f \end{pmatrix}\begin{pmatrix} a & b \\ c & d \end{pmatrix} = \begin{pmatrix} ea & eb \\ fc & fd \end{pmatrix}$

Note that each element of the first row of A is multiplied by e; each element of the second row of A is multiplied by f.

(*b*) $AD = \begin{pmatrix} a & b \\ c & d \end{pmatrix}\begin{pmatrix} e & 0 \\ 0 & f \end{pmatrix} = \begin{pmatrix} ea & fb \\ ec & fd \end{pmatrix}$

Note that each element of the first column of A is multiplied by e; each element of the second column of A is multiplied by f.

If all the diagonal elements of a diagonal matrix are equal, the diagonal matrix is called a *scalar matrix*. For example,

$$\begin{pmatrix} 2 & 0 \\ 0 & 2 \end{pmatrix} \quad \begin{pmatrix} -6 & 0 \\ 0 & -6 \end{pmatrix} \quad \begin{pmatrix} 3 & 0 & 0 \\ 0 & 3 & 0 \\ 0 & 0 & 3 \end{pmatrix}$$

are scalar matrices. Note that any null matrix of order n is also a scalar matrix.

Example 2

Let $A = \begin{pmatrix} a & b \\ c & d \end{pmatrix}$ be any square matrix of order 2; let $S = \begin{pmatrix} s & 0 \\ 0 & s \end{pmatrix}$ be any scalar matrix of order 2. Show that $SA = AS = sA$.

$$SA = \begin{pmatrix} s & 0 \\ 0 & s \end{pmatrix}\begin{pmatrix} a & b \\ c & d \end{pmatrix} = \begin{pmatrix} sa & sb \\ sc & sd \end{pmatrix} = s\begin{pmatrix} a & b \\ c & d \end{pmatrix}$$

$$AS = \begin{pmatrix} a & b \\ c & d \end{pmatrix}\begin{pmatrix} s & 0 \\ 0 & s \end{pmatrix} = \begin{pmatrix} sa & sb \\ sc & sd \end{pmatrix} = s\begin{pmatrix} a & b \\ c & d \end{pmatrix}$$

Thus, $SA = AS = sA$.

In general, the product of any matrix A and a conformable scalar matrix S with diagonal elements s is equivalent to the multiplication of A by the real number s. For example,

$$\begin{pmatrix} 3 & 0 \\ 0 & 3 \end{pmatrix}\begin{pmatrix} 2 & -1 & 5 \\ 4 & 0 & 2 \end{pmatrix} = \begin{pmatrix} 6 & -3 & 15 \\ 12 & 0 & 6 \end{pmatrix} = 3\begin{pmatrix} 2 & -1 & 5 \\ 4 & 0 & 2 \end{pmatrix}$$

A scalar matrix of order n whose diagonal elements are equal to 1 is called an *identity* or *unit matrix*. The matrices

$$\begin{pmatrix} 1 & 0 \\ 0 & 1 \end{pmatrix} \quad \text{and} \quad \begin{pmatrix} 1 & 0 & 0 \\ 0 & 1 & 0 \\ 0 & 0 & 1 \end{pmatrix}$$

are identity matrices of orders 2 and 3, respectively. An identity matrix, denoted by I, has the property that

$$AI = IA = A$$

where A is a square matrix and I is the identity matrix of the same order as A. Thus, the matrix I of order n is the *multiplicative identity element* for the set of all square matrices of order n. It should be noted that if A is a matrix of order m by n, the *premultiplication* identity matrix is of order m by m; the *postmultiplication* identity matrix is of order n by n. For example, if

$$A = \begin{pmatrix} 1 & 2 & 3 \\ 5 & 0 & 4 \end{pmatrix}$$

then

$$\begin{pmatrix} 1 & 0 \\ 0 & 1 \end{pmatrix}\begin{pmatrix} 1 & 2 & 3 \\ 5 & 0 & 4 \end{pmatrix} = \begin{pmatrix} 1 & 2 & 3 \\ 5 & 0 & 4 \end{pmatrix}$$

and

$$\begin{pmatrix} 1 & 2 & 3 \\ 5 & 0 & 4 \end{pmatrix}\begin{pmatrix} 1 & 0 & 0 \\ 0 & 1 & 0 \\ 0 & 0 & 1 \end{pmatrix} = \begin{pmatrix} 1 & 2 & 3 \\ 5 & 0 & 4 \end{pmatrix}$$

Sometimes we denote the identity matrix by $((\delta_{ij}))$, making use of the *Kronecker delta* δ_{ij}. This symbol is defined by

$$\delta_{ij} = \begin{cases} 1 & \text{if } i = j \\ 0 & \text{if } i \neq j \end{cases}$$

Example 3

Construct a matrix $((a_{ij}))$ of order 3 by 2, where $a_{ij} = 3i^2 + \delta_{ij}j$.

$$\begin{aligned} a_{11} &= 3(1)^2 + \delta_{11}(1) & a_{12} &= 3(1)^2 + \delta_{12}(2) \\ &= 3 + 1 = 4 & &= 3 + 0 = 3 \\ a_{21} &= 3(2)^2 + \delta_{21}(1) & a_{22} &= 3(2)^2 + \delta_{22}(2) \\ &= 12 + 0 = 12 & &= 12 + 2 = 14 \\ a_{31} &= 3(3)^2 + \delta_{31}(1) & a_{32} &= 3(3)^2 + \delta_{32}(2) \\ &= 27 + 0 = 27 & &= 27 + 0 = 27 \end{aligned}$$

Therefore,

$$((a_{ij})) = \begin{pmatrix} 4 & 3 \\ 12 & 14 \\ 27 & 27 \end{pmatrix}$$

Exercises

1. Write symbolically the diagonal elements of matrix $((a_{ij}))$ of order 3.
2. Show that the product of the diagonal matrices

$$\begin{pmatrix} a_{11} & 0 \\ 0 & a_{22} \end{pmatrix} \quad \text{and} \quad \begin{pmatrix} b_{11} & 0 \\ 0 & b_{22} \end{pmatrix}$$

is commutative.

3. Let $A = \begin{pmatrix} 3 & 0 \\ 5 & -1 \end{pmatrix}$ and $B = \begin{pmatrix} 1 & 1 \\ 1 & 0 \end{pmatrix}$. Find (*a*) $A - I$; (*b*) $B + 3I$.
4. Show that $AB = BA = I$, where

$$A = \begin{pmatrix} 2 & -3 \\ -1 & 2 \end{pmatrix} \quad \text{and} \quad B = \begin{pmatrix} 2 & 3 \\ 1 & 2 \end{pmatrix}$$

5. Show that $AB = BA = I$, where

$$A = \begin{pmatrix} x & -y \\ y & x \end{pmatrix} \quad \text{and} \quad B = \begin{pmatrix} x & y \\ -y & x \end{pmatrix}$$

provided that $x^2 + y^2 = 1$.

6. Show that $AB = BA = I$, where

$$A = \begin{pmatrix} 0 & 1 & 3 \\ -2 & 3 & 9 \\ 1 & -2 & -5 \end{pmatrix} \quad \text{and} \quad B = \begin{pmatrix} \frac{3}{2} & -\frac{1}{2} & 0 \\ -\frac{1}{2} & -\frac{3}{2} & -3 \\ \frac{1}{2} & \frac{1}{2} & 1 \end{pmatrix}$$

7. Let $A = \begin{pmatrix} 0 & 2 \\ 3 & -4 \end{pmatrix}$ and $B = \begin{pmatrix} 5 & 1 \\ 2 & 1 \end{pmatrix}$.

   (*a*) Premultiply B by A; (*b*) postmultiply B by A.

8. Let $A = \begin{pmatrix} 1 & 0 \\ 3 & 1 \\ -2 & 2 \end{pmatrix}$.

   (*a*) Find the premultiplication identity element.
   (*b*) Find the postmultiplication identity element.

In Exercises 9 through 14 determine whether each statement is true or false.

9. Every diagonal matrix is a scalar matrix.
10. Every scalar matrix is a diagonal matrix.
11. The matrix $\begin{pmatrix} 4 & 0 & 0 \\ 0 & 4 & 0 \end{pmatrix}$ is a diagonal matrix.

12. The product of two conformable diagonal matrices is a diagonal matrix.
13. The product of two conformable scalar matrices is a scalar matrix.
14. The matrix $\begin{pmatrix} 1 & 0 \\ 0 & 1 \\ 0 & 0 \end{pmatrix}$ is an identity matrix.
15. Find (*a*) δ_{34}; (*b*) δ_{77}; (*c*) δ_{12}; (*d*) δ_{nn}.
16. Construct a square matrix $((a_{ij}))$ of order 3, where $a_{ij} = 3i - \delta_{ij} j^2$.
17. Construct a matrix $((a_{ij}))$ of order 2 by 3, where $a_{ij} = i^2 + \delta_{ij} j$.
18. Write the matrix $((\delta_{ij} a_{ij}))$ of order 2.
19. Show that the matrix $A = \begin{pmatrix} 1 & 2 \\ 3 & 4 \end{pmatrix}$ satisfies the equation $A^2 - 5A - 2I = 0$.
20. Whenever the rows and columns of a given matrix A are interchanged, the resulting matrix is called the *transpose* of A and is denoted by A^T. Find A^T for

$$A = \begin{pmatrix} 2 & 3 & 5 \\ 6 & 1 & 1 \\ 2 & 4 & 0 \end{pmatrix}$$

21. Show that $(AB)^T = B^T A^T$, where $A = \begin{pmatrix} 5 & 0 \\ 3 & 1 \end{pmatrix}$ and $B = \begin{pmatrix} 2 & 2 \\ 1 & 0 \end{pmatrix}$. (See Exercise 20.)
22. A square real matrix A for which $A = A^T$ is called a *symmetric matrix*; a square real matrix A for which $A = -A^T$ is called a *skew-symmetric matrix*. Determine which of the following matrices are (*a*) symmetric; (*b*) skew-symmetric:

$$A = \begin{pmatrix} 1 & 0 & 4 \\ 0 & 2 & 5 \\ 4 & 5 & 3 \end{pmatrix} \qquad B = \begin{pmatrix} 1 & 3 & 4 \\ 2 & 0 & 2 \\ 4 & 3 & 1 \end{pmatrix} \qquad C = \begin{pmatrix} 0 & 0 & 0 \\ 0 & 0 & 0 \\ 0 & 0 & 0 \end{pmatrix}$$

$$D = \begin{pmatrix} 0 & 2 & -3 \\ 2 & 0 & 7 \\ -3 & 7 & 0 \end{pmatrix} \qquad E = \begin{pmatrix} 0 & 2 & -3 \\ -2 & 0 & 7 \\ 3 & -7 & 0 \end{pmatrix} \qquad F = \begin{pmatrix} 1 & 1 & 1 \\ 2 & 2 & 2 \\ 3 & 3 & 3 \end{pmatrix}$$

23. A road map of Florida has a mileage chart containing the following information:

Town	Miami	Orlando	Tampa	Tallahassee
Miami	0	229	264	476
Orlando	229	0	78	247
Tampa	264	78	0	253
Tallahassee	476	247	253	0

(*a*) What kind of matrix represents the data above? (See Exercise 22.)

(*b*) Represent the data above by deleting as much information as possible.

24. Suppose Art, Mike, Paul, and Joe played each other in tennis. The matrix

$$\begin{array}{c} \\ A \\ M \\ P \\ J \end{array}\begin{array}{c} \begin{array}{cccc} A & M & P & J \end{array} \\ \begin{pmatrix} 0 & 1 & 0 & 1 \\ 0 & 0 & 1 & 1 \\ 1 & 0 & 0 & 0 \\ 0 & 0 & 1 & 0 \end{pmatrix} \end{array}$$

indicates that Art beat Mike and Joe, Mike beat Paul and Joe, Paul beat Art, and Joe beat Paul.

(*a*) If we add the numbers in any particular row, what does the sum tell us?

(*b*) If we add the numbers in any particular column, what does the sum tell us?

(*c*) The matrix above is called a first-stage *dominance matrix*. For example, Art has two first-stage dominances over other players. Why? Mike has two first-stage dominances over other players. Why?

(*d*) Find $\begin{pmatrix} 0 & 1 & 0 & 1 \\ 0 & 0 & 1 & 1 \\ 1 & 0 & 0 & 0 \\ 0 & 0 & 1 & 0 \end{pmatrix}\begin{pmatrix} 0 & 1 & 0 & 1 \\ 0 & 0 & 1 & 1 \\ 1 & 0 & 0 & 0 \\ 0 & 0 & 1 & 0 \end{pmatrix}$.

The product matrix is called a second-stage dominance matrix. Which player has three second-stage dominances? Which player has two second-stage dominances?

(*e*) Who is the best tennis player in the group?

25. A tennis tournament among six girls, in which each girl was to play every other girl, was terminated prior to completion because of weather conditions. Suppose each girl had played four matches: Donna beat Ann and Edith; Edith beat Ann, Betty, and Teri; Teri beat Donna; Cathy beat Teri and Donna; Ann beat Cathy; and Betty beat Ann, Teri, and Cathy. (*a*) Construct first- and second-stage dominance matrices for the tournament. (*b*) Rank the girls in order of ability based upon the incomplete results of the tournament. (See Exercise 24.)

26. The Metro Conference Golf League consists of the Lakewater, Oak Lane, Seashore, and Moore Ridge teams. Each team plays every other team. Lakewater won all their matches, Oak Lane won none, and Seashore beat Moore Ridge and Oak Lane.

(*a*) Construct first- and second-stage dominance matrices.

(*b*) Find the total number of first- and second-stage dominances for each team. (See Exercise 24.)

27. An automobile manufacturer advises rotating tires every 6,000 miles as follows: left front tire to left rear; left rear tire to right front; right front tire to right rear; right rear tire to spare; spare tire to left front.

(*a*) Express this information in matrix form.

(*b*) Discuss the position of the original tires after two such rotations.

5.6

Multiplicative Inverse of a Matrix

Consider any square matrix A of order n. If a matrix B exists such that

$$AB = I \qquad \text{and} \qquad BA = I$$

then B is called the *multiplicative inverse* of matrix A. The multiplicative inverse of A is denoted by A^{-1}. Thus,

$$AA^{-1} = I \qquad \text{and} \qquad A^{-1}A = I$$

For example, if

$$A = \begin{pmatrix} 2 & 1 \\ 5 & 3 \end{pmatrix}$$

then

$$A^{-1} = \begin{pmatrix} 3 & -1 \\ -5 & 2 \end{pmatrix}$$

since

$$AA^{-1} = \begin{pmatrix} 2 & 1 \\ 5 & 3 \end{pmatrix}\begin{pmatrix} 3 & -1 \\ -5 & 2 \end{pmatrix} = \begin{pmatrix} 1 & 0 \\ 0 & 1 \end{pmatrix} = I$$

$$A^{-1}A = \begin{pmatrix} 3 & -1 \\ -5 & 2 \end{pmatrix}\begin{pmatrix} 2 & 1 \\ 5 & 3 \end{pmatrix} = \begin{pmatrix} 1 & 0 \\ 0 & 1 \end{pmatrix} = I$$

In this section we shall be concerned only with the multiplicative inverses of square matrices of order 2. Not every square matrix of order 2 has a multiplicative inverse; for example, the matrix A, where

$$A = \begin{pmatrix} 2 & 0 \\ 1 & 0 \end{pmatrix}$$

does not have a multiplicative inverse. If

$$\begin{pmatrix} a & b \\ c & d \end{pmatrix}$$

were the multiplicative inverse of A, then

$$\begin{pmatrix} 2 & 0 \\ 1 & 0 \end{pmatrix}\begin{pmatrix} a & b \\ c & d \end{pmatrix} = \begin{pmatrix} 1 & 0 \\ 0 & 1 \end{pmatrix}$$

$$\begin{pmatrix} 2a & 2b \\ a & b \end{pmatrix} = \begin{pmatrix} 1 & 0 \\ 0 & 1 \end{pmatrix}$$

The definition of equality of matrices requires that

$$\begin{aligned} 2a &= 1 \qquad & 2b &= 0 \\ a &= 0 \qquad & b &= 1 \end{aligned}$$

However, if $a = 0$, then $2a \neq 1$. Similarly, if $b = 1$, then $2b \neq 0$. Hence, values of a, b, c, and d do not exist such that

$$\begin{pmatrix} 2 & 0 \\ 1 & 0 \end{pmatrix}\begin{pmatrix} a & b \\ c & d \end{pmatrix} = \begin{pmatrix} 1 & 0 \\ 0 & 1 \end{pmatrix}$$

and the matrix A does not have a multiplicative inverse.

Now let us consider a general method for determining the multiplicative inverse of a matrix, if it exists. Let A be any square matrix of order 2:

$$A = \begin{pmatrix} a & b \\ c & d \end{pmatrix}$$

If A has a multiplicative inverse,

$$A^{-1} = \begin{pmatrix} w & x \\ y & z \end{pmatrix}$$

then

$$AA^{-1} = \begin{pmatrix} a & b \\ c & d \end{pmatrix}\begin{pmatrix} w & x \\ y & z \end{pmatrix} = \begin{pmatrix} 1 & 0 \\ 0 & 1 \end{pmatrix}$$

$$\begin{pmatrix} aw + by & ax + bz \\ cw + dy & cx + dz \end{pmatrix} = \begin{pmatrix} 1 & 0 \\ 0 & 1 \end{pmatrix}$$

The definition of equality of matrices requires that

$$\begin{cases} aw + by = 1 \\ cw + dy = 0 \end{cases} \quad \text{and} \quad \begin{cases} ax + bz = 0 \\ cx + dz = 1 \end{cases}$$

This pair of systems of equations places conditions upon w, x, y, and z in terms of a, b, c, and d. The first system of linear equations is equivalent to

$$\begin{cases} adw + bdy = d \\ bcw + bdy = 0 \end{cases}$$

Subtracting equals from equals, it follows that

$$adw - bcw = d$$
$$(ad - bc)w = d$$

and

$$w = \frac{d}{ad - bc}$$

Substituting $d/(ad - bc)$ for w in either equation of the system, we obtain

$$y = \frac{-c}{ad - bc}$$

Similarly, using the second system of linear equations, it can be shown that

$$x = \frac{-b}{ad - bc} \quad \text{and} \quad z = \frac{a}{ad - bc}$$

Hence the multiplicative inverse A^{-1} is

$$\begin{pmatrix} \dfrac{d}{ad-bc} & \dfrac{-b}{ad-bc} \\ \dfrac{-c}{ad-bc} & \dfrac{a}{ad-bc} \end{pmatrix}$$

provided that $ad - bc \neq 0$. (Why?) That is,

$$AA^{-1} = \begin{pmatrix} a & b \\ c & d \end{pmatrix} \begin{pmatrix} \dfrac{d}{ad-bc} & \dfrac{-b}{ad-bc} \\ \dfrac{-c}{ad-bc} & \dfrac{a}{ad-bc} \end{pmatrix} = \begin{pmatrix} 1 & 0 \\ 0 & 1 \end{pmatrix}$$

$$A^{-1}A = \begin{pmatrix} \dfrac{d}{ad-bc} & \dfrac{-b}{ad-bc} \\ \dfrac{-c}{ad-bc} & \dfrac{a}{ad-bc} \end{pmatrix} \begin{pmatrix} a & b \\ c & d \end{pmatrix} = \begin{pmatrix} 1 & 0 \\ 0 & 1 \end{pmatrix}$$

The method of obtaining A^{-1} indicates that it is unique when it exists.

Example 1

Find the multiplicative inverse of A, if it exists, where

$$A = \begin{pmatrix} 4 & 1 \\ 6 & 2 \end{pmatrix}$$

Let $a = 4$, $b = 1$, $c = 6$, and $d = 2$. Then

$$ad - bc = 4 \cdot 2 - 1 \cdot 6 = 2$$

Since $ad - bc \neq 0$, the multiplicative inverse of A exists:

$$A^{-1} = \begin{pmatrix} \dfrac{d}{ad-bc} & \dfrac{-b}{ad-bc} \\ \dfrac{-c}{ad-bc} & \dfrac{a}{ad-bc} \end{pmatrix} = \begin{pmatrix} \dfrac{2}{2} & -\dfrac{1}{2} \\ -\dfrac{6}{2} & \dfrac{4}{2} \end{pmatrix} = \begin{pmatrix} 1 & -\dfrac{1}{2} \\ -3 & 2 \end{pmatrix}$$

Check:

$$AA^{-1} = \begin{pmatrix} 4 & 1 \\ 6 & 2 \end{pmatrix} \begin{pmatrix} 1 & -\frac{1}{2} \\ -3 & 2 \end{pmatrix} = \begin{pmatrix} 1 & 0 \\ 0 & 1 \end{pmatrix}$$

$$A^{-1}A = \begin{pmatrix} 1 & -\frac{1}{2} \\ -3 & 2 \end{pmatrix} \begin{pmatrix} 4 & 1 \\ 6 & 2 \end{pmatrix} = \begin{pmatrix} 1 & 0 \\ 0 & 1 \end{pmatrix}$$

Example 2

Find the multiplicative inverse of A, if it exists, where

$$A = \begin{pmatrix} 10 & 5 \\ 2 & 1 \end{pmatrix}$$

Let $a = 10$, $b = 5$, $c = 2$, and $d = 1$. Then

$$ad - bc = 10 \cdot 1 - 5 \cdot 2 = 0$$

Since $ad - bc = 0$, the multiplicative inverse of A does not exist.

The real number $ad - bc$ associated with the matrix

$$A = \begin{pmatrix} a & b \\ c & d \end{pmatrix}$$

is called the *determinant of matrix A* and is denoted by

$$\begin{vmatrix} a & b \\ c & d \end{vmatrix}$$

that is,

$$\begin{vmatrix} a & b \\ c & d \end{vmatrix} = ad - bc$$

The determinant of a square matrix A is also denoted by det A. Hence, we may state that a matrix A has a multiplicative inverse if and only if det $A \neq 0$.

Exercises

1. Verify that $\begin{pmatrix} 5 & 3 \\ 3 & 2 \end{pmatrix}$ is the multiplicative inverse of $\begin{pmatrix} 2 & -3 \\ -3 & 5 \end{pmatrix}$.

In Exercises 2 through 9 find the multiplicative inverse, if it exists, of each matrix.

2. $\begin{pmatrix} 3 & 1 \\ 4 & 2 \end{pmatrix}$
3. $\begin{pmatrix} 3 & 6 \\ 1 & 2 \end{pmatrix}$
4. $\begin{pmatrix} 8 & 5 \\ 12 & 8 \end{pmatrix}$
5. $\begin{pmatrix} 5 & 4 \\ 3 & 2 \end{pmatrix}$

6. $\begin{pmatrix} 1 & 3 \\ 0 & 0 \end{pmatrix}$

7. $\begin{pmatrix} 4 & 0 \\ 0 & 1 \end{pmatrix}$

8. $\begin{pmatrix} \frac{4}{5} & -\frac{3}{5} \\ \frac{3}{5} & \frac{4}{5} \end{pmatrix}$

9. $\begin{pmatrix} a & b \\ -b & a \end{pmatrix}$ where $ab \neq 0$

10. Let $A = \begin{pmatrix} 9 & 3 \\ 2 & 1 \end{pmatrix}$. Show that $(A^{-1})^{-1} = A$.

11. Let $A = \begin{pmatrix} 2 & 5 \\ 1 & 3 \end{pmatrix}$. Show that $(A^{-1})^T = (A^T)^{-1}$; that is, the transpose of A^{-1} equals the inverse of A^T.

12. Let $A = \begin{pmatrix} 2 & 3 \\ 1 & 2 \end{pmatrix}$ and $B = \begin{pmatrix} 2 & 1 \\ 4 & 3 \end{pmatrix}$. Show that $(AB)^{-1} = B^{-1}A^{-1}$.

13. If $AB = AC$ and A^{-1} exists, show that $B = C$.

14. Show that the multiplication of the matrix $\begin{pmatrix} a & b \\ c & d \end{pmatrix}$ and its inverse

$$\begin{pmatrix} \dfrac{d}{ad - bc} & \dfrac{-b}{ad - bc} \\ \dfrac{-c}{ad - bc} & \dfrac{a}{ad - bc} \end{pmatrix}$$ is commutative.

15. Find det A, where $A = \begin{pmatrix} 1 & 4 \\ 5 & 2 \end{pmatrix}$.

16. Show that det AB = det $A \cdot$ det B, where

$$A = \begin{pmatrix} 1 & 2 \\ 3 & 4 \end{pmatrix} \quad \text{and} \quad B = \begin{pmatrix} 2 & 1 \\ 0 & 1 \end{pmatrix}$$

17. Verify that $\begin{pmatrix} 1 & 6 & 6 \\ 0 & 2 & 12 \\ 0 & 0 & 3 \end{pmatrix}$ is the multiplicative inverse of $\begin{pmatrix} 1 & -3 & 10 \\ 0 & \frac{1}{2} & -2 \\ 0 & 0 & \frac{1}{3} \end{pmatrix}$.

18. Verify that $\begin{pmatrix} \frac{3}{2} & -\frac{1}{2} & 0 \\ -\frac{1}{2} & -\frac{3}{2} & -3 \\ \frac{1}{2} & \frac{1}{2} & 1 \end{pmatrix}$ is the multiplicative inverse of

$$\begin{pmatrix} 0 & 1 & 3 \\ -2 & 3 & 9 \\ 1 & -2 & -5 \end{pmatrix}$$

19. Consider sending a secret message by replacing each letter with a numerical value: $A = 1$, $B = 2$, $\ldots$, $Z = 26$. Arrange the successive coded letters into groups of two letters placed in column form:

$$\begin{pmatrix} \text{1st letter} & \text{3d letter} & \text{5th letter} & \cdots \\ \text{2d letter} & \text{4th letter} & \text{6th letter} & \cdots \end{pmatrix}$$

So that a given letter is not represented by the same numerical value throughout the entire message (the frequencies of letters make it easy to break the code), we multiply the message by the following square matrix of order 2:

$$\begin{pmatrix} 2 & 1 \\ 3 & 2 \end{pmatrix}$$

The message received appears as

$$\begin{pmatrix} 65 & 66 & 20 & 34 & 45 & 22 & 64 \\ 105 & 111 & 39 & 63 & 72 & 37 & 108 \end{pmatrix}$$

Decode the message.

5.7

Matrices and Systems of Equations

The concept of a multiplicative inverse of a matrix can be used to solve systems of linear equations. The following examples illustrate the procedure.

Example 1

Solve the system of linear equations

$$\begin{cases} 3x + y = 5 \\ 5x + 2y = 8 \end{cases}$$

The system of linear equations may be expressed in matrix form as

$$\begin{pmatrix} 3 & 1 \\ 5 & 2 \end{pmatrix}\begin{pmatrix} x \\ y \end{pmatrix} = \begin{pmatrix} 5 \\ 8 \end{pmatrix}$$

Since the determinant of the matrix $\begin{pmatrix} 3 & 1 \\ 5 & 2 \end{pmatrix}$ is not equal to zero, the multiplicative inverse of the matrix exists. Let $a = 3$, $b = 1$, $c = 5$, and $d = 2$. The multiplicative inverse of $\begin{pmatrix} 3 & 1 \\ 5 & 2 \end{pmatrix}$ is $\begin{pmatrix} 2 & -1 \\ -5 & 3 \end{pmatrix}$. If we premultiply both sides of the matrix equation representing the system of linear equations by this multiplicative inverse, we obtain

$$\begin{pmatrix} 2 & -1 \\ -5 & 3 \end{pmatrix}\begin{pmatrix} 3 & 1 \\ 5 & 2 \end{pmatrix}\begin{pmatrix} x \\ y \end{pmatrix} = \begin{pmatrix} 2 & -1 \\ -5 & 3 \end{pmatrix}\begin{pmatrix} 5 \\ 8 \end{pmatrix}$$

$$\begin{pmatrix} 1 & 0 \\ 0 & 1 \end{pmatrix}\begin{pmatrix} x \\ y \end{pmatrix} = \begin{pmatrix} 2 \\ -1 \end{pmatrix}$$

$$\begin{pmatrix} x \\ y \end{pmatrix} = \begin{pmatrix} 2 \\ -1 \end{pmatrix}$$

Hence, by the definition of equality of matrices, $x = 2$ and $y = -1$. Therefore, the solution of the given system of linear equations is the ordered pair of real numbers $(2, -1)$.

Example 2

Solve the system of linear equations

$$\begin{cases} 3x + 2y = 9 \\ 5x + 4y = 17 \end{cases}$$

The system of linear equations may be expressed in matrix form as

$$\begin{pmatrix} 3 & 2 \\ 5 & 4 \end{pmatrix}\begin{pmatrix} x \\ y \end{pmatrix} = \begin{pmatrix} 9 \\ 17 \end{pmatrix}$$

Since the multiplicative inverse of $\begin{pmatrix} 3 & 2 \\ 5 & 4 \end{pmatrix}$ is $\begin{pmatrix} 2 & -1 \\ -\frac{5}{2} & \frac{3}{2} \end{pmatrix}$, then

$$\begin{pmatrix} 2 & -1 \\ -\frac{5}{2} & \frac{3}{2} \end{pmatrix}\begin{pmatrix} 3 & 2 \\ 5 & 4 \end{pmatrix}\begin{pmatrix} x \\ y \end{pmatrix} = \begin{pmatrix} 2 & -1 \\ -\frac{5}{2} & \frac{3}{2} \end{pmatrix}\begin{pmatrix} 9 \\ 17 \end{pmatrix}$$

$$\begin{pmatrix} 1 & 0 \\ 0 & 1 \end{pmatrix}\begin{pmatrix} x \\ y \end{pmatrix} = \begin{pmatrix} 1 \\ 3 \end{pmatrix}$$

$$\begin{pmatrix} x \\ y \end{pmatrix} = \begin{pmatrix} 1 \\ 3 \end{pmatrix}$$

Hence, $x = 1$ and $y = 3$. The solution of the given system of linear equations is the ordered pair of real numbers $(1, 3)$.

Note that every system of two linear equations in two variables

$$\begin{cases} ax + by = e \\ cx + dy = f \end{cases}$$

can be expressed in the matrix form

$$\begin{pmatrix} a & b \\ c & d \end{pmatrix}\begin{pmatrix} x \\ y \end{pmatrix} = \begin{pmatrix} e \\ f \end{pmatrix}$$

If the multiplicative inverse of $\begin{pmatrix} a & b \\ c & d \end{pmatrix}$ exists, then

$$\begin{pmatrix} \frac{d}{ad-bc} & \frac{-b}{ad-bc} \\ \frac{-c}{ad-bc} & \frac{a}{ad-bc} \end{pmatrix}\begin{pmatrix} a & b \\ c & d \end{pmatrix}\begin{pmatrix} x \\ y \end{pmatrix} = \begin{pmatrix} \frac{d}{ad-bc} & \frac{-b}{ad-bc} \\ \frac{-c}{ad-bc} & \frac{a}{ad-bc} \end{pmatrix}\begin{pmatrix} e \\ f \end{pmatrix}$$

$$\begin{pmatrix} 1 & 0 \\ 0 & 1 \end{pmatrix}\begin{pmatrix} x \\ y \end{pmatrix} = \begin{pmatrix} \frac{d}{ad-bc} & \frac{-b}{ad-bc} \\ \frac{-c}{ad-bc} & \frac{a}{ad-bc} \end{pmatrix}\begin{pmatrix} e \\ f \end{pmatrix}$$

$$\begin{pmatrix} x \\ y \end{pmatrix} = \begin{pmatrix} \frac{d}{ad-bc} & \frac{-b}{ad-bc} \\ \frac{-c}{ad-bc} & \frac{a}{ad-bc} \end{pmatrix}\begin{pmatrix} e \\ f \end{pmatrix}$$

Thus, this last matrix equation represents the solution of the system. For example, consider the system of linear equations

$$\begin{cases} 2x - 3y = 4 \\ x + y = 7 \end{cases}$$

The solution can be expressed by the matrix equation

$$\begin{pmatrix} x \\ y \end{pmatrix} = \begin{pmatrix} 2 & -3 \\ 1 & 1 \end{pmatrix}^{-1}\begin{pmatrix} 4 \\ 7 \end{pmatrix}$$

where $\begin{pmatrix} 2 & -3 \\ 1 & 1 \end{pmatrix}^{-1}$ denotes the multiplicative inverse of $\begin{pmatrix} 2 & -3 \\ 1 & 1 \end{pmatrix}$. Since

$$\begin{pmatrix} 2 & -3 \\ 1 & 1 \end{pmatrix}^{-1} = \begin{pmatrix} \frac{1}{5} & \frac{3}{5} \\ -\frac{1}{5} & \frac{2}{5} \end{pmatrix}$$

then

$$\begin{pmatrix} x \\ y \end{pmatrix} = \begin{pmatrix} \frac{1}{5} & \frac{3}{5} \\ -\frac{1}{5} & \frac{2}{5} \end{pmatrix}\begin{pmatrix} 4 \\ 7 \end{pmatrix} = \begin{pmatrix} 5 \\ 2 \end{pmatrix}$$

Hence, $x = 5$ and $y = 2$; that is, (5, 2) is the solution of the system of linear equations.

Exercises

In Exercises 1 through 8 solve each system of linear equations.

1. $\begin{cases} 2x - y = -3 \\ x + 2y = 6 \end{cases}$

2. $\begin{cases} 2x + y = 5 \\ x + 3y = 10 \end{cases}$

3. $\begin{cases} x - 2y = 5 \\ 3x - y = 5 \end{cases}$

4. $\begin{cases} 2x + 3y = 5 \\ x + 2y = 4 \end{cases}$

5. $\begin{cases} x + 3y = 7 \\ 3x + y = 5 \end{cases}$

6. $\begin{cases} 2x + y = 4 \\ 3x + 4y = 1 \end{cases}$

7. $\begin{cases} 5x - 2y = 12 \\ x + 2y = 0 \end{cases}$

8. $\begin{cases} x + 2y = 1 \\ 3x + 4y = 3 \end{cases}$

9. Solve the system of linear equations

$$\begin{cases} x + 2y + 3z = 1 \\ x + 3y + 5z = 2 \\ x + 5y + 12z = -2 \end{cases}$$

by using the fact that the multiplicative inverse of

$$\begin{pmatrix} 1 & 2 & 3 \\ 1 & 3 & 5 \\ 1 & 5 & 12 \end{pmatrix} \quad \text{is} \quad \begin{pmatrix} \frac{11}{3} & -3 & \frac{1}{3} \\ -\frac{7}{3} & 3 & -\frac{2}{3} \\ \frac{2}{3} & -1 & \frac{1}{3} \end{pmatrix}$$

10. A championship high school basketball game was attended by 3,512 people. An adult admission ticket cost \$1.25; a student admission ticket was \$0.75. How many tickets of each kind were sold if the total receipts for the game were \$3,014?

11. State University's baseball team played 42 games last season. The team won six more than twice the number of games lost. Determine the baseball team's record for last season.

12. State University's football stadium has a seating capacity of 44,000. Reserved seat tickets cost \$6.00; general admission tickets sell for \$2.50. The total receipts possible for one game are \$215,000. Determine the number of reserved seats.

13. Two machines can produce a certain number of items in 5 hours. The same number of items can be produced if machine *A*, which is fully automatic, runs for 3 hours and machine *B*, which needs an operator, runs for 10 hours. Find the time for machine *A* to produce the items alone.

14. A dairy mixes milk consisting of 5 percent butterfat with milk consisting of 8 percent butterfat to obtain 150 pounds of milk consisting of 6 percent butterfat. Determine how many pounds of each type were used in the mixture.

15. An investment of \$4,500 in stocks and \$3,000 in a savings and loan institution earned a total interest of \$382.50. If the interest rates had been interchanged, the interest earned would have been \$367.50. Find the interest rate on each investment.
16. A chemist mixes a 50% sulfuric acid solution with a 10% sulfuric acid solution to obtain 120 cubic centimeters of a 25% sulfuric acid solution. Find how much of each solution he used.
17. A garden nursery offers a 20 percent discount on evergreens when ten or more are purchased and a 10 percent discount on grass seed when 100 pounds or more are purchased. The cost of 12 evergreens and 150 pounds of grass seed is \$129.90; the cost of 8 evergreens and 60 pounds of grass seed is \$82. Determine the price of each evergreen and each pound of grass seed.

5.8

Elementary Row Transformations

Consider a system of three linear equations in three variables x, y, and z:

$$\begin{cases} a_{11}x + a_{12}y + a_{13}z = c_1 \\ a_{21}x + a_{22}y + a_{23}z = c_2 \\ a_{31}x + a_{32}y + a_{33}z = c_3 \end{cases}$$

The system can be expressed in matrix form as

$$\begin{pmatrix} a_{11} & a_{12} & a_{13} \\ a_{21} & a_{22} & a_{23} \\ a_{31} & a_{32} & a_{33} \end{pmatrix}\begin{pmatrix} x \\ y \\ z \end{pmatrix} = \begin{pmatrix} c_1 \\ c_2 \\ c_3 \end{pmatrix}$$

In addition to the *matrix of coefficients*

$$\begin{pmatrix} a_{11} & a_{12} & a_{13} \\ a_{21} & a_{22} & a_{23} \\ a_{31} & a_{32} & a_{33} \end{pmatrix}$$

we shall define the matrix composed of elements a_{ij} with constants c_1, c_2, and c_3 included as an additional column to be the *augmented matrix* of the system; i.e.,

$$\begin{pmatrix} a_{11} & a_{12} & a_{13} & c_1 \\ a_{21} & a_{22} & a_{23} & c_2 \\ a_{31} & a_{32} & a_{33} & c_3 \end{pmatrix}$$

is the augmented matrix of the system of linear equations.

For finding solutions, if they exist, to a system of linear equations, three simple operations may be employed to transform the system to an *equivalent system* (a system with the same solution set). An equivalent system is obtained if

1. Two equations are interchanged
2. The terms of any equation are multiplied by a nonzero real number
3. A nonzero real multiple of the terms of one equation is added to the corresponding terms of another equation of the system

One method of solution of the system consists of performing these three operations until the system can be represented in matrix form by

$$\begin{pmatrix} 1 & 0 & 0 \\ 0 & 1 & 0 \\ 0 & 0 & 1 \end{pmatrix}\begin{pmatrix} x \\ y \\ z \end{pmatrix} = \begin{pmatrix} k_1 \\ k_2 \\ k_3 \end{pmatrix}$$

which indicates k_1, k_2, and k_3 as the solution of the system. The augmented matrix of the system now appears in the form

$$\begin{pmatrix} 1 & 0 & 0 & k_1 \\ 0 & 1 & 0 & k_2 \\ 0 & 0 & 1 & k_3 \end{pmatrix}$$

Therefore, the solution of the system may be obtained by considering only the augmented matrix in an attempt to reduce it to this *diagonal form* by performing three *elementary row transformations* on the matrix as follows:

1. The interchange of two rows
2. Multiplication of the elements of any row by a nonzero real number
3. Addition of a nonzero real multiple of the elements of any row onto the corresponding elements of another row

Example 1

Solve the system of linear equations by performing the elementary row transformations on the augmented matrix:

$$\begin{cases} x + y + z = 5 \\ 3x - 2y - z = 3 \\ 2x + 3y + 5z = 8 \end{cases}$$

The augmented matrix of the system is

$$\begin{pmatrix} 1 & 1 & 1 & 5 \\ 3 & -2 & -1 & 3 \\ 2 & 3 & 5 & 8 \end{pmatrix}$$

Multiply the elements of row one by -3 and add the results to the corresponding elements of row two:

$$\begin{pmatrix} 1 & 1 & 1 & 5 \\ 0 & -5 & -4 & -12 \\ 2 & 3 & 5 & 8 \end{pmatrix}$$

Multiply the elements of row one by -2 and add the results to the corresponding elements of row three:

$$\begin{pmatrix} 1 & 1 & 1 & 5 \\ 0 & -5 & -4 & -12 \\ 0 & 1 & 3 & -2 \end{pmatrix}$$

Interchange rows two and three:

$$\begin{pmatrix} 1 & 1 & 1 & 5 \\ 0 & 1 & 3 & -2 \\ 0 & -5 & -4 & -12 \end{pmatrix}$$

Multiply the elements of row two by -1 and add the results to the corresponding elements of row one:

$$\begin{pmatrix} 1 & 0 & -2 & 7 \\ 0 & 1 & 3 & -2 \\ 0 & -5 & -4 & -12 \end{pmatrix}$$

Multiply the elements of row two by 5 and add the results to the corresponding elements of row three:

$$\begin{pmatrix} 1 & 0 & -2 & 7 \\ 0 & 1 & 3 & -2 \\ 0 & 0 & 11 & -22 \end{pmatrix}$$

Multiply the elements of row three by $\frac{1}{11}$:

$$\begin{pmatrix} 1 & 0 & -2 & 7 \\ 0 & 1 & 3 & -2 \\ 0 & 0 & 1 & -2 \end{pmatrix}$$

Multiply the elements of row three by 2 and add the results to the corresponding elements of row one:

$$\begin{pmatrix} 1 & 0 & 0 & 3 \\ 0 & 1 & 3 & -2 \\ 0 & 0 & 1 & -2 \end{pmatrix}$$

Multiply the elements of row three by -3 and add the results to the corresponding elements of row two:

$$\begin{pmatrix} 1 & 0 & 0 & 3 \\ 0 & 1 & 0 & 4 \\ 0 & 0 & 1 & -2 \end{pmatrix}$$

The transformed augmented matrix represents the equivalent system of linear equations

$$\begin{pmatrix} 1 & 0 & 0 \\ 0 & 1 & 0 \\ 0 & 0 & 1 \end{pmatrix} \begin{pmatrix} x \\ y \\ z \end{pmatrix} = \begin{pmatrix} 3 \\ 4 \\ -2 \end{pmatrix}$$

that is,

$$\begin{pmatrix} x \\ y \\ z \end{pmatrix} = \begin{pmatrix} 3 \\ 4 \\ -2 \end{pmatrix}$$

Thus, $x = 3$, $y = 4$, and $z = -2$. The solution of the given system of linear equations is the ordered triple of real numbers $(3, 4, -2)$.

Example 2

Solve the system of linear equations

$$\begin{cases} x + y + z = 4 \\ 2x + 3y + 3z = 11 \\ x - 3y + 4z = 13 \end{cases}$$

The augmented matrix is

$$\begin{pmatrix} 1 & 1 & 1 & 4 \\ 2 & 3 & 3 & 11 \\ 1 & -3 & 4 & 13 \end{pmatrix}$$

If we proceed in the manner described in Example 1, the augmented matrix takes on the following forms:

$$\begin{pmatrix} 1 & 1 & 1 & 4 \\ \mathbf{0} & \mathbf{1} & \mathbf{1} & \mathbf{3} \\ 1 & -3 & 4 & 13 \end{pmatrix}$$

$$\begin{pmatrix} 1 & 1 & 1 & 4 \\ 0 & 1 & 1 & 3 \\ \mathbf{0} & \mathbf{-4} & \mathbf{3} & \mathbf{9} \end{pmatrix}$$

$$\begin{pmatrix} \mathbf{1} & \mathbf{0} & \mathbf{0} & \mathbf{1} \\ 0 & 1 & 1 & 3 \\ 0 & -4 & 3 & 9 \end{pmatrix}$$

$$\begin{pmatrix} 1 & 0 & 0 & 1 \\ 0 & 1 & 1 & 3 \\ \mathbf{0} & \mathbf{0} & \mathbf{7} & \mathbf{21} \end{pmatrix}$$

$$\begin{pmatrix} 1 & 0 & 0 & 1 \\ 0 & 1 & 1 & 3 \\ \mathbf{0} & \mathbf{0} & \mathbf{1} & \mathbf{3} \end{pmatrix}$$

$$\begin{pmatrix} 1 & 0 & 0 & 1 \\ \mathbf{0} & \mathbf{1} & \mathbf{0} & \mathbf{0} \\ 0 & 0 & 1 & 3 \end{pmatrix}$$

Thus, $x = 1$, $y = 0$, and $z = 3$ and the solution is the ordered triple of real numbers (1, 0, 3).

It may not be always possible to reduce the augmented matrix of a system of linear equations to a diagonal form by means of the elementary row transformations. In that event, a unique solution may not exist, as illustrated in Example 3.

Example 3

Solve the system of linear equations

$$\begin{cases} x + 2y - z = 1 \\ x + 3y - z = 0 \\ x - \ \ y - z = -2 \end{cases}$$

The augmented matrix is

$$\begin{pmatrix} 1 & 2 & -1 & 1 \\ 1 & 3 & -1 & 0 \\ 1 & -1 & -1 & -2 \end{pmatrix}$$

Multiply the elements of row one by -1 and add the results to the corresponding elements of row two:

$$\begin{pmatrix} 1 & 2 & -1 & 1 \\ 0 & 1 & 0 & -1 \\ 1 & -1 & -1 & -2 \end{pmatrix}$$

Multiply the elements of row one by -1 and add the results to the corresponding elements of row three:

$$\begin{pmatrix} 1 & 2 & -1 & 1 \\ 0 & 1 & 0 & -1 \\ 0 & -3 & 0 & -3 \end{pmatrix}$$

Multiply the elements of row two by -2 and add the results to the corresponding elements of row one:

$$\begin{pmatrix} 1 & 0 & -1 & 3 \\ 0 & 1 & 0 & -1 \\ 0 & -3 & 0 & -3 \end{pmatrix}$$

Multiply the elements of row two by 3 and add the results to the corresponding elements of row three:

$$\begin{pmatrix} 1 & 0 & -1 & 3 \\ 0 & 1 & 0 & -1 \\ 0 & 0 & 0 & -6 \end{pmatrix}$$

The transformed augmented matrix represents the system of linear equations

$$\begin{cases} x - z = 3 \\ \quad\; y = -1 \\ \quad\; 0 = -6 \end{cases}$$

Since no values of x, y, and z exist such that $0 = -6$, the system is *inconsistent*; that is, no solution exists. If at least one solution of a system of linear equations exists, the system is said to be *consistent*.

Exercises

In Exercises 1 and 2 find a system of linear equations whose augmented matrix is given.

1. $\begin{pmatrix} 3 & 5 & -3 & 2 \\ 2 & -1 & 1 & 3 \\ 1 & -3 & 1 & -1 \end{pmatrix}$

2. $\begin{pmatrix} 1 & 1 & 2 & 3 \\ 3 & 0 & 2 & 2 \\ -1 & 2 & 2 & 4 \end{pmatrix}$

In Exercises 3 through 10 solve each system of linear equations by performing the elementary row transformations on the augmented matrix.

3. $\begin{cases} x - 2y + z = 3 \\ 2x - 5y + 6z = 20 \\ x + y - 2z = -3 \end{cases}$

4. $\begin{cases} 2x - 4y - 5z = 21 \\ 4x - y + 5z = 4 \\ 2x - 11y - z = 2 \end{cases}$

5. $\begin{cases} x - y + z = 1 \\ -x + y + z = 1 \\ x + y + z = 3 \end{cases}$

6. $\begin{cases} x - y + 3z = 0 \\ 3x + 2y - z = 15 \\ -3x + 2y + 2z = 6 \end{cases}$

7. $\begin{cases} 3x + 5z = 4 \\ 7y + 4z = 8 \\ 5x + 4y = -10 \end{cases}$

8. $\begin{cases} x + 2y + 3z = 14 \\ 2x + 3y + 4z = 20 \\ 3x + 4y + 4z = 23 \end{cases}$

9. $\begin{cases} w + x + y - z = 0 \\ 2w - 3x - 2y + 2z = 19 \\ w + 3x + 2y - 2z = -4 \\ 3w + x - y + z = 22 \end{cases}$

10. $\begin{cases} w + x + y + z = 8 \\ w - x - z = -2 \\ x + y + 2z = 9 \\ w + 2y = 6 \end{cases}$

11. Consider the system of linear equations in x, y, and z:

$$\begin{cases} x + 2y + 3z = 4 \\ 2x + 5y + 8z = 11 \\ 3x + 6y + tz = r \end{cases}$$

(*a*) Determine t and r so that the system has no solutions.
(*b*) Determine t and r so that the system has exactly one solution.
(*c*) Determine t and r so that the system has more than one solution.

12. What is the maximum number of elementary row transformations necessary to transform a square matrix of order n to diagonal form?

13. (*a*) Show that the first elementary row transformation on a square matrix A of order 3 is equivalent to a premultiplication of matrix A by one of the following matrices:

$$\begin{pmatrix} 0 & 1 & 0 \\ 1 & 0 & 0 \\ 0 & 0 & 1 \end{pmatrix} \quad \begin{pmatrix} 0 & 0 & 1 \\ 0 & 1 & 0 \\ 1 & 0 & 0 \end{pmatrix} \quad \begin{pmatrix} 1 & 0 & 0 \\ 0 & 0 & 1 \\ 0 & 1 & 0 \end{pmatrix}$$

(*b*) Show that the second elementary row transformation is equivalent to a premultiplication of matrix A by one of the following matrices:

$$\begin{pmatrix} k & 0 & 0 \\ 0 & 1 & 0 \\ 0 & 0 & 1 \end{pmatrix} \quad \begin{pmatrix} 1 & 0 & 0 \\ 0 & k & 0 \\ 0 & 0 & 1 \end{pmatrix} \quad \begin{pmatrix} 1 & 0 & 0 \\ 0 & 1 & 0 \\ 0 & 0 & k \end{pmatrix}$$

where k is a nonzero real number.

(*c*) Show that the third elementary row transformation is equivalent to a premultiplication of matrix A by one of the following matrices:

$$\begin{pmatrix} 1 & k & 0 \\ 0 & 1 & 0 \\ 0 & 0 & 1 \end{pmatrix} \quad \begin{pmatrix} 1 & 0 & k \\ 0 & 1 & 0 \\ 0 & 0 & 1 \end{pmatrix} \quad \begin{pmatrix} 1 & 0 & 0 \\ k & 1 & 0 \\ 0 & 0 & 1 \end{pmatrix}$$

$$\begin{pmatrix} 1 & 0 & 0 \\ 0 & 1 & k \\ 0 & 0 & 1 \end{pmatrix} \quad \begin{pmatrix} 1 & 0 & 0 \\ 0 & 1 & 0 \\ k & 0 & 1 \end{pmatrix} \quad \begin{pmatrix} 1 & 0 & 0 \\ 0 & 1 & 0 \\ 0 & k & 1 \end{pmatrix}$$

(These twelve matrices are called *elementary row transformation matrices.*)

14. Find the product of the matrices

$$\begin{pmatrix} 0 & 1 & 0 \\ 1 & 0 & 0 \\ 0 & 0 & 1 \end{pmatrix} \begin{pmatrix} 6 & -2 & 3 & 1 \\ -2 & 0 & 1 & 2 \\ 2 & 3 & 4 & -4 \end{pmatrix}$$

without carrying out the multiplication.

15. Find all the elementary row transformation matrices of order 2.

16. Express the inverse of $\begin{pmatrix} 6 & 8 \\ 1 & 4 \end{pmatrix}$ as the product of a set of elementary row transformation matrices.

17. A trucking company owns three types of trucks that are equipped to haul three different items per load according to the following table:

Item	Type 1	Type 2	Type 3
A	2	1	1
B	4	0	2
C	2	6	1

How many trucks of each type are needed to haul exactly 8 units of item A, 10 units of item B, and 23 units of item C?

18. Suppose the trucking company in Exercise 17 discontinues hauling units of item B and can now haul units of the other items according to the following table:

Item	Type 1	Type 2	Type 3
A	2	1	1
C	4	6	2

How many trucks of each type are needed to haul exactly 4 units of item A and 16 units of item C?

19. A chemist has three solutions containing nitric acid; the first contains 30% acid, the second 50%, and the third 10%. If he needs to obtain a mixture of 25 gallons containing 40% nitric acid, using twice as much of the 50% solution as the 30% solution, how many gallons of each solution should he use?

5.9

Determinants

As noted in Sec. 5.6, for each square matrix $A = ((a_{ij}))$ of order 2 there exists a rule (function) that associates with the matrix a unique real number. The rule is called the *determinant* of the matrix and is denoted either by det A or

$$\begin{vmatrix} a_{11} & a_{12} \\ a_{21} & a_{22} \end{vmatrix}$$

The determinant function assigns to each square matrix of order 2 the real number $a_{11}a_{22} - a_{12}a_{21}$, called the *value of the determinant*; that is, det $A = a_{11}a_{22} - a_{12}a_{21}$. We sometimes speak of the "value of the determinant" as simply the "determinant."

The value of the determinant of a matrix of order 2 may be remembered by the schematic diagram

$$\begin{vmatrix} a_{11} & a_{12} \\ a_{21} & a_{22} \end{vmatrix}\begin{matrix} -a_{12}a_{21} \\ +a_{11}a_{22} \end{matrix} = a_{11}a_{22} - a_{12}a_{21}$$

Note that the value of the determinant is equal to the product of the elements in the upper left-hand and lower right-hand corners minus the product of the elements in the upper right-hand and lower left-hand corners.

Example 1

Find the value of each determinant of order 2:

(*a*) $\begin{vmatrix} 5 & 1 \\ 3 & 2 \end{vmatrix}$

(*b*) $\begin{vmatrix} x & 1 \\ 7 & 3 \end{vmatrix}$

(*a*) $\begin{vmatrix} 5 & 1 \\ 3 & 2 \end{vmatrix}\begin{matrix} -(1)(3) \\ +(5)(2) \end{matrix} = (5)(2) - (1)(3) = 10 - 3 = 7$

(*b*) $\begin{vmatrix} x & 1 \\ 7 & 3 \end{vmatrix}\begin{matrix} -(1)(7) \\ +(x)(3) \end{matrix} = 3x - 7$

Example 2

Find x such that

$$\begin{vmatrix} 2 & 5 \\ 4 & x - 1 \end{vmatrix} = 0$$

Since $\begin{vmatrix} 2 & 5 \\ 4 & x - 1 \end{vmatrix} = 2(x - 1) - 5 \cdot 4 = 2x - 22$, then

$$2x - 22 = 0$$
$$x = 11$$

A *determinant* of a matrix $A = ((a_{ij}))$ of order 3 is a function of the square matrix and is denoted by det A or

$$\begin{vmatrix} a_{11} & a_{12} & a_{13} \\ a_{21} & a_{22} & a_{23} \\ a_{31} & a_{32} & a_{33} \end{vmatrix}$$

The *value of the determinant* of A is defined by

$$a_{11}a_{22}a_{33} + a_{12}a_{23}a_{31} + a_{13}a_{21}a_{32} - a_{13}a_{22}a_{31} - a_{11}a_{23}a_{32} - a_{12}a_{21}a_{33}$$

The value of the determinant of order 3 may be remembered by use of a particular scheme similar to that used for determinants of order 2:

$$-a_{13}a_{22}a_{31} - a_{11}a_{23}a_{32} - a_{12}a_{21}a_{33}$$

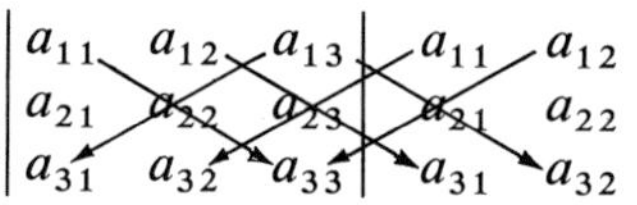

$$+ a_{11}a_{22}a_{33} + a_{12}a_{23}a_{31} + a_{13}a_{21}a_{32}$$

In the above schematic diagram, the first two columns of the determinant are repeated at its right. The products of the three elements along the arrows running downward to the right are noted, as well as the negatives of the products of the three elements along the arrows running downward to the left. The algebraic sum of these six products is the value of the determinant.

Example 3

Find the value of the determinant

$$\begin{vmatrix} 2 & 3 & 1 \\ 0 & -2 & 4 \\ 5 & -1 & 7 \end{vmatrix}$$

Using the schematic diagram for values of determinants of order 3, we obtain

$$-(1)(-2)(5) - (2)(4)(-1) - (3)(0)(7)$$

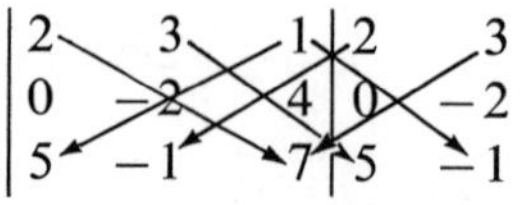

$$+(2)(-2)(7) + (3)(4)(5) + (1)(0)(-1)$$

Hence, the value of the determinant is

$$(2)(-2)(7) + (3)(4)(5) + (1)(0)(-1) \\ -(1)(-2)(5) - (2)(4)(-1) - (3)(0)(7)$$

that is, 50.

Exercises

In Exercises 1 through 4 find the value of each second-order determinant.

1. $\begin{vmatrix} 2 & 5 \\ 1 & 3 \end{vmatrix}$

2. $\begin{vmatrix} -3 & 1 \\ 4 & 2 \end{vmatrix}$

3. $\begin{vmatrix} 0.2 & 0.3 \\ 0.5 & 0.8 \end{vmatrix}$

4. $\begin{vmatrix} a & 1 \\ b & 1 \end{vmatrix}$

In Exercises 5 through 8 find the solution of each equation.

5. $\begin{vmatrix} x & 1 \\ 6 & 3 \end{vmatrix} = 0$

6. $\begin{vmatrix} 5 & x \\ 3 & x \end{vmatrix} = 10$

7. $\begin{vmatrix} x-1 & x \\ x & x \end{vmatrix} = 3$

8. $\begin{vmatrix} x-2 & x+3 \\ 2 & 4 \end{vmatrix} = 2$

In Exercises 9 through 12 show that each statement is true.

9. $\begin{vmatrix} a & c \\ b & d \end{vmatrix} = \begin{vmatrix} a & b \\ c & d \end{vmatrix}$

10. $\begin{vmatrix} a & c \\ b & d \end{vmatrix} = -\begin{vmatrix} b & d \\ a & c \end{vmatrix}$

11. $\begin{vmatrix} a & c \\ 0 & 0 \end{vmatrix} = 0$

12. $\begin{vmatrix} ka & kc \\ b & d \end{vmatrix} = k\begin{vmatrix} a & c \\ b & d \end{vmatrix}$

In Exercises 13 through 16 find the value of each third-order determinant.

13. $\begin{vmatrix} 1 & 0 & 1 \\ -2 & 3 & 2 \\ 1 & 0 & 1 \end{vmatrix}$

14. $\begin{vmatrix} 2 & -2 & 1 \\ 3 & 1 & 3 \\ -1 & 0 & 1 \end{vmatrix}$

15. $\begin{vmatrix} 2 & 3 & 1 \\ 1 & 4 & -3 \\ -1 & 2 & 0 \end{vmatrix}$

16. $\begin{vmatrix} 5 & 1 & 1 \\ 1 & 5 & 1 \\ 1 & 1 & 5 \end{vmatrix}$

17. Show that

$$\begin{vmatrix} a+b & a & a \\ a & a+b & a \\ a & a & a+b \end{vmatrix} = b^2(3a+b)$$

18. Show that $\det AB = \det A \cdot \det B$, where

$$A = \begin{vmatrix} 1 & 2 & 1 \\ 0 & 1 & 0 \\ -1 & 2 & 0 \end{vmatrix} \quad \text{and} \quad B = \begin{vmatrix} 1 & 1 & 0 \\ 2 & 0 & 1 \\ 3 & 1 & 4 \end{vmatrix}$$

19. Show that the equation

$$\begin{vmatrix} 5-x & 1 \\ 2 & 3-x \end{vmatrix} = 0$$

is satisfied if x is replaced by the matrix

$$A = \begin{pmatrix} 5 & 1 \\ 2 & 3 \end{pmatrix}$$

and each real number n is replaced by nI, where I is the identity matrix of order 2.

20. If each element of a second-order determinant is either 0 or 1, what is the probability that the value of the determinant is positive?

5.10

Properties of Determinants

In general, a *determinant* of a matrix $A = ((a_{ij}))$ of order n is a function of the square matrix and is denoted by $\det A$ or

$$\begin{vmatrix} a_{11} & a_{12} & \cdots & a_{1n} \\ a_{21} & a_{22} & \cdots & a_{2n} \\ \cdots & \cdots & \cdots & \cdots \\ a_{n1} & a_{n2} & \cdots & a_{nn} \end{vmatrix}$$

The *value of the determinant* of A is defined as the sum of $n!$ distinct terms of the form

$$(-1)^k a_{1i_1} a_{2i_2} a_{3i_3} \cdots a_{ni_n}$$

where the second subscripts $i_1, i_2, i_3, \ldots, i_n$ are equal to $1, 2, 3, \ldots, n$ taken in some order; each term contains one and only one element from each row and one and only one element from each column; and the exponent k represents the number of interchanges of two elements necessary for the second subscripts to be placed in the order 1, 2, 3, . . . , n. For example, consider the term containing $a_{13}a_{22}a_{35}a_{41}a_{54}$ in the evaluation of a determinant of order 5. The term has associated with it the factor $(-1)^3$ because three interchanges of two elements are necessary for the second subscripts to be placed in the order 1, 2, 3, 4, 5:

$$\begin{aligned} a_{13}a_{22}a_{35}a_{41}a_{54} &= a_{41}a_{22}a_{35}a_{13}a_{54} \\ &= a_{41}a_{22}a_{13}a_{35}a_{54} \\ &= a_{41}a_{22}a_{13}a_{54}a_{35} \end{aligned}$$

The value of determinants of order $n \geq 4$ are seldom obtained by the use of the definition because the labor of computation is almost prohibitive. A number of theorems that enable us to compute the value of a determinant with less labor exist. The theorems in this section will be stated and proved for determinants of order 3. However, each theorem can be extended to determinants of order n.

Theorem 5.1 The value of a determinant remains unchanged if corresponding rows and columns are interchanged; that is,

$$\begin{vmatrix} a_{11} & a_{12} & a_{13} \\ a_{21} & a_{22} & a_{23} \\ a_{31} & a_{32} & a_{33} \end{vmatrix} = \begin{vmatrix} a_{11} & a_{21} & a_{31} \\ a_{12} & a_{22} & a_{32} \\ a_{13} & a_{23} & a_{33} \end{vmatrix}$$

Proof: By the definition of the value of a determinant of order 3,

$$\begin{vmatrix} a_{11} & a_{12} & a_{13} \\ a_{21} & a_{22} & a_{23} \\ a_{31} & a_{32} & a_{33} \end{vmatrix} = a_{11}a_{22}a_{33} + a_{12}a_{23}a_{31} + a_{13}a_{21}a_{32} - a_{13}a_{22}a_{31} - a_{11}a_{23}a_{32} - a_{12}a_{21}a_{33}$$

$$\begin{vmatrix} a_{11} & a_{21} & a_{31} \\ a_{12} & a_{22} & a_{32} \\ a_{13} & a_{23} & a_{33} \end{vmatrix} = a_{11}a_{22}a_{33} + a_{21}a_{32}a_{13} + a_{31}a_{12}a_{23} - a_{31}a_{22}a_{13} - a_{11}a_{32}a_{23} - a_{21}a_{12}a_{33}$$

Hence by a simple rearrangement of factors and terms, we have

$$\begin{vmatrix} a_{11} & a_{12} & a_{13} \\ a_{21} & a_{22} & a_{23} \\ a_{31} & a_{32} & a_{33} \end{vmatrix} = \begin{vmatrix} a_{11} & a_{21} & a_{31} \\ a_{12} & a_{22} & a_{32} \\ a_{13} & a_{23} & a_{33} \end{vmatrix}$$

As a result of Theorem 5.1, for every theorem concerning the rows of a determinant there exists an associated corollary concerning the columns of the determinant. The associated corollary will not be stated in each instance with respect to the theorems presented in this section.

Theorem 5.2 If any two rows of a determinant are interchanged, then the value of the determinant is changed by a factor (-1). For example,

$$\begin{vmatrix} a_{21} & a_{22} & a_{23} \\ a_{11} & a_{12} & a_{13} \\ a_{31} & a_{32} & a_{33} \end{vmatrix} = - \begin{vmatrix} a_{11} & a_{12} & a_{13} \\ a_{21} & a_{22} & a_{23} \\ a_{31} & a_{32} & a_{33} \end{vmatrix}$$

The proof of Theorem 5.2 is left to the reader as Exercise 12.

Theorem 5.3 If every element of a row of a determinant is multiplied by a factor t, then the value of the determinant is multiplied by t.

Proof: Let every element of the ith row of a determinant be multiplied by a factor t. By the definition of the value of a determinant of order 3, every term contains one and only one element from each row as a factor. Therefore, every term will contain the additional factor t. Hence the value of the determinant will be multiplied by t. For example,

$$\begin{vmatrix} a_{11} & a_{12} & a_{13} \\ ta_{21} & ta_{22} & ta_{23} \\ a_{31} & a_{32} & a_{33} \end{vmatrix} = t\begin{vmatrix} a_{11} & a_{12} & a_{13} \\ a_{21} & a_{22} & a_{23} \\ a_{31} & a_{32} & a_{33} \end{vmatrix}$$

Theorem 5.4 The value of a determinant remains unchanged if a constant multiple of each of the elements of a row is added to the corresponding elements of another row.

For example,

$$\begin{vmatrix} a_{11} & a_{12} & a_{13} \\ a_{21} & a_{22} & a_{23} \\ a_{31} & a_{32} & a_{33} \end{vmatrix} = \begin{vmatrix} a_{11} + ta_{31} & a_{12} + ta_{32} & a_{13} + ta_{33} \\ a_{21} & a_{22} & a_{23} \\ a_{31} & a_{32} & a_{33} \end{vmatrix}$$

The proof of Theorem 5.4 is left to the reader as Exercise 13.

Consider the definition of the value of a determinant of order 3. By a rearrangement of terms, we have

$$\begin{vmatrix} a_{11} & a_{12} & a_{13} \\ a_{21} & a_{22} & a_{23} \\ a_{31} & a_{32} & a_{33} \end{vmatrix} = a_{11}(a_{22}a_{33} - a_{23}a_{32}) - a_{12}(a_{21}a_{33} - a_{23}a_{31}) + a_{13}(a_{21}a_{32} - a_{22}a_{31})$$

Using the definition of the value of a determinant of order 2, we have

$$\begin{vmatrix} a_{11} & a_{12} & a_{13} \\ a_{21} & a_{22} & a_{23} \\ a_{31} & a_{32} & a_{33} \end{vmatrix} = a_{11}\begin{vmatrix} a_{22} & a_{23} \\ a_{32} & a_{33} \end{vmatrix} - a_{12}\begin{vmatrix} a_{21} & a_{23} \\ a_{31} & a_{33} \end{vmatrix} + a_{13}\begin{vmatrix} a_{21} & a_{22} \\ a_{31} & a_{32} \end{vmatrix}$$

This is a special case of Theorem 5.5.

Note that each of the three determinants of order 2 may be obtained from the determinant of order 3 by deleting certain rows and columns. The determinant $\begin{vmatrix} a_{22} & a_{23} \\ a_{32} & a_{33} \end{vmatrix}$ may be obtained by deleting the row and column containing the element a_{11}; the determinant $\begin{vmatrix} a_{21} & a_{23} \\ a_{31} & a_{33} \end{vmatrix}$ may be obtained by deleting the row and column containing the element a_{12}; the determinant $\begin{vmatrix} a_{21} & a_{22} \\ a_{31} & a_{32} \end{vmatrix}$ may be obtained by deleting the row and column containing the element a_{13}. The *cofactor* of any element a_{ij}

of a determinant $\det((a_{ij}))$ is defined as the product of $(-1)^{i+j}$ and the determinant obtained by deleting the elements of the ith row and the jth column of $\det((a_{ij}))$.

Theorem 5.5 The value of a determinant is equal to the sum of the products of the elements of any row and their cofactors.

The proof of Theorem 5.5 is left to the reader as Exercise 14.

Example

Use Theorem 5.5 to find the value of the determinant

$$\begin{vmatrix} 3 & 0 & 5 \\ 1 & -1 & 2 \\ 8 & 3 & 4 \end{vmatrix}$$

By Theorem 5.5, using the elements of row one and their cofactors, we have

$$\begin{vmatrix} 3 & 0 & 5 \\ 1 & -1 & 2 \\ 8 & 3 & 4 \end{vmatrix} = 3\begin{vmatrix} -1 & 2 \\ 3 & 4 \end{vmatrix} - 0\begin{vmatrix} 1 & 2 \\ 8 & 4 \end{vmatrix} + 5\begin{vmatrix} 1 & -1 \\ 8 & 3 \end{vmatrix}$$

$$= (3)(-10) + (5)(11)$$
$$= 25$$

Exercises

1. True or false: $\det A = \det A^T$?

Let $\begin{vmatrix} a & b & c \\ d & e & f \\ g & h & j \end{vmatrix} = 4$. In Exercises 2 through 7 find the value of each determinant by using the theorems of this section.

2. $\begin{vmatrix} g & h & j \\ d & e & f \\ a & b & c \end{vmatrix}$

3. $\begin{vmatrix} a & g & d \\ b & h & e \\ c & j & f \end{vmatrix}$

4. $\begin{vmatrix} a & 3b & 2c \\ d & 3e & 2f \\ g & 3h & 2j \end{vmatrix}$

5. $\begin{vmatrix} a+2d & b+2e & c+2f \\ d & e & f \\ g & h & j \end{vmatrix}$

6. $\begin{vmatrix} 0 & 0 & 1 & 0 \\ a & b & 2 & c \\ d & e & 3 & f \\ g & h & 4 & j \end{vmatrix}$

7. $\begin{vmatrix} 3 & 0 & 0 & 0 \\ 3 & a & b & c \\ 3 & d & e & f \\ 3 & g & h & j \end{vmatrix}$

8. Evaluate the following by using the cofactors of the elements of the second row:

$$\begin{vmatrix} 2 & 3 & -1 \\ -2 & 1 & 0 \\ 1 & 3 & 1 \end{vmatrix}$$

9. Find the value of

$$\begin{vmatrix} 1 & 1 & 1 \\ a & b & c \\ b+c & c+a & a+b \end{vmatrix}$$

10. Prove that if every element of a row of a determinant is zero, then the value of the determinant is zero.

11. Prove that the value of the determinant of a matrix of order n is zero if two rows are identical.

12. Prove Theorem 5.2.

13. Prove Theorem 5.4.

14. Prove Theorem 5.5.

15. Show that the sum of the products of the elements of the second row and the cofactors of the corresponding elements of either the first or the third row is zero for the matrix

$$\begin{pmatrix} 3 & 4 & 0 \\ -1 & 2 & 1 \\ 2 & 1 & -2 \end{pmatrix}$$

16. Prove that for any determinant of a matrix of order 3 the sum of the products of the elements of any row and the cofactors of the corresponding elements of another row is zero. (*Note:* This result is true for the determinant of a matrix of any order n.)

17. Find $\det((ka_{ij}))$ if $((a_{ij}))$ is of order 4 and $\det((a_{ij})) = m$.

18. Find $\det((a_{ij}))$ if matrix $((a_{ij}))$ is of order n and $a_{ij} = 0$ for every pair of values i and j such that $i > j$.

19. Show that

$$\begin{vmatrix} a & b & m & n \\ c & d & r & s \\ 0 & 0 & e & f \\ 0 & 0 & g & h \end{vmatrix} = \begin{vmatrix} a & b \\ c & d \end{vmatrix} \cdot \begin{vmatrix} e & f \\ g & h \end{vmatrix}$$

In Exercises 20 and 21 find the value of each determinant.

20. $\begin{vmatrix} 3 & 1 & 0 & 1 \\ 1 & 0 & 2 & 1 \\ 0 & 1 & 2 & 1 \\ 1 & 2 & 0 & 0 \end{vmatrix}$

21. $\begin{vmatrix} 2 & -1 & 2 & 3 \\ 1 & 1 & 0 & 2 \\ 2 & 5 & 6 & 3 \\ 2 & 2 & 7 & 1 \end{vmatrix}$

5.11

More About Matrix Inverses

It is convenient to denote the cofactor of the element a_{ij} of a square matrix $A = ((a_{ij}))$ by the symbol A_{ij}. For example, the cofactor of a_{12} in

$$A = \begin{pmatrix} a_{11} & a_{12} & a_{13} \\ a_{21} & a_{22} & a_{23} \\ a_{31} & a_{32} & a_{33} \end{pmatrix}$$

is denoted by A_{12}, and

$$\begin{aligned} A_{12} &= (-1)^{1+2} \begin{vmatrix} a_{21} & a_{23} \\ a_{31} & a_{33} \end{vmatrix} \\ &= -a_{21}a_{33} + a_{23}a_{31} \end{aligned}$$

Let $A = ((a_{ij}))$ be any square matrix of order 3. From our study of determinants we know that the value of any determinant is equal to the sum of the products of the elements of any row and their cofactors. Hence

$$\begin{aligned} \det A &= a_{11}A_{11} + a_{12}A_{12} + a_{13}A_{13} \\ \det A &= a_{21}A_{21} + a_{22}A_{22} + a_{23}A_{23} \\ \det A &= a_{31}A_{31} + a_{32}A_{32} + a_{33}A_{33} \end{aligned}$$

Recall from Exercise 11 of Sec. 5.10 that if two rows of a determinant are identical, then the value of the determinant is 0. Thus the sum of the products of the elements of any row with the cofactors of the corresponding elements of another row is 0 because the sum represents the evaluation of a determinant in which two rows are identical:

$$\begin{aligned} a_{11}A_{21} + a_{12}A_{22} + a_{13}A_{23} &= 0 \\ a_{11}A_{31} + a_{12}A_{32} + a_{13}A_{33} &= 0 \\ a_{21}A_{11} + a_{22}A_{12} + a_{23}A_{13} &= 0 \\ a_{21}A_{31} + a_{22}A_{32} + a_{23}A_{33} &= 0 \\ a_{31}A_{11} + a_{32}A_{12} + a_{33}A_{13} &= 0 \\ a_{31}A_{21} + a_{32}A_{22} + a_{33}A_{23} &= 0 \end{aligned}$$

We may express these nine equations by the single matrix equation

$$\begin{pmatrix} a_{11} & a_{12} & a_{13} \\ a_{21} & a_{22} & a_{23} \\ a_{31} & a_{32} & a_{33} \end{pmatrix} \begin{pmatrix} A_{11} & A_{21} & A_{31} \\ A_{12} & A_{22} & A_{32} \\ A_{13} & A_{23} & A_{33} \end{pmatrix} = \det A \begin{pmatrix} 1 & 0 & 0 \\ 0 & 1 & 0 \\ 0 & 0 & 1 \end{pmatrix}$$

This equation can be expressed in more concise notation as

$$((a_{ij}))((A_{ij}))^T = \det A \cdot I$$

Note that the matrix $((A_{ij}))^T$ is the transpose of the matrix obtained by replacing each element of $((a_{ij}))$ by its cofactor. Since $A = ((a_{ij}))$, the matrix equation can also be written as

$$A((A_{ij}))^T = \det A \cdot I$$

whereby

$$A \frac{((A_{ij}))^T}{\det A} = I$$

provided that $\det A \neq 0$. Thus, since $AA^{-1} = I$,

$$A^{-1} = \frac{((A_{ij}))^T}{\det A}$$

provided that the determinant of matrix A is nonzero.

It should be mentioned that we have shown only that $((A_{ij}))^T/\det A$ is the right-hand inverse of matrix A. However, for square matrices the right-hand inverse is the inverse; that is,

$$\frac{((A_{ij}))^T}{\det A} A = A \frac{((A_{ij}))^T}{\det A} = I$$

Example 1

Determine A^{-1}, if it exists, where

$$A = \begin{pmatrix} 0 & -2 & 1 \\ 1 & 3 & -2 \\ 3 & 9 & -5 \end{pmatrix}$$

Since $\det A = \begin{vmatrix} 0 & -2 & 1 \\ 1 & 3 & -2 \\ 3 & 9 & -5 \end{vmatrix} = 2$, the multiplicative inverse of matrix A exists. The cofactors of the elements of A are

$$A_{11} = \begin{vmatrix} 3 & -2 \\ 9 & -5 \end{vmatrix} \qquad A_{12} = -\begin{vmatrix} 1 & -2 \\ 3 & -5 \end{vmatrix} \qquad A_{13} = \begin{vmatrix} 1 & 3 \\ 3 & 9 \end{vmatrix}$$

$$A_{21} = -\begin{vmatrix} -2 & 1 \\ 9 & -5 \end{vmatrix} \qquad A_{22} = \begin{vmatrix} 0 & 1 \\ 3 & -5 \end{vmatrix} \qquad A_{23} = -\begin{vmatrix} 0 & -2 \\ 3 & 9 \end{vmatrix}$$

$$A_{31} = \begin{vmatrix} -2 & 1 \\ 3 & -2 \end{vmatrix} \qquad A_{32} = -\begin{vmatrix} 0 & 1 \\ 1 & -2 \end{vmatrix} \qquad A_{33} = \begin{vmatrix} 0 & -2 \\ 1 & 3 \end{vmatrix}$$

We now determine the matrix of cofactors, which is

$$((A_{ij})) = \begin{pmatrix} A_{11} & A_{12} & A_{13} \\ A_{21} & A_{22} & A_{23} \\ A_{31} & A_{32} & A_{33} \end{pmatrix} = \begin{pmatrix} 3 & -1 & 0 \\ -1 & -3 & -6 \\ 1 & 1 & 2 \end{pmatrix}$$

The transpose of the matrix of cofactors is

$$((A_{ij}))^T = \begin{pmatrix} 3 & -1 & 1 \\ -1 & -3 & 1 \\ 0 & -6 & 2 \end{pmatrix}$$

Hence

$$A^{-1} = \frac{((A_{ij}))^T}{\det A} = \begin{pmatrix} \frac{3}{2} & -\frac{1}{2} & \frac{1}{2} \\ -\frac{1}{2} & -\frac{3}{2} & \frac{1}{2} \\ 0 & -3 & 1 \end{pmatrix}$$

Check:

$$AA^{-1} = \begin{pmatrix} 0 & -2 & 1 \\ 1 & 3 & -2 \\ 3 & 9 & -5 \end{pmatrix} \begin{pmatrix} \frac{3}{2} & -\frac{1}{2} & \frac{1}{2} \\ -\frac{1}{2} & -\frac{3}{2} & \frac{1}{2} \\ 0 & -3 & 1 \end{pmatrix} = \begin{pmatrix} 1 & 0 & 0 \\ 0 & 1 & 0 \\ 0 & 0 & 1 \end{pmatrix}$$

Example 2

Show that the *multiplicative inverse of a matrix*, if it exists, is unique.

Let A be a square matrix with a multiplicative inverse A^{-1}. Let B be any other multiplicative inverse of A. Since $AA^{-1} = I$ and $AB = I$, it follows that

$$AA^{-1} = AB$$

By the associative property of matrix multiplication,

$$\begin{aligned} A^{-1}(AA^{-1}) &= A^{-1}(AB) \\ (A^{-1}A)A^{-1} &= (A^{-1}A)B \\ IA^{-1} &= IB \\ A^{-1} &= B \end{aligned}$$

Hence, if it exists, the multiplicative inverse of a matrix is unique.

Exercises

In Exercises 1 through 8 find the multiplicative inverse, if it exists, of each matrix.

1. $\begin{pmatrix} 1 & 2 & 3 \\ 2 & 4 & 5 \\ 3 & 5 & 6 \end{pmatrix}$

2. $\begin{pmatrix} 3 & 0 & 1 \\ -1 & -1 & 0 \\ 2 & 0 & 0 \end{pmatrix}$

3. $\begin{pmatrix} 2 & 5 & 8 \\ 1 & 3 & 5 \\ 0 & 2 & 10 \end{pmatrix}$

4. $\begin{pmatrix} 1 & 6 & 6 \\ 0 & 2 & 12 \\ 0 & 0 & 3 \end{pmatrix}$

5. $\begin{pmatrix} \frac{2}{3} & -\frac{2}{3} & \frac{1}{3} \\ \frac{1}{3} & \frac{2}{3} & \frac{2}{3} \\ \frac{2}{3} & \frac{1}{3} & -\frac{2}{3} \end{pmatrix}$

6. $\begin{pmatrix} 2 & 3 & 4 \\ 0 & 5 & 6 \\ 0 & 0 & 1 \end{pmatrix}$

7. $\begin{pmatrix} 2 & 1 & 1 \\ 2 & 0 & 2 \\ 0 & 1 & -3 \end{pmatrix}$

8. $\begin{pmatrix} 0 & a & b \\ -a & 0 & c \\ -b & -c & 0 \end{pmatrix}$

9. Show that

$$\begin{pmatrix} a & 0 & 0 \\ 0 & b & 0 \\ 0 & 0 & c \end{pmatrix}^{-1} = \begin{pmatrix} \frac{1}{a} & 0 & 0 \\ 0 & \frac{1}{b} & 0 \\ 0 & 0 & \frac{1}{c} \end{pmatrix}$$

10. Let $A = \begin{pmatrix} 7 & -3 & -3 \\ -2 & 2 & 0 \\ -1 & 0 & 1 \end{pmatrix}$. Show that $(A^{-1})^T = (A^T)^{-1}$.

11. Let $A = \begin{pmatrix} 2 & 0 & 2 \\ 2 & 0 & 1 \\ 0 & 1 & 0 \end{pmatrix}$ and $B = \begin{pmatrix} 3 & 0 & 1 \\ -1 & -1 & 0 \\ 2 & 0 & 0 \end{pmatrix}$. Show that

$$(AB)^{-1} = B^{-1}A^{-1}$$

12. Decode the following message if $A = 1$, $B = 2$, ..., $Z = 26$ and the matrix $\begin{pmatrix} 1 & 2 & 3 \\ 2 & 4 & 5 \\ 3 & 5 & 6 \end{pmatrix}$ was used to code the information:

$$\begin{pmatrix} 57 & 120 & 18 & 109 \\ 107 & 216 & 35 & 194 \\ 141 & 268 & 44 & 241 \end{pmatrix}$$

(See Exercise 19 of Sec. 5.6.)

5.12

Probability Matrices

An experiment in which the outcomes of each trial depend at most on the outcomes of the preceding trial is called a *Markov chain*; the outcomes are called *states*. In this section we shall describe the process of a Markov chain by means of *probability matrices*. A probability matrix of order n is a matrix $((a_{ij}))$ whose elements a_{ij} satisfy the following conditions:

1. $0 \leq a_{ij} \leq 1$.
2. $a_{i1} + a_{i2} + \cdots + a_{in} = 1$ for each $i = 1, 2, \ldots, n$.

As an example of a Markov chain, suppose that from past experience the weather bureau predicts that if the weather is good today, then the probability that it will be good tomorrow is $\frac{2}{3}$; if it is raining today, then the probability that the weather will be good tomorrow is $\frac{1}{2}$. The weather on any given day is dependent upon the weather of the preceding day. If we designate "good weather" by G and "rain" by R, the possible states of this Markov chain are G and R. The *transition probabilities*, that is, the probabilities of going from one state to another, can be represented geometrically by the arrow diagram in Fig. 5.1.

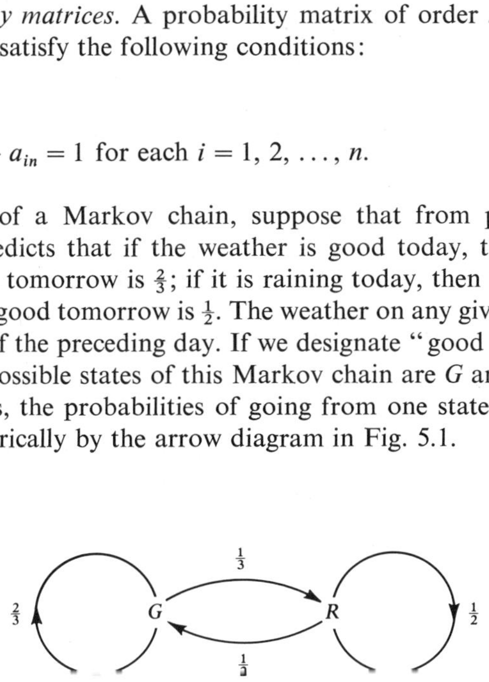

Figure 5.1

The probability matrix for this Markov chain is

$$\begin{array}{cc} & \overbrace{\text{Good} \quad \text{Rain}}^{\text{Tomorrow}} \\ \text{Today}\begin{cases}\text{Good} \\ \text{Rain}\end{cases} & \begin{pmatrix} \frac{2}{3} & \frac{1}{3} \\ \frac{1}{2} & \frac{1}{2} \end{pmatrix} \end{array}$$

We can use the probabilities in Fig. 5.1 to find the probability that the weather is good or it rains on any future day. For example, if the weather is good today, what is the probability that the weather will be good the day after tomorrow?

Now, the weather will be good the day after tomorrow if one of the two following sequences of events occurs:

$$G \to G \to G \qquad \text{or} \qquad G \to R \to G$$

The probability that one of these two sequences of events occurs is given by

$$\tfrac{2}{3} \cdot \tfrac{2}{3} + \tfrac{1}{3} \cdot \tfrac{1}{2}$$

Note that this result is the element in the top left-hand corner of the square of the probability matrix for this Markov chain:

$$\begin{pmatrix} \frac{2}{3} & \frac{1}{3} \\ \frac{1}{2} & \frac{1}{2} \end{pmatrix}\begin{pmatrix} \frac{2}{3} & \frac{1}{3} \\ \frac{1}{2} & \frac{1}{2} \end{pmatrix} = \begin{pmatrix} \frac{2}{3} \cdot \frac{2}{3} + \frac{1}{3} \cdot \frac{1}{2} & \cdots \\ \cdots & \cdots \end{pmatrix}$$

The probability that it will rain the day after tomorrow if the weather is good today can be found by considering the following two sequences of events:

$$G \to G \to R \qquad \text{or} \qquad G \to R \to R$$

The probability that one of these two sequences of events occurs is given by

$$\tfrac{2}{3} \cdot \tfrac{1}{3} + \tfrac{1}{3} \cdot \tfrac{1}{2}$$

This result is the element in the top right-hand corner of the square of the probability matrix for this Markov chain:

$$\begin{pmatrix} \frac{2}{3} & \frac{1}{3} \\ \frac{1}{2} & \frac{1}{2} \end{pmatrix}\begin{pmatrix} \frac{2}{3} & \frac{1}{3} \\ \frac{1}{2} & \frac{1}{2} \end{pmatrix} = \begin{pmatrix} \cdots & \frac{2}{3} \cdot \frac{1}{3} + \frac{1}{3} \cdot \frac{1}{2} \\ \cdots & \cdots \end{pmatrix}$$

The probability that the weather will be good the day after tomorrow if it is raining today can be found by considering the following two sequences of events:

$$R \to G \to G \qquad \text{or} \qquad R \to R \to G$$

The probability that one of these two sequences of events occurs is given by

$$\tfrac{1}{2} \cdot \tfrac{2}{3} + \tfrac{1}{2} \cdot \tfrac{1}{2}$$

This result is the element in the bottom left-hand corner of the square of the probability matrix for this Markov chain:

$$\begin{pmatrix} \frac{2}{3} & \frac{1}{3} \\ \frac{1}{2} & \frac{1}{2} \end{pmatrix}\begin{pmatrix} \frac{2}{3} & \frac{1}{3} \\ \frac{1}{2} & \frac{1}{2} \end{pmatrix} = \begin{pmatrix} \cdots & \cdots \\ \frac{1}{2} \cdot \frac{2}{3} + \frac{1}{2} \cdot \frac{1}{2} & \cdots \end{pmatrix}$$

Finally, the probability that it will rain the day after tomorrow if it is raining today can be found by considering the following two sequences of events:

$$R \to G \to R \qquad \text{or} \qquad R \to R \to R$$

The probability that one of these two sequences of events occurs is given by

$$\tfrac{1}{2}\cdot\tfrac{1}{3}+\tfrac{1}{2}\cdot\tfrac{1}{2}$$

This result is the element in the bottom right-hand corner of the square of the probability matrix for this Markov chain:

$$\begin{pmatrix}\frac{2}{3} & \frac{1}{3}\\ \frac{1}{2} & \frac{1}{2}\end{pmatrix}\begin{pmatrix}\frac{2}{3} & \frac{1}{3}\\ \frac{1}{2} & \frac{1}{2}\end{pmatrix}=\begin{pmatrix}\cdots & \cdots\\ \cdots & \frac{1}{2}\cdot\frac{1}{3}+\frac{1}{2}\cdot\frac{1}{2}\end{pmatrix}$$

Thus we have

$$\begin{array}{ll} & \overbrace{\begin{array}{cc}\text{Good} & \text{Rain}\end{array}}^{\text{Day after tomorrow}} \\ \text{Today}\begin{cases}\text{Good}\\ \text{Rain}\end{cases} & \begin{pmatrix}\frac{11}{18} & \frac{7}{18}\\ \frac{7}{12} & \frac{5}{12}\end{pmatrix}\end{array}$$

It is important to note that this is also a probability matrix since

1. $0\le\frac{11}{18}\le 1 \qquad 0\le\frac{7}{18}\le 1 \qquad 0\le\frac{7}{12}\le 1 \qquad 0\le\frac{5}{12}\le 1$
2. $\frac{11}{18}+\frac{7}{18}=1 \qquad \frac{7}{12}+\frac{5}{12}=1$

The above analysis suggests that squaring a probability matrix for a Markov chain yields a matrix whose elements are the probabilities that an experiment now in some state S_i will be in state S_j after the second trial. In general, the element a_{ij} of the nth power of a probability matrix for a Markov chain represents the probability that an experiment in state S_i will be in state S_j after the nth trial.

Example 1

Use the preceding data and determine the probability matrix for the weather conditions for
(*a*) 3 days' time; (*b*) 4 days' time.

(*a*) $$\begin{pmatrix}\frac{2}{3} & \frac{1}{3}\\ \frac{1}{2} & \frac{1}{2}\end{pmatrix}^3=\begin{pmatrix}\frac{11}{18} & \frac{7}{18}\\ \frac{7}{12} & \frac{5}{12}\end{pmatrix}\begin{pmatrix}\frac{2}{3} & \frac{1}{3}\\ \frac{1}{2} & \frac{1}{2}\end{pmatrix}=\begin{pmatrix}\frac{65}{108} & \frac{43}{108}\\ \frac{43}{72} & \frac{29}{72}\end{pmatrix}$$

that is,

$$\begin{array}{ll} & \overbrace{\begin{array}{cc}\text{Good} & \text{Rain}\end{array}}^{\text{3 days from today}} \\ \text{Today}\begin{cases}\text{Good}\\ \text{Rain}\end{cases} & \begin{pmatrix}0.602 & 0.398\\ 0.597 & 0.403\end{pmatrix}\end{array}$$

(*b*) $$\begin{pmatrix}\frac{2}{3} & \frac{1}{3}\\ \frac{1}{2} & \frac{1}{2}\end{pmatrix}^4=\begin{pmatrix}\frac{65}{108} & \frac{43}{108}\\ \frac{43}{72} & \frac{29}{72}\end{pmatrix}\begin{pmatrix}\frac{2}{3} & \frac{1}{3}\\ \frac{1}{2} & \frac{1}{2}\end{pmatrix}=\begin{pmatrix}\frac{389}{648} & \frac{259}{648}\\ \frac{259}{432} & \frac{173}{432}\end{pmatrix}$$

that is,

$$\text{Today}\begin{cases}\text{Good} \\ \text{Rain}\end{cases}\overset{\overbrace{\text{Good} \qquad \text{Rain}}^{\text{4 days from today}}}{\begin{pmatrix} 0.6003 & 0.3997 \\ 0.5995 & 0.4005 \end{pmatrix}}$$

The results of Example 1 indicate that the nth power of the original probability matrix tends to the matrix

$$\begin{pmatrix} 0.6 & 0.4 \\ 0.6 & 0.4 \end{pmatrix}$$

as n increases. This shows that in the long run 0.6 of the days the weather will be good and 0.4 of the days it will rain.

Example 2

If a customer buys Brand X this week, the probability that he will buy Brand X next week is $\frac{4}{5}$. If he buys Brand Y this week, the probability that he will buy Brand X next week is $\frac{2}{5}$.

(*a*) Find the probability matrix for next week's purchase.
(*b*) Estimate the probabilities for his purchases after 1 year.

(*a*)

$$\text{This week}\begin{cases}\text{Brand } X \\ \text{Brand } Y\end{cases}\overset{\overbrace{\text{Brand } X \quad \text{Brand } Y}^{\text{Next week}}}{\begin{pmatrix} \frac{4}{5} & \frac{1}{5} \\ \frac{2}{5} & \frac{3}{5} \end{pmatrix}}$$

(*b*)
$$\begin{pmatrix} \frac{4}{5} & \frac{1}{5} \\ \frac{2}{5} & \frac{3}{5} \end{pmatrix}^2 = \begin{pmatrix} \frac{18}{25} & \frac{7}{25} \\ \frac{14}{25} & \frac{11}{25} \end{pmatrix} = \begin{pmatrix} 0.72 & 0.28 \\ 0.56 & 0.44 \end{pmatrix}$$

$$\begin{pmatrix} \frac{4}{5} & \frac{1}{5} \\ \frac{2}{5} & \frac{3}{5} \end{pmatrix}^3 = \begin{pmatrix} \frac{86}{125} & \frac{39}{125} \\ \frac{78}{125} & \frac{47}{125} \end{pmatrix} = \begin{pmatrix} 0.688 & 0.312 \\ 0.624 & 0.376 \end{pmatrix}$$

$$\begin{pmatrix} \frac{4}{5} & \frac{1}{5} \\ \frac{2}{3} & \frac{3}{5} \end{pmatrix}^4 = \begin{pmatrix} \frac{422}{625} & \frac{203}{625} \\ \frac{406}{625} & \frac{219}{625} \end{pmatrix} = \begin{pmatrix} 0.675 & 0.325 \\ 0.650 & 0.350 \end{pmatrix}$$

$$\begin{pmatrix} \frac{4}{5} & \frac{1}{5} \\ \frac{2}{5} & \frac{3}{5} \end{pmatrix}^5 = \begin{pmatrix} \frac{2094}{3125} & \frac{1031}{3125} \\ \frac{2062}{3125} & \frac{1063}{3125} \end{pmatrix} = \begin{pmatrix} 0.670 & 0.330 \\ 0.660 & 0.340 \end{pmatrix}$$

At this point it appears that the nth power of the original probability matrix tends to the matrix

$$\begin{pmatrix} \frac{2}{3} & \frac{1}{3} \\ \frac{2}{3} & \frac{1}{3} \end{pmatrix}$$

that is,

$$\begin{pmatrix} \frac{4}{5} & \frac{1}{5} \\ \frac{2}{5} & \frac{3}{5} \end{pmatrix}^n \to \begin{pmatrix} \frac{2}{3} & \frac{1}{3} \\ \frac{2}{3} & \frac{1}{3} \end{pmatrix}$$

as n increases. Note that

$$\begin{pmatrix} \frac{2}{3} & \frac{1}{3} \\ \frac{2}{3} & \frac{1}{3} \end{pmatrix}\begin{pmatrix} \frac{4}{5} & \frac{1}{5} \\ \frac{2}{5} & \frac{3}{5} \end{pmatrix} = \begin{pmatrix} \frac{2}{3} & \frac{1}{3} \\ \frac{2}{3} & \frac{1}{3} \end{pmatrix}$$

which verifies this fact. Thus, in the long run Brand X will have $\frac{2}{3}$ of the market; Brand Y will have $\frac{1}{3}$ of the market.

Exercises

1. If a textbook salesman is in Orlando today, he will definitely go to Miami tomorrow. However, if he is in Miami today, the probability that he will stay tomorrow is $\frac{4}{5}$.
   (*a*) Find the probability matrix for his journeys.
   (*b*) Determine the proportion of his time spent in Orlando and Miami.
2. Of the people on North Beach Island who vote Conservative in an election, $\frac{9}{10}$ will vote Conservative again in the next election. Of those who vote Liberal, only $\frac{2}{3}$ will vote Liberal again in the next election. What percentage of the future votes can the Conservative Party expect to obtain?
3. By the end of each day, $\frac{1}{10}$ of the liquid present in a container at the beginning of the day is changed into vapor; $\frac{4}{5}$ of the vapor present at the beginning of the day is changed into liquid. What proportion of liquid and vapor will ultimately be obtained?
4. A message consisting of the digits 0 and 1 is received and transmitted by radio through several relay stations under less than ideal conditions. The probability of a digit being changed is $\frac{1}{10}$.
   (*a*) Find the probability matrix for this situation.
   (*b*) What is the probability of a digit being changed after many transmissions?
5. If our baseball team's star pitcher wins, then the probability that he wins next time is $\frac{3}{4}$; if he loses, then the probability that he wins next time is $\frac{1}{2}$. What fraction of the games that he pitches this year should the manager expect him to win?
6. Only two brands of frozen crabmeat are available to shoppers of a particular town. During the past year $\frac{2}{5}$ of the customers buying Crabby Crabmeat changed to buying the competitor's brand; $\frac{5}{8}$ of the competitor's customers changed to buying Crabby Crabmeat. What percentage of the market can the Crabby Crabmeat Company expect to obtain in the long run if the products are not changed?

7. The following table indicates the educational background of parents and their children:

		Highest educational level of children, %		
		College	High School	Elementary School
Highest educational level of parents, %	College	75	20	5
	High School	50	40	10
	Elementary School	25	50	25

(*a*) Find the probability matrix for this situation.
(*b*) What is the probability that the grandchild of an elementary school graduate is a high school graduate?
(*c*) What is the probability that the grandchild of an elementary school graduate is a college graduate?
(*d*) What is the probability that the grandchild of a college graduate is a college graduate?

Keyed Summary

In this chapter we studied matrices, operations defined on matrices, and special matrices. The concepts and methods presented are part of the body of knowledge called the *algebra of matrices.*

5.1 Matrices

Rectangular arrays of elements are called matrices. A matrix of m rows and n columns is a matrix of order m by n. If the number of rows is equal to the number of columns, then the matrix is called a square matrix. If the elements of a matrix are real numbers or real variables, then the matrix is called a real matrix.

Two matrices A and B are equal if and only if they are of the same order and corresponding elements are equal.

Any matrix, all of whose elements are zero, is called a null or zero matrix.

5.2 *Addition and Subtraction of Matrices*

If A and B are matrices of the order m by n, then the sum $A + B$ is defined as a third matrix C of the same order such that each element of C is the sum of the corresponding elements of A and B.

Matrix addition is commutative and associative.

The null matrix of order m by n is the additive identity element for the set of all matrices of order m by n.

If A and B are matrices of the order m by n, then the difference $A - B$ is defined as a third matrix C of the same order such that each element of C is the difference of the corresponding elements of A and B.

5.3 *Multiplication of a Matrix by a Real Number*

Given any matrix A and a real number t, then the product tA is a matrix whose elements are the products of t and the corresponding elements of A.

Let r and s represent any two real numbers, and let A and B represent any two matrices of the same order:

1. $(r + s)A = rA + sA$
2. $r(A + B) = rA + rB$
3. $r(sA) = (rs)A$

5.4 *Matrix Multiplication*

Let A be a matrix of order m by n; let B be a matrix of order n by r. The product AB is defined as a matrix C of order m by r such that the element in the ith row and jth column of C is equal to the sum of the products of the elements in the ith row of A and the corresponding elements in the jth column of B.

When AB exists, the matrices A and B are said to be conformable.

Matrix multiplication is associative and distributive with respect to addition. Matrix multiplication is not commutative. The cancellation property of multiplication is not valid for matrix multiplication.

5.5 Special Matrices

A square matrix whose nondiagonal elements are all equal to zero is called a diagonal matrix.

If all the diagonal elements of a diagonal matrix are equal, the diagonal matrix is called a scalar matrix.

A scalar matrix of order n, whose diagonal elements are equal to 1, is called an identity or unit matrix. An identity matrix, denoted by I, has the property that $AI = IA = A$, where A is a square matrix of the same order. The matrix I of order n is the multiplicative identity element for the set of all square matrices of order n.

Whenever the rows and columns of a given matrix A are interchanged, the resulting matrix is called the transpose of the original matrix A and is denoted by A^T.

5.6 Multiplicative Inverse of a Matrix

The multiplicative inverse of A, denoted by A^{-1}, is such that $AA^{-1} = A^{-1}A = I$.

The multiplicative inverse of $\begin{pmatrix} a & b \\ c & d \end{pmatrix}$ is the matrix

$$\begin{pmatrix} \dfrac{d}{ad - bc} & \dfrac{-b}{ad - bc} \\[2ex] \dfrac{-c}{ad - bc} & \dfrac{a}{ad - bc} \end{pmatrix}$$

provided $ad - bc \neq 0$.

The determinant $\begin{vmatrix} a & b \\ c & d \end{vmatrix}$ is called the determinant of the matrix $\begin{pmatrix} a & b \\ c & d \end{pmatrix}$.

5.7 Matrices and Systems of Equations

The solution of the system of two linear equations in two variables

$$\begin{cases} ax + by = e \\ cx + dy = f \end{cases}$$

can be expressed in matrix form by

$$\begin{pmatrix} x \\ y \end{pmatrix} = \begin{pmatrix} \dfrac{d}{ad - bc} & \dfrac{-b}{ad - bc} \\[2ex] \dfrac{-c}{ad - bc} & \dfrac{a}{ad - bc} \end{pmatrix} \begin{pmatrix} e \\ f \end{pmatrix}$$

provided $ad - bc \neq 0$.

5.8 Elementary Row Transformations

The solution, if it exists, of a system of n linear equations in n variables can be obtained by performing the elementary row transformations on the augmented matrix of the system. The elementary row transformations are

1. The interchange of two rows
2. Multiplication of the elements of any row by a nonzero real number
3. Addition of a nonzero real multiple of the elements of any row onto the corresponding elements of another row

5.9 Determinants

The determinant of a square matrix $A = ((a_{ij}))$ of order 2 is denoted either by det A or

$$\begin{vmatrix} a_{11} & a_{12} \\ a_{21} & a_{22} \end{vmatrix}$$

The value of the determinant is $a_{11}a_{22} - a_{12}a_{21}$ and may be remembered by the schematic diagram:

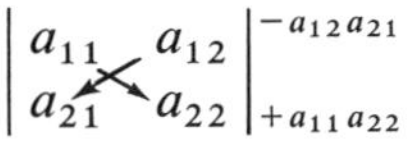

The determinant of a square matrix $A = ((a_{ij}))$ of order 3 is denoted by det A or

$$\begin{vmatrix} a_{11} & a_{12} & a_{13} \\ a_{21} & a_{22} & a_{23} \\ a_{31} & a_{32} & a_{33} \end{vmatrix}$$

The value of the determinant is $a_{11}a_{22}a_{33} + a_{12}a_{23}a_{31} + a_{13}a_{21}a_{32} - a_{13}a_{22}a_{31} - a_{11}a_{23}a_{32} - a_{12}a_{21}a_{33}$ and may be remembered by the schematic diagram:

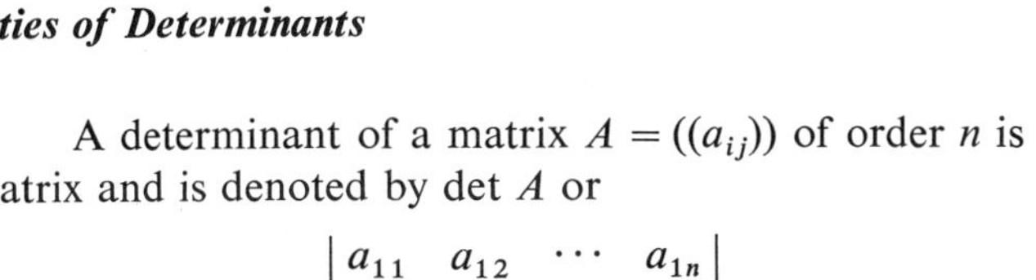

5.10 Properties of Determinants

A determinant of a matrix $A = ((a_{ij}))$ of order n is a function of the square matrix and is denoted by det A or

$$\begin{vmatrix} a_{11} & a_{12} & \cdots & a_{1n} \\ a_{21} & a_{22} & \cdots & a_{2n} \\ \cdots & \cdots & \cdots & \cdots \\ a_{n1} & a_{n2} & \cdots & a_{nn} \end{vmatrix}$$

The value of the determinant of A is defined as the sum of $n!$ distinct terms of the form

$$(-1)^k a_{1i_1} a_{2i_2} a_{3i_3} \cdots a_{ni_n}$$

where the second subscripts $i_1, i_2, i_3, \ldots, i_n$ are equal to 1, 2, 3, ..., n taken in some order; each term contains one and only one element from each row and one and only one element from each column; and the exponent k represents the number of interchanges of two elements necessary for the second subscripts to be placed in the order 1, 2, 3, ..., n.

The value of a determinant remains unchanged if corresponding rows and columns are interchanged.

If any two rows of a determinant are interchanged, then the value of the determinant is changed by a factor (-1).

If every element of a row of a determinant is multiplied by a factor t, then the value of the determinant is multiplied by t.

The value of a determinant remains unchanged if a constant multiple of each of the elements of a row is added to the corresponding elements of another row.

The cofactor of any element a_{ij} of a determinant det $((a_{ij}))$ is defined as the product of $(-1)^{i+j}$ and the determinant obtained by deleting the elements of the ith row and the jth column of det $((a_{ij}))$.

The value of a determinant is equal to the sum of the products of the elements of any row and their cofactors.

5.11 *More About Matrix Inverses*

The cofactor of the element a_{ij} of a square matrix $A = ((a_{ij}))$ is denoted by the symbol A_{ij}.

The inverse A^{-1} of a square matrix A is given by the expression

$$\frac{((A_{ij}))^T}{\det A}$$

provided that the determinant of matrix A is nonzero.

5.12 *Probability Matrices*

An experiment in which the outcomes of each trial depend at most on the outcomes of the preceding trial is called a Markov chain; the outcomes are called states.

The process of a Markov chain can be described by means of probability matrices. A probability matrix of order n is a matrix $((a_{ij}))$ whose elements a_{ij} satisfy the following conditions:

1. $0 \le a_{ij} \le 1$
2. $a_{i1} + a_{i2} + \cdots + a_{in} = 1$ for each $i = 1, 2, \ldots, n$

Building Your Mathematical Vocabulary

Addition of matrices (5.2)
Additive identity element (5.2)
Augmented matrix (5.8)
Cofactor (5.10, 5.11)
Column matrix (5.2)
Conformable (5.2, 5.4)
Consistent (5.8)
Determinant of a matrix (5.6, 5.9, 5.10)
Diagonal elements (5.5)
Diagonal matrix (5.5)
Dominance matrix (5.5)
Elementary row transformation matrices (5.8)
Elementary row transformations (5.8)
Equality of matrices (5.1)
Equivalent systems (5.8)
Fibonacci sequence (5.4)
Identity matrix (5.5)
Inconsistent (5.8)
Kronecker delta (5.5)
Main diagonal (5.5)
Markov chain (5.12)
Matrices (5.1)
Matrix (5.1)
Matrix of coefficients (5.8)
Multiplication of a matrix by a real number (5.3)
Multiplication of matrices (5.4)
Multiplicative identity element (5.5)
Multiplicative inverse of a matrix (5.6, 5.11)
Null matrix (5.1)
Order of a matrix (5.1)
Pauli matrices (5.4)
Postmultiplication (5.5)
Premultiplication (5.5)
Probability matrices (5.12)
Real matrix (5.1)
Rectangular array (5.1)
Row matrix (5.3)
Scalar matrix (5.5)
Skew-symmetric matrix (5.5)
Square matrix (5.1)
States (5.12)
Subtraction of matrices (5.2)
Symmetric matrix (5.5)
Transition probabilities (5.12)
Transpose (5.5)
Unit matrix (5.5)
Value of a determinant (5.9, 5.10)
Zero matrix (5.1)

6

Linear Programming and Game Theory

6.1

Order Relations of Numbers

The set R of real numbers can be illustrated graphically by means of a number line. Recall that every line is a set of points; a *number line* is a representation of numbers by points on a line.

To represent the real numbers on a line, first select an arbitrary point A as the origin and then select a unit length. Now mark off unit segments to the right and left of the origin. The origin point corresponds to the whole number 0. End points of the unit segments to the right of the origin point correspond to the positive integers 1, 2, 3, ..., respectively; end points to the left correspond to the negative integers -1, -2, -3, ... A real number associated with a point is called the *coordinate* of the point; the point is called the *graph* of the number. The set of points on the number line and the set R of real numbers are in a *one-to-one correspondence*; that is, each point is the graph of exactly one real number, and each real number is the coordinate of exactly one point. The number line extends without bound in both directions.

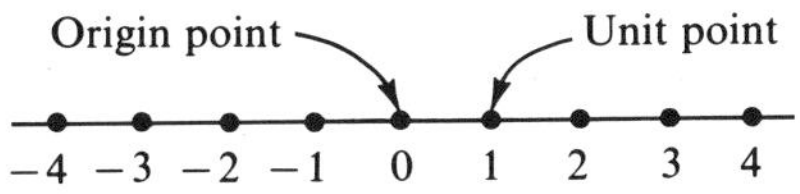

Figure 6.1

Given any two real numbers a and b, it is always possible to determine if a is less than b, a is equal to b, or a is greater than b. Whenever this is possible for a set of numbers in general, the set is *ordered*. The set of real numbers is an ordered set of numbers.

If $b - a$ is a positive number, then a is less than b and we write $a < b$.
If $b - a$ is zero, then a is equal to b and we write $a = b$.
If $b - a$ is a negative number, then a is greater than b and we write $a > b$.

Note that if $a > b$, then $b < a$.

As a matter of convenience, the compound statement $a < b$ or $a = b$ is written $a \leq b$. Similarly, the compound statement $a > b$ or $a = b$ is written $a \geq b$.

The relations of equality and inequality can be illustrated geometrically by means of a number line. If two numbers are equal, then they both are coordinates of the same point on the number line. However, if two numbers are unequal, then they are coordinates of two distinct points. Let the unit point of a number line be to the right of the origin. Then any point P with coordinate a is to the left of point Q with coordinate b if and only if $a < b$, as shown in Fig. 6.2. Any point P with

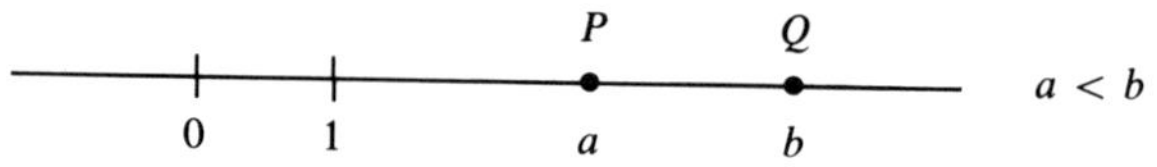

Figure 6.2

coordinate a coincides with point Q with coordinate b if and only if $a = b$, as shown in Fig. 6.3. Any point P with coordinate a is to the right of point Q with coordinate b if and only if $a > b$, as shown in Fig. 6.4.

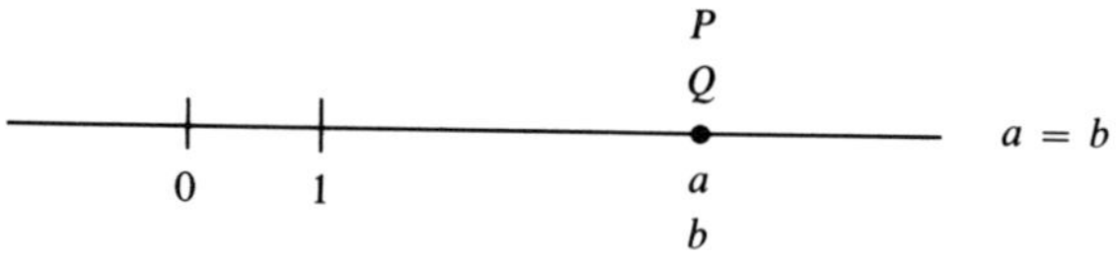

Figure 6.3

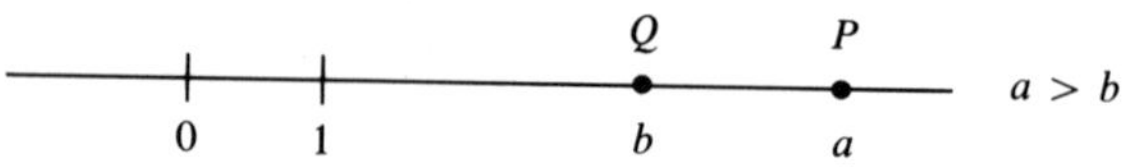

Figure 6.4

Example 1

Graph the set of real numbers less than 3.

The point whose coordinate is 3 separates the points on a number line into three sets: (*a*) the set of points whose coordinates are less than 3; (*b*) the set consisting of the point whose coordinate is 3; (*c*) the set of points whose coordinates are greater than 3. The sets of points in (*a*) and (*c*) are examples of *half lines*. A half line may be described as consisting of the points to the left or to the right of a particular point on a line.

The graph of the set of real numbers less than 3 is the half line shown in Fig. 6.5. We indicate that the point whose coordinate is 3 does not belong to the graph by use of a hollow dot.

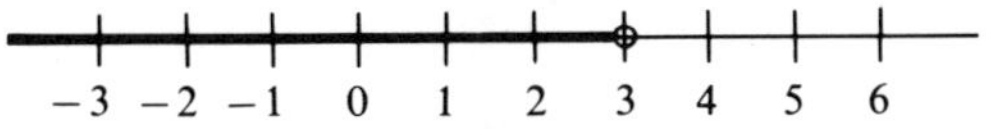

Figure 6.5

Example 2

Graph the set of real numbers greater than or equal to 1.

The graph of the set of real numbers greater than or equal to 1 is the ray indicated in Fig. 6.6. A *ray* is the union of a point on

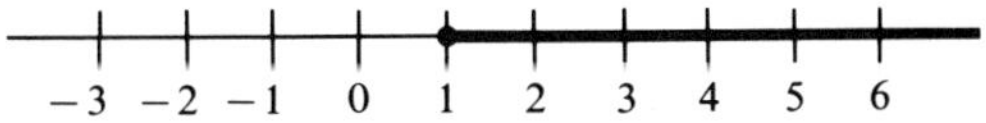

Figure 6.6

a line and one of the two half lines determined by that point. We indicate that the point whose coordinate is 1 does belong to the graph by the use of a solid dot.

In algebra a *conditional statement* is an equality or inequality involving at least one variable. For example,

$$x + 1 = 4 \qquad x - 2 > 0 \qquad x + 5 \leq 8$$

are conditional statements in x. The set of values of the variable x for which a conditional statement is true is called the *solution set* of the conditional statement. Each conditional statement in one variable has a graph on a number line. The *graph of a conditional statement* is the set of points whose coordinates are in the solution set. For example, the graph in Fig. 6.5 is the graph of the conditional statement $x < 3$; the graph in Fig. 6.6 is the graph of the conditional statement $x \geq 1$.

Conditional statements of inequality such as

$$2x + 5 > 0 \qquad x - 1 < 0 \qquad 4x - 3 \leq 0 \qquad 2x + 4 \geq 0$$

are examples of *linear inequalities in x*. Any conditional statement of inequality that can be expressed in one of the forms

$$ax + b > 0 \qquad ax + b < 0 \qquad ax + b \geq 0 \qquad \text{or} \qquad ax + b \leq 0$$

where a and b are real numbers and $a \neq 0$, is called a *linear inequality in one variable.* The graphs of the solution sets of the linear inequalities $ax + b > 0$ and $ax + b < 0$ are half lines; the graphs of $ax + b \geq 0$ and $ax + b \leq 0$ are rays. For example:

$2x + 5 > 0$

Figure 6.7

$x - 1 < 0$

Figure 6.8

$4x - 3 \leq 0$

Figure 6.9

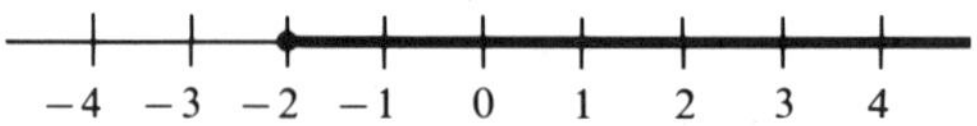

$2x + 4 \geq 0$

Figure 6.10

It is often convenient to write $a < x < b$ to denote the numbers x simultaneously greater than a and less than b. A similar interpretation is placed upon $a \leq x < b$, $a < x \leq b$, and $a \leq x \leq b$.

Example 3

Graph the solution set of the compound conditional statement $2 \le x \le 4$.

The graph of the solution set of $x \ge 2$ is shown in Fig. 6.11.

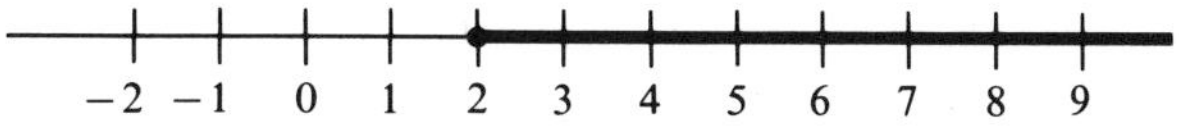

Figure 6.11

The graph of the solution set of $x \le 4$ is shown in Fig. 6.12.

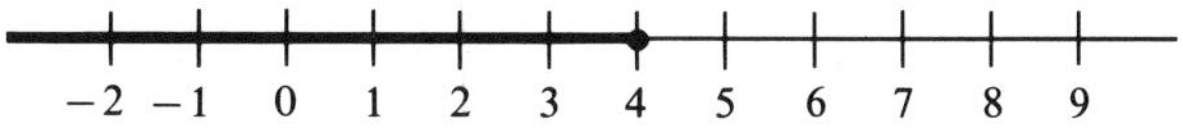

Figure 6.12

The set of points that the two graphs have in common represents the graph of the solution set of the compound statement $2 \le x \le 4$. The graph of the solution set is shown in Fig. 6.13. The graph

Figure 6.13

is the *line segment* whose end points have coordinates 2 and 4, respectively.

The order relations < and > have a number of other important properties with which you should be familiar. We shall illustrate each of these properties with the "less than" relation. Similar statements are valid for the "greater than" relation. Let a, b, and c represent any real numbers.

Transitive property: If $a < b$ and $b < c$, then $a < c$. For example, since $1 < 4$ and $4 < 6$, then $1 < 6$.

Addition property: If $a < b$, then $a + c < b + c$. For example, since $3 < 7$, then $3 + 4 < 7 + 4$.

Multiplication property: If $a < b$ and $c > 0$, then $ac < bc$; if $a < b$ and $c < 0$, then $ac > bc$. For example, since $2 < 5$ and $3 > 0$, then $2 \cdot 3 < 5 \cdot 3$; since $4 < 6$ and $-2 < 0$, then $4 \cdot (-2) > 6 \cdot (-2)$.

Exercises

1. Write statements (*a*) and (*b*) in notational form: (*a*) x is less than y; (*b*) x is greater than or equal to 4.

In Exercises 2 through 21 graph the solution set of each conditional statement.

2. $x \geq 3$
3. $x > 1$
4. $x < 2$
5. $x \leq 2$
6. $x > 2$
7. $x \geq 2$
8. $x \leq 4$
9. $0 \leq x \leq 3$
10. $1 < x < 5$
11. $1 \leq x < 5$
12. $1 < x \leq 5$
13. $1 \leq x \leq 5$
14. $x - 3 < 0$
15. $3x + 1 \leq 0$
16. $3x - 12 > 0$
17. $5x + 10 \geq 0$
18. $3x + 7 < 28$
19. $5x + 1 > 2(1 + 2x)$
20. $4 + 2x \geq 7 + x$
21. $4(x + 1) \leq x + 10$
22. Write the conditional statements with graphs (*a*) to (*d*):

(*a*) 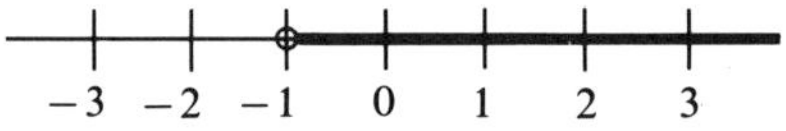

(*b*) 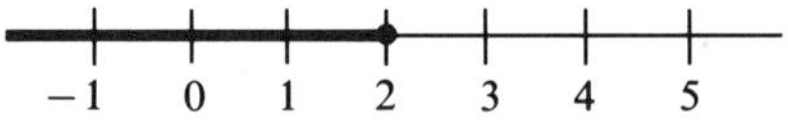

(*c*) 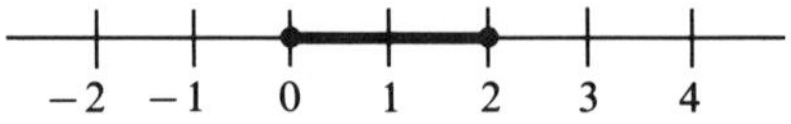

(*d*) 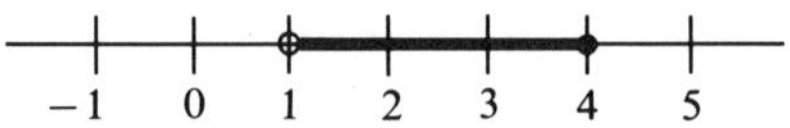

23. Express inequalities (*a*) and (*b*) in the form $a \leq x \leq b$: (*a*) $2 \leq x - 3 \leq 5$; (*b*) $-6 \leq 2x \leq 4$.

In Exercises 24 through 28 determine whether each statement is true or false.

24. If $a < b$, then $a - 2 < b - 2$.
25. If $a > b$, then $a + 7 > b + 7$.

26. If $a < b$, then $-2a < -2b$.
27. If $a < b$, then $3a < 3b$.
28. If $a > 2$ and $2 > b$, then $a > b$.

6.2

Graphs of Linear Equations

One of the most important Cartesian sets is the set $R \times R$, where R is the set of real numbers. The Cartesian set $R \times R$ contains infinitely many ordered pairs of real numbers. A graph of the Cartesian set $R \times R$ consists of all the points of the plane and is called the *Cartesian plane.*

To locate the point associated with a particular element of the set $R \times R$, a horizontal reference line called the *x axis* and a vertical reference line called the *y axis* are identified on the plane. These *coordinate axes* are number lines with the point of intersection of the two lines chosen as the origin in each case. Each element (a, b) contained in the Cartesian set $R \times R$ corresponds to the point of intersection of two lines: one line is parallel to the y axis and passes through the point on the x axis associated with the real number a; the second line is parallel to the x axis and passes through the point on the y axis associated with the real number b, as shown in Fig. 6.14. The numbers a and b are called the *x coordinate* and the

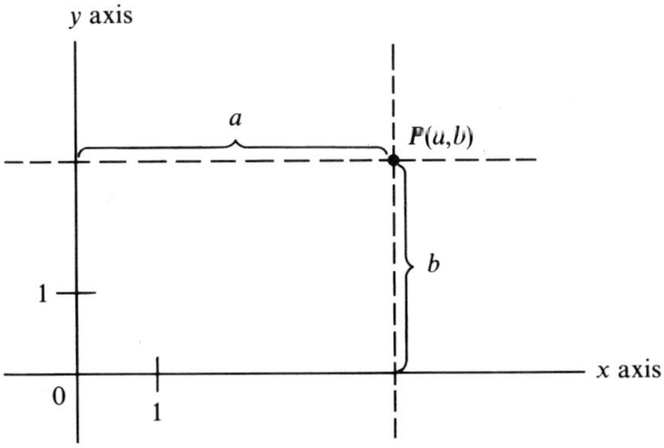

Figure 6.14

y coordinate of the point, respectively. We often denote a point on the plane by a single capital letter. For example, we may denote the point whose coordinates are (a, b) by the letter P and write $P(a, b)$.

Recall that any statement of equality which can be expressed in the form

$$ax + by + c = 0$$

where a, b, and c are real numbers and a and b are not both zero, is called a *linear equation* in the two variables x and y. The solution set of a linear equation x and y is the set of ordered pairs of real numbers (x, y) that make the statement true. For example, consider the linear equation $3x + 2y - 6 = 0$. By considering real values of x and determining corresponding real values of y that make the conditional statement true, we can obtain members of the solution set such as

$$(4, -3) \qquad (2, 0) \qquad (0, 3) \qquad (-2, 6) \qquad (-4, 9)$$

The graph of this partial solution is shown in Fig. 6.15. Generally the solution set of a linear equation in two variables consists of infinitely many ordered pairs of real numbers. The graph of the partial solution set of $3x + 2y - 6 = 0$ suggests that the graph of the solution set is a line, as shown in Fig. 6.16. We shall assume that the graph of any linear equation in two variables is a straight line.

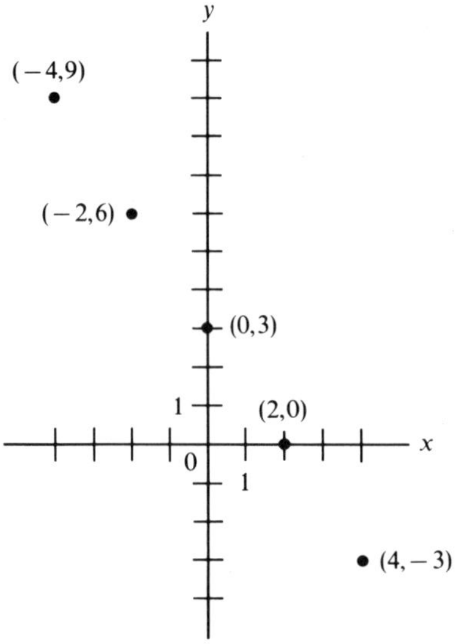

Figure 6.15

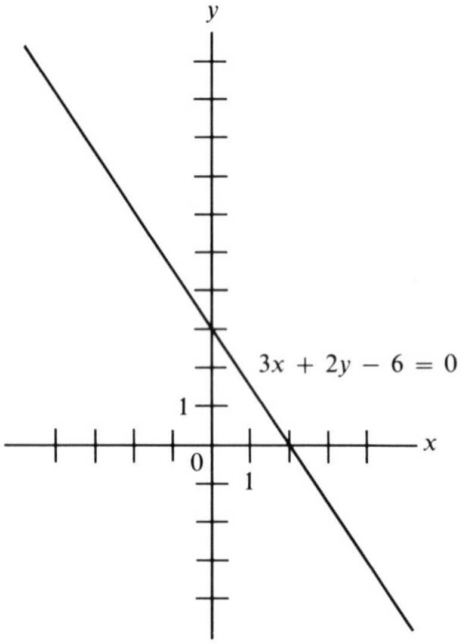

Figure 6.16

Note that one and only one straight line can be drawn through two points. Therefore, we need to locate only two points on the graph of a linear equation to draw its graph.

Example 1

Graph the linear equation

$$2x - y + 1 = 0$$

Two ordered pairs of real numbers that satisfy the equation may be obtained by assigning values to x and determining the associated values of y. If $x = 0$, then $y = 1$; if $x = 3$, then $y = 7$. Thus, the straight line containing the points (0, 1) and (3, 7) is the graph of the linear equation $2x - y + 1 = 0$, as shown in Fig. 6.17.

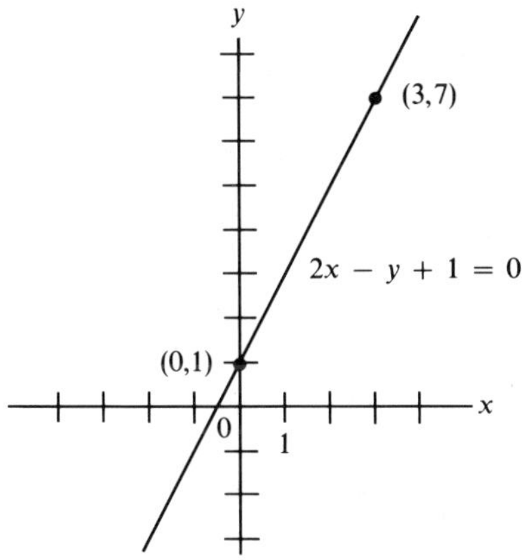

Figure 6.17

Consider the graph of the linear equation $ax + by + c = 0$, where $b = 0$. Then $a \neq 0$ and

$$x = -\frac{c}{a}$$

a constant. The graph of this equation is a line parallel to the y axis. If $c = 0$, the line is the y axis. If $a = 0$, then $b \neq 0$ and

$$y = -\frac{c}{b}$$

a constant. The graph of this equation is a line parallel to the x axis. If $c = 0$ in this case, the line is the x axis.

Example 2

Graph

(*a*) $2x - 3 = 0$; (*b*) $y - 4 = 0$.

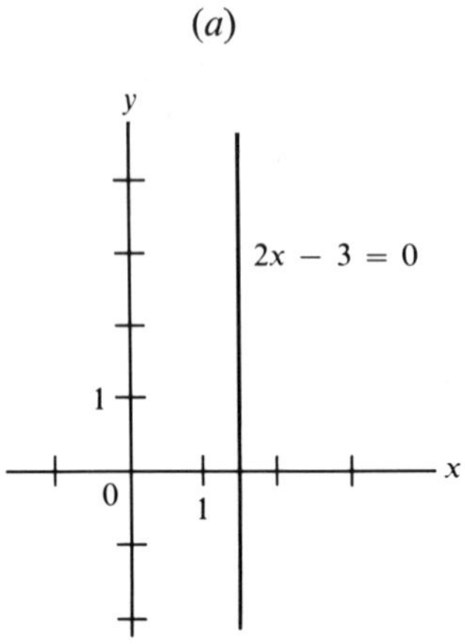

Figure 6.18

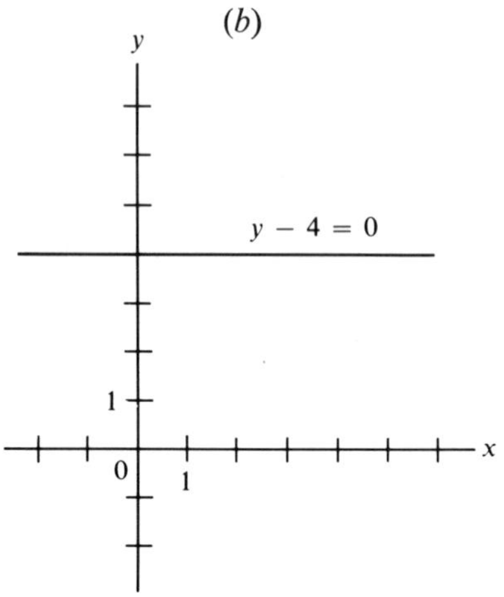

Figure 6.19

Let $A(x_1, y_1)$ and $B(x_2, y_2)$ be any two distinct points on the line $ax + by + c = 0$, except $x_1 \neq x_2$, as shown in Fig. 6.20. Then

$$ax_2 + by_2 + c = 0$$
$$ax_1 + by_1 + c = 0$$

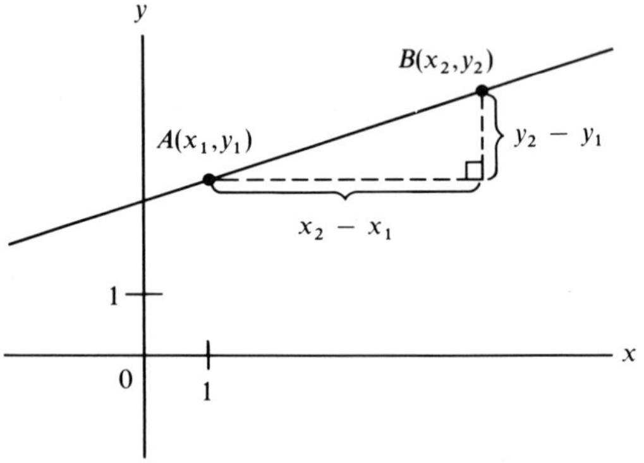

Figure 6.20

Subtracting equals from equals, we may obtain

$$(ax_2 + by_2 + c) - (ax_1 + by_1 + c) = 0$$

whereby

$$a(x_2 - x_1) + b(y_2 - y_1) = 0$$
$$b(y_2 - y_1) = -a(x_2 - x_1)$$
$$\frac{y_2 - y_1}{x_2 - x_1} = -\frac{a}{b}$$

Because a and b are constants, the ratio $(y_2 - y_1)/(x_2 - x_1)$ is always constant for any two points A and B. If we consider a movement from point A to point B, the expression $x_2 - x_1$ represents the change in x; the expression $y_2 - y_1$ represents the change in y. The ratio of the change in y to the change in x is called the *slope* of line AB. In general, the slope of a line is constant, provided $x_2 - x_1 \neq 0$.

Note that the slope of a line parallel to or coincident with the x axis is defined as zero since $y_2 - y_1 = 0$; the slope of a line parallel to or coincident with the y axis is undefined.

Example 3

Find the slope of the line whose equation is

$$2x - y + 1 = 0$$

Choose any two points on the graph of the linear equation $2x - y + 1 = 0$ shown in Fig. 6.21. For example, $A(1, 3)$ and $B(3, 7)$. Then

$$\frac{y_2 - y_1}{x_2 - x_1} = \frac{7 - 3}{3 - 1} = \frac{4}{2} = 2$$

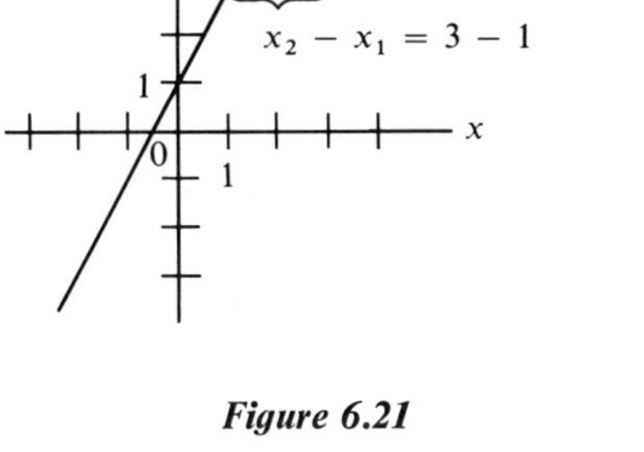

Figure 6.21

Hence, the slope of the line $2x - y + 1 = 0$ is 2. We may interpret this to mean that if we move from any point to another point on the graph, the ratio of the change in the y coordinate to the change in the x coordinate is 2.

The slope of the line could have been determined also by considering the equation of the line as being of the form $ax + by + c = 0$, where $a = 2$, $b = -1$, and $c = 1$. By definition,

$$\text{Slope of line } AB = -\frac{a}{b} = -\frac{2}{-1} = 2$$

The slope of a line is dependent upon the coefficients of x and y in the equation $ax + by + c = 0$.

A set of lines with a common property is called a *family of lines.*

Example 4

Graph the lines whose equations are of the form

$$2x + 5y + c = 0$$

where $c = -10, -5, 0, 5$.

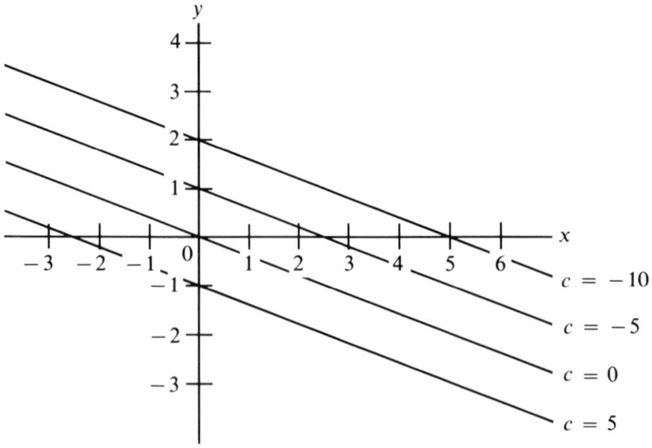

Figure 6.22

The property the lines have in common is that they have the same slope; the set of lines represents a family of parallel lines.

Exercises

1. Find the ordered pairs of real numbers associated with each of the points A, B, C, D, E, and F on the Cartesian plane in the figure:

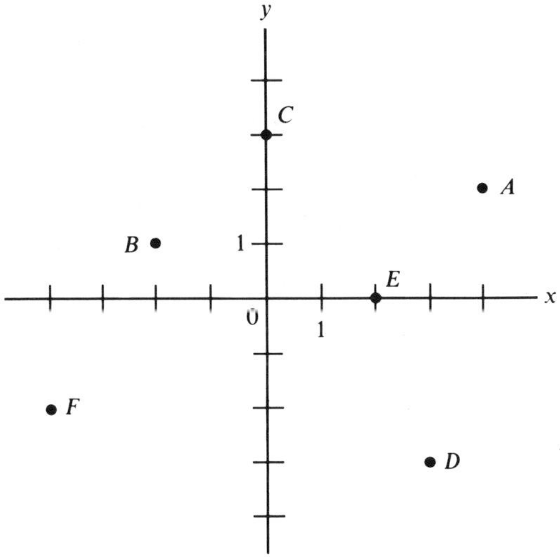

In Exercises 2 through 9 graph each linear equation.

2. $2x - y - 3 = 0$
3. $x + y = 0$
4. $3x - y + 5 = 0$
5. $2x + 5y - 10 = 0$
6. $2y - 5 = 0$
7. $x + 2 = 0$
8. $7x - 3y + 21 = 0$
9. $4x + 3y - 12 = 0$
10. Graph linear equations $2x - y + 1 = 0$, $2x - y + 3 = 0$, and $2x - y + 5 = 0$ on the same Cartesian plane. What property do the lines have in common?
11. Determine a conditional statement whose graph is the (*a*) x axis; (*b*) y axis.
12. Find the equation of the line parallel to the x axis and containing the point (*a*) (2, 3); (*b*) (5, 1).
13. Find the equation of the line parallel to the y axis and containing the point (*a*) (2, 3); (*b*) (5, 1).
14. Find the slope, if it exists, of the line containing the pair of points (*a*) $A(3, 1)$ and $B(5, 9)$; (*b*) $A(2, 1)$ and $B(7, 1)$.
15. Find the slope, if it exists, of the line whose equation is (*a*) $6x + 3y - 8 = 0$; (*b*) $3x - y + 4 = 0$.

16. Show that the equation $ax + by + c = 0$ can be written in the form $y = mx + k$, provided $b \neq 0$. Discuss m.
17. Graph the lines whose equations are of the form $y = mx + 2$, where $m = -\frac{1}{2}, -1, -2, \frac{1}{2}, 1, 2$.
18. Graph the lines whose equations are of the form $2x - 3y + c = 0$, where $c = -6, -3, 0, 3, 6$.

6.3

Linear Inequalities in Two Variables

Conditional statements of inequality such as

$$x - y < 0 \qquad x - 2y + 4 > 0 \qquad 3x - y \leq 0 \qquad 2x + 7 \geq 0$$

are examples of linear inequalities in x and y. Any conditional statement of inequality that can be expressed in one of the forms

$$\begin{array}{ccc} ax + by + c > 0 & & ax + by + c < 0 \\ ax + by + c \geq 0 & \text{or} & ax + by + c \leq 0 \end{array}$$

where a, b, and c are real numbers and a and b are not both zero, is called a *linear inequality in the two variables x and y.*

The solution set of a linear inequality in x and y is the set of ordered pairs of real numbers (x, y) that make the conditional statement true. In general, there are infinitely many members of the solution set. Therefore, the solution set is best represented graphically.

Every straight line $ax + by + c = 0$ in a plane separates the plane into three disjoint sets of points:

1. The set of points either "above" or "to the left of" the line
2. The set of points on the line
3. The set of points either "below" or "to the right of" the line

The sets of points described in (1) and (3) are called *half planes*; the line is the *boundary* of each half plane. The points on the line do not belong to either half plane determined by the line.

Example 1

Graph the set of points whose coordinates satisfy linear inequalities (*a*) and (*b*):
(*a*) $x < 3$; (*b*) $y \geq 2$.

(*a*) The shaded half plane to the left of the line $x = 3$, as shown in Fig. 6.23, is the graph of the linear inequality $x < 3$. (*b*) The shaded half plane above the line $y = 2$ and the boundary line itself, as shown in Fig. 6.24, is the graph of the linear inequality $y \geq 2$.

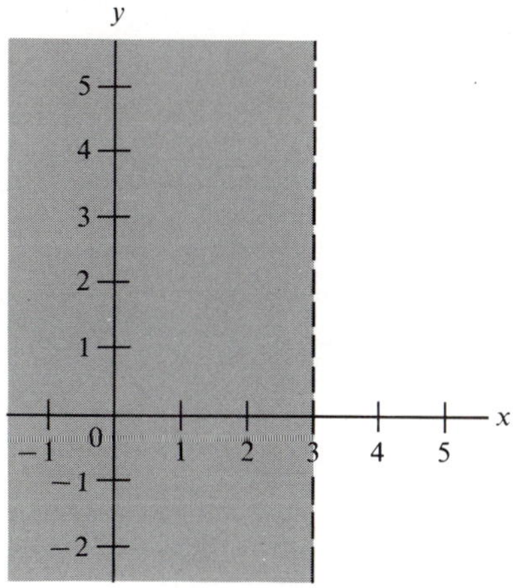

Figure 6.23

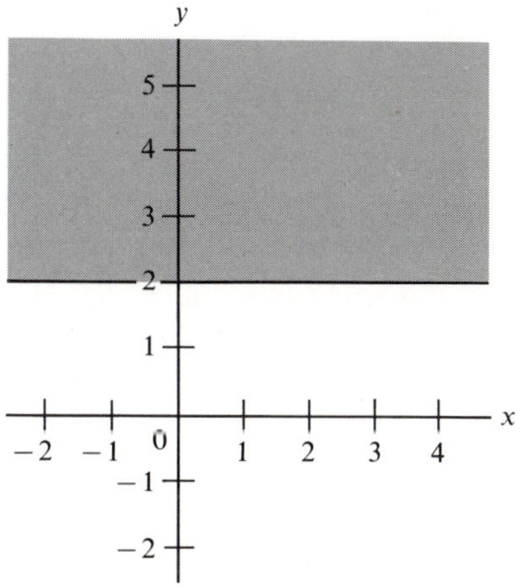

Figure 6.24

Note that if the boundary line is not part of the graph of a linear inequality in two variables, the line is drawn dashed. If the boundary line is part of the graph of the linear inequality, the line is drawn solid.

Example 2

Graph the set of points whose coordinates satisfy the linear inequality $2x + 3y - 12 > 0$.

The graph of the related linear equation $2x + 3y - 12 > 0$ is shown as a dashed line in Fig. 6.25. The coordinates of every point on the line make the linear expression $2x + 3y - 12$ identically equal to 0. Let $P(x, y)$ be any point on that line. Consider a vertical line l that contains the point P. Every point on line l has the same x coordinate as P. However, each point A on line l above point P has a y coordinate greater than the y coordinate of P. The coordinates of the points on l above P make the linear expression $2x + 3y - 12$ greater than 0. In a similar manner, the coordinates of each point B on l below P make the linear expression $2x + 3y - 12$ less than 0. Hence the graph of the set of points whose coordinates satisfy the linear inequality $2x + 3y - 12 > 0$ is the shaded half plane to which A belongs. The line $2x + 3y$

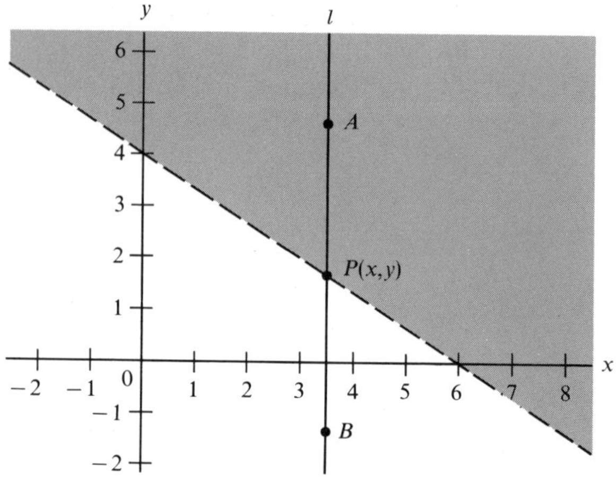

Figure 6.25

$-12 = 0$ is shown as a dashed line in Fig. 6.25 because the points on the line do not satisfy the linear inequality; that is, the boundary $2x + 3y - 12 = 0$ does not belong to the graph of the linear inequality $2x + 3y - 12 > 0$.

In general, the graph of a linear inequality $ax + by + c > 0$ or $ax + by + c < 0$ is a half plane determined by the line $ax + by + c = 0$. The set of points whose coordinates satisfy the inequality $ax + by + c \geq 0$ is equal to the set of points whose coordinates satisfy the linear inequality $ax + by + c > 0$ or the linear equation $ax + by + c = 0$. The graph of a linear inequality $ax + by + c \geq 0$ or $ax + by + c \leq 0$ consists of the set of points of a half plane determined by the line $ax + by + c = 0$ and the points on the line.

If $b > 0$***, the graph of the linear inequality*** $ax + by + c > 0$ ***is the half plane above the line*** $ax + by + c = 0$***; the graph of the linear inequality*** $ax + by + c < 0$ ***is the half plane below the line.***

To prove the above statement, let $P_0(x_0, y_0)$ be any point on the line $ax + by + c = 0$. Then

$$ax_0 + by_0 + c = 0$$

Let $P(x_0, y)$ be any point on a vertical line l that contains $P_0(x_0, y_0)$. If

$$ax_0 + by + c > 0$$

then by subtraction

$$\begin{aligned} (ax_0 + by + c) - (ax_0 + by_0 + c) &> 0 \\ by - by_0 &> 0 \\ b(y - y_0) &> 0 \end{aligned}$$

If $b > 0$, then $y - y_0 > 0$; that is, $y > y_0$. Hence the point $P(x_0, y)$ is above the line $ax + by + c = 0$.

Conversely, if $P(x_0, y)$ is a point on l above the line $ax + by + c = 0$, then $y > y_0$. Since $b > 0$, then $by > by_0$ and

$$ax_0 + by + c > ax_0 + by_0 + c$$
$$ax_0 + by + c > 0$$

Therefore, the points above the line $ax + by + c = 0$ satisfy the linear inequality $ax + by + c > 0$.

Now, only those points whose coordinates satisfy the linear equation $ax + by + c = 0$ lie on the line. If $b > 0$, those points and only those points whose coordinates satisfy the linear inequality $ax + by + c > 0$ lie in the half plane above the line. Every point in the plane lies on the line $ax + by + c = 0$, in the half plane above the line, or in the half plane below the line. Furthermore, every point in the plane has coordinates that satisfy the linear equation $ax + by + c = 0$, the linear inequality $ax + by + c > 0$, or the linear inequality $ax + by + c < 0$. Hence it follows that those points and only those points whose coordinates satisfy the linear inequality $ax + by + c < 0$ lie in the half plane below the line.

Example 1

Graph the linear inequality $3x + 5y + 15 \leq 0$.

Since $b > 0$, the graph of the linear inequality $3x + 5y + 15 < 0$ is the half plane in Fig. 6.26 below the line $3x + 5y + 15 = 0$. The line $3x + 5y + 15 = 0$ is shown in the figure as a solid line because its points belong to the graph of $3x + 5y + 15 \leq 0$.

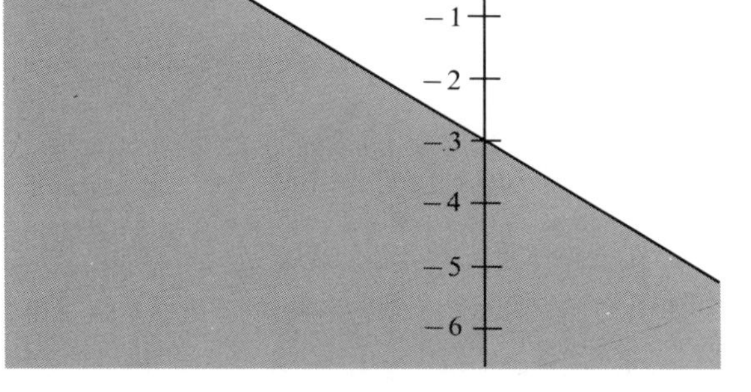

Figure 6.26

Exercises

In Exercises 1 through 16 graph each linear inequality.

1. $x \leq 2$
2. $x > 3$
3. $y > 1$
4. $y \leq 4$
5. $y - x < 0$
6. $x + y + 1 \geq 0$
7. $2x - y + 4 \geq 0$
8. $3x + 4y + 12 < 0$
9. $2x + y - 2 \leq 0$
10. $x + 3y - 6 < 0$
11. $x + 3y - 6 > 0$
12. $x + 3y - 6 \leq 0$
13. $y \geq x + 1$
14. $x + y < 5$
15. $2x + 3y > 6$
16. $x + 2y \geq 4$
17. Show that if $b = 0$ and $a > 0$, then the graph of the linear inequality $ax + by + c > 0$ is the half plane to the right of the line $ax + by + c = 0$; the graph of the linear inequality $ax + by + c < 0$ is the half plane to the left of the line.

6.4

Systems of Linear Inequalities

A compound statement consisting of the conjunction of two or more linear inequalities is called a *system of linear inequalities*. For example,

$$\begin{cases} x + 2y - 4 \leq 0 \\ x - y + 1 > 0 \end{cases}$$

is a system of linear inequalities whose solution set may be expressed as

$$\{(x, y) \mid x + 2y - 4 \leq 0\} \cap \{(x, y) \mid x - y + 1 > 0\}$$

We have already noted that each line separates the plane into two half planes. To solve a system of linear inequalities, we graph each linear inequality on the same coordinate plane. The intersection of the graphs (half planes with or without boundaries) is the graph of the solution set. The graph of the solution set of the system of linear inequalities given above is shown in Fig. 6.27 as a cross-hatched region.

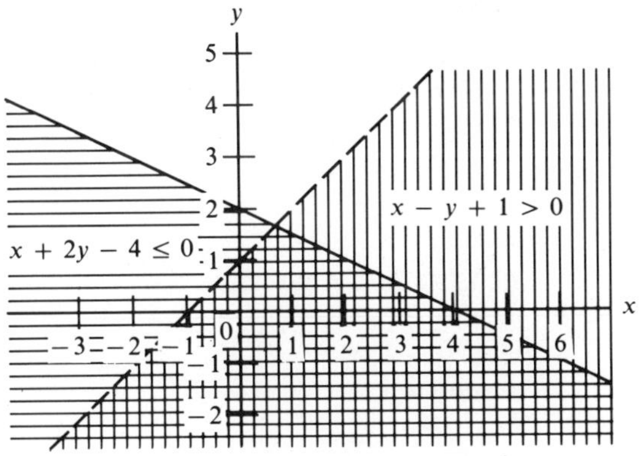

Figure 6.27

Example 1

Solve graphically the system of linear inequalities

$$\begin{cases} x + 2 \geq 0 \\ x - 2 \leq 0 \\ y - x \geq 0 \end{cases}$$

The graph of $x + 2 \geq 0$ consists of the points to the right of or on the line $x + 2 = 0$; the graph of $x - 2 \leq 0$ consists of the points to the left of or on the line $x - 2 = 0$. The intersection of these two half planes consists of the points in the region shaded by the horizontal lines in Fig. 6.28, including the points on the lines $x + 2 = 0$ and $x - 2 = 0$.

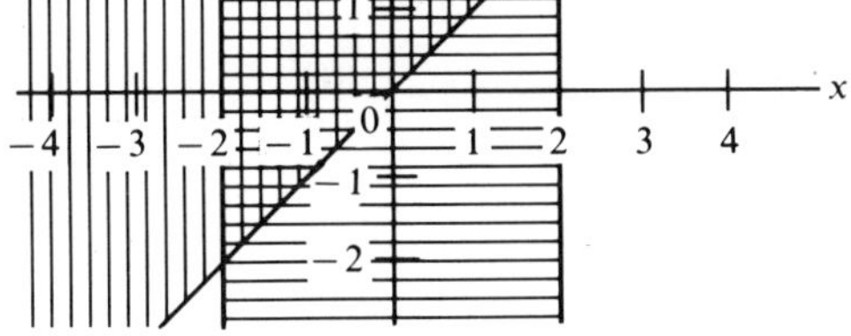

Figure 6.28

The points in the region shaded by the vertical lines represent the graph of $y - x \geq 0$.

Hence, the graph of the solution set of the system

$$\begin{cases} x + 2 \geq 0 \\ x - 2 \leq 0 \\ y - x \geq 0 \end{cases}$$

consists of the points in the cross-hatched region in Fig. 6.28.

Example 2

Solve graphically the system of linear inequalities

$$\begin{cases} x + y - 5 \leq 0 \\ \qquad\quad x \geq 0 \\ \quad\; y - 1 \geq 0 \end{cases}$$

The graph of the solution set consists of the set of points of the triangle with vertices at (0, 5), (0, 1), and (4, 1) and the interior points of the triangle, as indicated by the shaded region in Fig. 6.29.

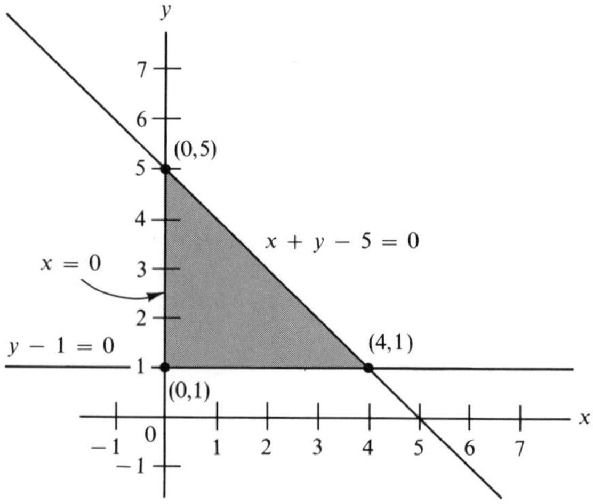

Figure 6.29

In Examples 1 and 2 the solution sets are the intersections of three half planes with boundaries. The intersection of n half planes with boundaries is called a *polygonal convex set.* In Example 1, the polygonal convex set had infinite area; in Example 2, the polygonal convex set had finite area. In general, a polygonal convex set with finite area consists of a polygon and its interior points. Hereafter we shall refer to a polygonal convex set with finite area as a *convex polygon.*

Note that a convex polygon formed by the intersection of n half planes has n sides and n vertices. The n vertices may be determined by solving pairs of equations representing the boundaries of the half planes, as illustrated in Example 3.

Example 3

Graph the solution set of the system of linear inequalities

$$\begin{cases} x + y - 6 \leq 0 \\ x - 1 \geq 0 \\ 3x + y - 12 \leq 0 \\ y \geq 0 \end{cases}$$

The shaded area in Fig. 6.30 represents the solution set of the system of linear inequalities:

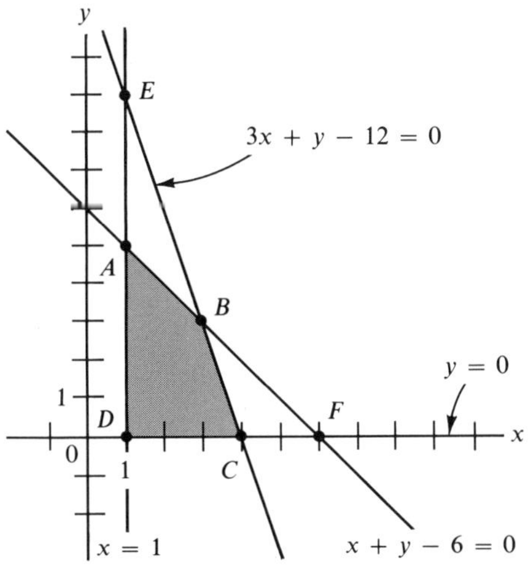

Figure 6.30

Note that the convex polygon determined by the four half planes and their boundaries is a quadrilateral with vertices A, B, C, D and its interior. To find the vertices we may consider pairs of the inequalities of the system, treat them as equalities, and solve simultaneously:

$$A(1, 5) \text{ is the solution of } \begin{cases} x + y - 6 = 0 \\ x - 1 = 0 \end{cases}$$

$$B(3, 3) \text{ is the solution of } \begin{cases} x + y - 6 = 0 \\ 3x + y - 12 = 0 \end{cases}$$

$$C(4, 0) \text{ is the solution of } \begin{cases} 3x + y - 12 = 0 \\ y = 0 \end{cases}$$

$$D(1, 0) \text{ is the solution of } \begin{cases} x - 1 = 0 \\ y = 0 \end{cases}$$

Although there are ${}_4C_2$, that is, six, different ways of selecting two of the four boundary lines, the solutions of the remaining two systems of linear equations do not satisfy the given system of linear inequalities. Hence, their graphs do not belong to the convex polygon:

$$E(1, 9) \text{ is the solution of } \begin{cases} 3x + y - 12 = 0 \\ x - 1 = 0 \end{cases}$$

$$\text{but does not satisfy} \quad x + y - 6 \leq 0$$

$$F(6, 0) \text{ is the solution of } \begin{cases} x + y - 6 = 0 \\ y = 0 \end{cases}$$

$$\text{but does not satisfy} \quad 3x + y - 12 \leq 0$$

Exercises

In Exercises 1 through 16 graph the solution set of each system of linear inequalities.

1. $\begin{cases} x < 2 \\ y > 1 \end{cases}$

2. $\begin{cases} x - 3 \leq 0 \\ y + 1 > 0 \end{cases}$

3. $\begin{cases} x \geq 1 \\ x + y \geq 2 \end{cases}$

4. $\begin{cases} x \geq 0 \\ y \geq 0 \\ x + y \leq 3 \end{cases}$

5. $\begin{cases} x - 3y - 3 \leq 0 \\ 2x + 3y - 6 \geq 0 \end{cases}$

6. $\begin{cases} 2x + 5y \geq 10 \\ x - 3y \leq 6 \end{cases}$

7. $\begin{cases} x + y \leq 1 \\ x - y \geq 0 \\ y \geq 0 \end{cases}$

8. $\begin{cases} x \geq 0 \\ y \geq 0 \\ 2x + 3y \leq 6 \\ 2x + y \leq 4 \end{cases}$

9. $\begin{cases} x + y \geq 1 \\ x - y \geq -1 \\ x \leq 1 \end{cases}$

10. $\begin{cases} x + y \geq 1 \\ x - y \geq -1 \\ x + 2y \leq 4 \\ y \geq 0 \end{cases}$

11. $\begin{cases} 3x + y \leq 3 \\ x - y \geq 1 \\ x - 2y \leq 4 \end{cases}$

12. $\begin{cases} x + y - 1 \leq 0 \\ x - y + 1 \leq 0 \\ y \geq 0 \end{cases}$

13. $\begin{cases} x + 2y - 2 \geq 0 \\ x - 2y - 2 \leq 0 \\ 2x - y - 2 \geq 0 \end{cases}$

14. $\begin{cases} 4x + 3y - 12 \leq 0 \\ 2x - 3y + 6 \geq 0 \\ x + y \geq 0 \end{cases}$

15. $\begin{cases} x \geq 0 \\ y \geq 0 \\ x + y \geq 1 \\ 2x + y \leq 6 \end{cases}$

16. $\begin{cases} 2y - x + 4 \leq 0 \\ x - y - 4 \leq 0 \\ x - 1 \geq 0 \\ y + 1 \geq 0 \end{cases}$

In Exercises 17 through 20 write a system of linear inequalities for the following graphs:

17.

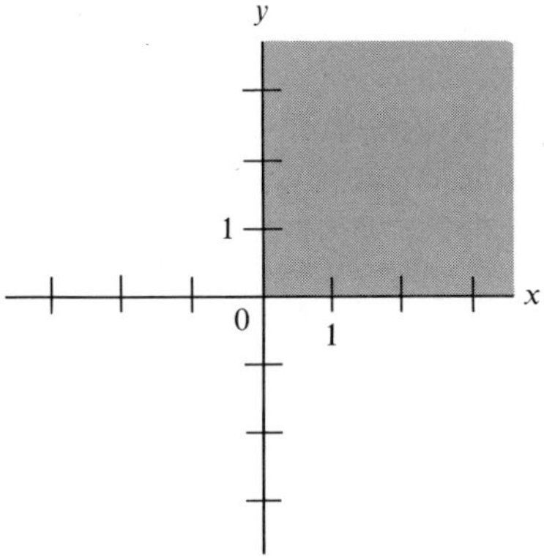

18.

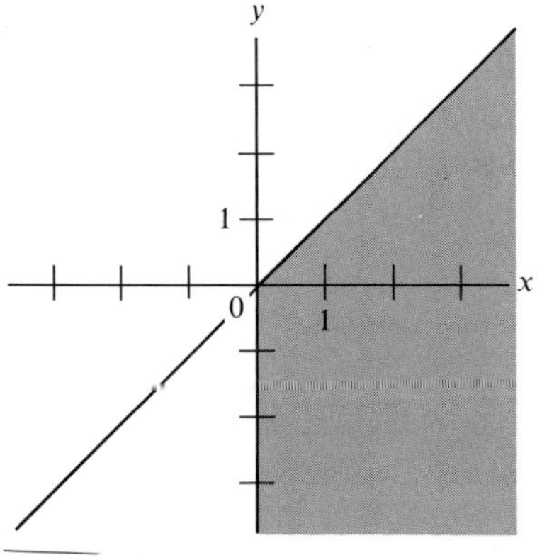

19.

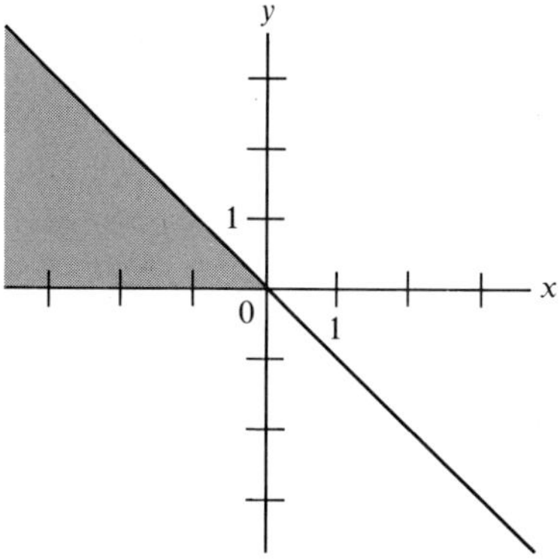

20.

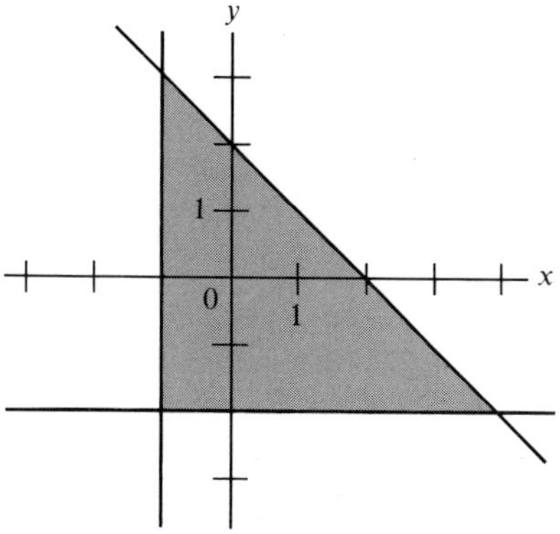

6.5

Convex Sets and Applications

In this section we introduce the topic of *linear programming.* Linear programming is the study of problems in which several conditions have to be satisfied simultaneously. Each condition may be expressed by means of a linear inequality. The intersection of the solution sets of the given linear inequalities represents the *feasible solutions.* Subject to these conditions, the maximum or minimum value of a linear function must be found.[1] That is, an *optimal solution* among the feasible solutions is desired.

By graphing each linear inequality, a polygonal convex set representing the feasible solution set can be obtained. Each point of this set represents a possible solution. A linear function defined over the convex set must be evaluated, in some manner, for each point of the feasible solution set.

The following examples illustrate the fundamental idea of the linear programming problem.

Example 1

A road contractor has eight 6-ton trucks, five 10-ton trucks, and ten available drivers. He has contracted to deliver a minimum of 432 tons of road gravel from pit to roadsite daily. The 6-ton trucks can make seven trips per day and the 10-ton trucks can make six trips per day. How should the contractor dispatch his trucks to (*a*) deliver the maximum tonnage per day; (*b*) use the fewest number of drivers; (*c*) use the trucks at minimum cost if a 6-ton truck costs \$8/day to run and a 10-ton truck costs \$10/day?

Let x represent the number of 6-ton trucks used by the contractor; let y represent the number of 10-ton trucks. The conditions placed on the number of trucks and drivers used based on those available can be expressed by the inequalities

$$x \le 8 \qquad y \le 5 \qquad x + y \le 10$$

Each day a 6-ton truck can deliver 42 tons; a 10-ton truck can deliver 60 tons. Therefore, to fulfill the contract of a minimum of 432 tons of road gravel per day,

$$42x + 60y \ge 432$$

---

[1] The feasible solution set is the domain of the linear function.

This linear inequality can be simplified and is equivalent to

$$7x + 10y \geq 72$$

The graphs of the four linear inequalities placing conditions on x and y are shown in Fig. 6.31. The intersection of these graphs is indicated by the shaded convex polygon in the figure. Because only a countable number of each type of truck can be used, the

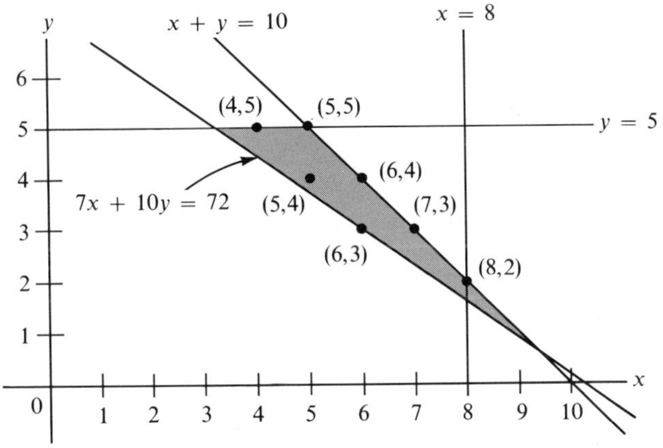

Figure 6.31

feasible solutions are denoted by the points contained in the convex polygon that have nonnegative integral coordinates. From the graph it is evident that the points

$$(4, 5) \qquad (5, 4) \qquad (5, 5) \qquad (6, 3) \qquad (6, 4) \qquad (7, 3) \qquad (8, 2)$$

belong to the feasible solution set.

To find the optimal solution(s), we next consider the contractor's policy stated in (*a*) to (*c*):

(*a*) The maximum tonnage that can be delivered per day can be found by evaluating the linear function $42x + 60y$ for each point of the feasible solution set. As indicated by the following table, using five 6-ton trucks and five 10-ton trucks gives a maximum tonnage of 510 tons. However, this requires all ten available drivers.

x	4	5	5	6	6	7	8
y	5	4	5	3	4	3	2
$42x + 60y$	468	450	510	432	492	474	456

(*b*) The least number of drivers that could be used is nine, and the required tonnage can be delivered by using the feasible solutions (4, 5), (5, 4), or (6, 3). However, the cost will not be the same in each case.

(*c*) The dollar cost of using x 6-ton trucks and y 10-ton trucks is expressed by the linear function

$$8x + 10y$$

The minimum cost is \$78 when the feasible solution (6, 3) is used. Thus, the optimal solution subject to the contractor's policy is to use six 6-ton trucks and three 10-ton trucks.

Example 2

A small factory produces baseball bats and tennis rackets. The production of a baseball bat requires 2 hours of machine time and 1 hour of a craftsman's time; a tennis racket requires 1 hour of machine time and 3 hours of a craftsman's time. There are 28 hours of machine time and 24 hours of craftsmen's time each day. The profits on a baseball bat and on a tennis racket are \$2 and \$3, respectively. Find the number of baseball bats and tennis rackets that should be produced daily for maximum profit.

Let x represent the number of baseball bats produced daily; let y represent the number of tennis rackets produced. The greatest number of baseball bats that can be manufactured is 14 because of the machine-time restriction; the greatest number of tennis rackets that can be manufactured is 8 because of the restriction on the craftsmen's time. Therefore,

$$0 \leq x \leq 14 \qquad \text{and} \qquad 0 \leq y \leq 8$$

Furthermore, the restrictions on machine time and craftsmen's time require that

$$2x + y \leq 28 \qquad \text{and} \qquad x + 3y \leq 24$$

The shaded convex polygon in Fig. 6.32 represents the intersection of the graphs of the linear inequalities placing conditions on x and y. Because only a countable number of each type of sporting equipment can be maufactured, the feasible solutions are denoted by the points contained in the convex polygon that have nonnegative integral coordinates. From the graph it is evident that many points belong to the feasible solution set.

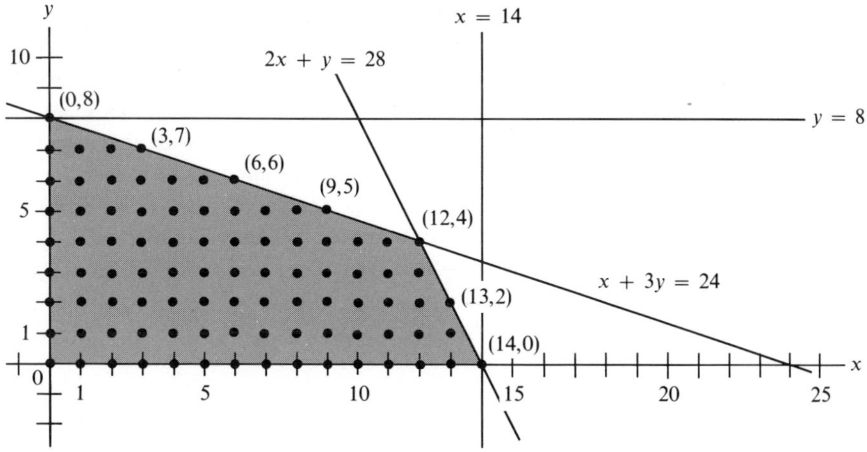

Figure 6.32

The profit in producing x baseball bats and y tennis rackets is expressed by $2x + 3y$. Consider the points in the feasible solution set for some constant x. For example, consider the points (10, 0), (10, 1), (10, 2), (10, 3), and (10, 4). It is obvious that among these points the profit is maximized for (10, 4). In general, for a constant x, the profit $2x + 3y$ will be maximized when y assumes its greatest value. Hence, in this example, we need to consider only the points (0, 8), (1, 7), (2, 7), (3, 7), (4, 6), (5, 6), (6, 6), (7, 5), (8, 5), (9, 5), (10, 4), (11, 4), (12, 4), (13, 2), and (14, 0) to obtain an optimal solution, that is, to maximize profit. Furthermore, for a constant y, the profit $2x + 3y$ will be maximized when x assumes its greatest value. Hence, we may further restrict our attention to the points

(0, 8) (3, 7) (6, 6) (9, 5) (12, 4) (13, 2) (14, 0)

The maximum profit can be found by evaluating the linear function $2x + 3y$ for each of these points. As indicated by the following table, producing 12 baseball bats and 4 tennis rackets daily gives a maximum profit:

x	0	3	6	9	12	13	14
y	8	7	6	5	4	2	0
$2x + 3y$	24	27	30	33	36	32	28

There are a number of typical problems in business and industry in which one is required to maximize or minimize a linear function subject to constraints expressed as linear inequalities. For example, given quantities of raw materials used in the manufacture of certain products, how can the manufacturer decide on the mix that will maximize profits or minimize costs? Given a variety of foods with known caloric, vitamin, and mineral content, how much of each food must one consume to simultaneously satisfy certain nutritional requirements and minimize daily caloric intake? Given a set of production sites and a set of markets for a product, how can a company minimize the cost of shipment to satisfy the contractual demands of the markets? The problems in this section and subsequent sections, although artificial in character, represent the nature of this type of linear programming problem.

Exercises

1. A manufacturer makes two products A and B on two machines. The first machine produces 1 unit each of A and B per hour; the second machine produces 2 units of A only per hour. A contract calls for the manufacturer to make 12 units of A and 6 units of B. Let x represent the number of hours that the first machine operates; let y represent the number of hours that the second machine operates. Sketch a polygonal convex set that contains the possible number of hours the manufacturer can operate each machine to meet the requirements of the contract. (Note that only points of the polygonal convex set with non-negative integral coordinates can be solutions.)
2. The cost of operating the first machine in Exercise 1 is \$6/hour; the cost of operating the second machine is \$4/hour. How many hours should the manufacturer operate each machine to meet the requirements of the contract at the minimum cost?
3. A film company produces two types of short films: documentaries and cartoons. A documentary takes 80 man-hours to make and costs \$3,000; a cartoon takes 50 man-hours and costs \$5,000. There are 400 man-hours available per week, and the weekly budget is \$30,000. Sketch a polygonal convex set that contains the possible number of each type of film which the company can produce weekly.
4. Assume that the profit on each type of film in Exercise 3 is the same. Find the number of each type of film the company should produce to maximize its profit.
5. In Exercise 3, assume that the profit on a cartoon is twice the profit on a documentary film. Find the number of each type of film the company should produce to maximize their profit.
6. A distributor agrees to purchase a minimum of 20 large bulldozers and small tractors from the manufacturer each month. He must purchase at least 10

bulldozers. The distributor makes a profit of \$400 on each bulldozer sold and a profit of \$250 on each tractor sold. He desires to make a total profit of at least \$7,000 each month. Sketch the polygonal convex set that contains the possible number of bulldozers and tractors the distributor can purchase.

7. The distributor in Exercise 6 must store the bulldozers and tractors prior to their sale to his customers. In the storage garage, a bulldozer requires twice the space required of a tractor. How many bulldozers and tractors should he purchase to minimize the space required in the storage garage?

6.6

Linear Functions Over Convex Polygons

In each of the linear programming problems of Sec. 6.5, we were required to find the maximum or minimum value of the linear function

$$f(x, y) = ax + by + c \qquad (a,\ b,\ \text{and } c \text{ are real numbers})$$

where x and y were subject to the following conditions:

$$\begin{cases} a_{11}x + a_{12}y + a_{13} \le 0 \\ a_{21}x + a_{22}y + a_{23} \le 0 \\ a_{31}x + a_{32}y + a_{33} \le 0 \\ \vdots \\ a_{n1}x + a_{n2}y + a_{n3} \le 0 \end{cases}$$

Each a_{ij} for $1 \le i \le n$ and $1 \le j \le 3$ represents a real number. Since the sense of an inequality can be changed by multiplying the inequality by -1, there is no loss in generality in assuming each inequality above to be of the "less than" form. Hereafter we shall refer to this type of problem as that of finding the maximum or minimum value of a linear function over a polygonal convex set.

In this section we shall examine the problem of finding the maximum or minimum value of a linear function over a polygonal convex set by the graphical method. In particular, we shall be concerned with the situation in which the polygonal convex set is a convex polygon.

Example 1

Find the maximum and minimum values of the function

$$f(x, y) = x - y + 5$$

over the convex polygon defined by the following system of linear inequalities:

$$\begin{cases} x - 2 \geq 0 \\ y - 1 \geq 0 \\ x - 2y - 2 \leq 0 \\ x + 4y - 14 \leq 0 \end{cases}$$

The graph of the solution set of the system of linear inequalities is shown in Fig. 6.33:

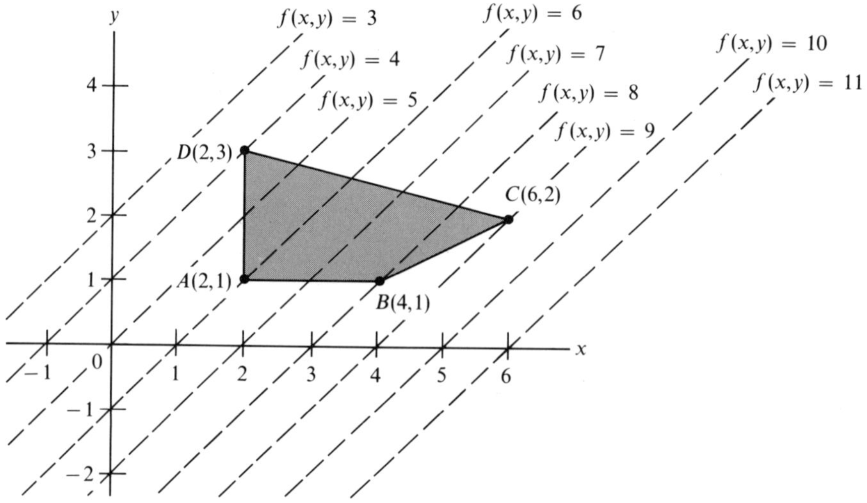

Figure 6.33

The convex polygon is a quadrilateral with vertices at $A(2, 1)$, $B(4, 1)$, $C(6, 2)$, and $D(2, 3)$ and its interior. Remember that each point of the convex polygon represents a feasible solution.

Consider a family of lines in the plane, each of the form

$$f(x, y) = x - y + 5$$

We treat $f(x, y)$ as a parameter. For different assigned values of $f(x, y)$, we obtain different members of the family of parallel lines. The ordered pairs of numbers (x, y) that we seek are the coordinates of those points of the convex polygon which yield optimal solutions—a maximum value of $f(x, y)$ and a minimum value of $f(x, y)$. In Fig. 6.33, several members of the family of parallel lines are shown for different values of $f(x, y)$. Beginning with large values of $f(x, y)$, the first member of the family of parallel lines that intersects the convex polygon does so at point C. Therefore, the linear function $f(x, y) = x - y + 5$ has a maximum value at point C of the convex polygon. The maximum value of $f(x, y)$ is

$$f(6, 2) = 6 - 2 + 5 = 9$$

Similarly, beginning with small values of $f(x, y)$, the first member of the family of parallel lines that intersects the convex polygon does so at point D. Therefore, the linear function $f(x, y) = x - y + 5$ has a minimum value at point D of the convex polygon. The minimum value of $f(x, y)$ is

$$f(2, 3) = 2 - 3 + 5 = 4$$

A graphical method of solution for a linear programming problem is suggested by Example 1. The graphical method of solution may be described as follows:

To find the maximum (minimum) value of a linear function $f(x, y) = ax + by + c$ over a convex polygon,

1. ***Sketch the convex polygon that represents the set of feasible solutions.***
2. ***Graph the line $f(x, y) = ax + by + c$ for some convenient value of $f(x, y)$.***
3. ***Use a straight edge to sketch lines parallel to the line in part (2), moving the straight edge in the direction of increasing (decreasing) values of $f(x, y)$.***
4. ***Locate the last point of the convex polygon that intersects one of the lines in part (3).***
5. ***The coordinates of the point determined in part (4) represent the values of x and y, respectively, for which $f(x, y)$ is a maximum (minimum).***

Example 2

Use the graphical method of solution to find the maximum and minimum values of the function $f(x, y) = 2x + 3y$ over the polygonal convex set defined by the following system of linear inequalities:

$$\begin{cases} x - 1 \geq 0 \\ x - 4 \leq 0 \\ x - 3y + 2 \leq 0 \end{cases}$$

The graph of the solution set of the system of linear inequalities is the polygonal convex set shaded in Fig. 6.34. Care should be taken to note that the set of feasible solutions is not represented by a convex polygon. The graphical method of solution is nonetheless applicable. However, in such cases, maximum or minimum values of the linear function may not exist.

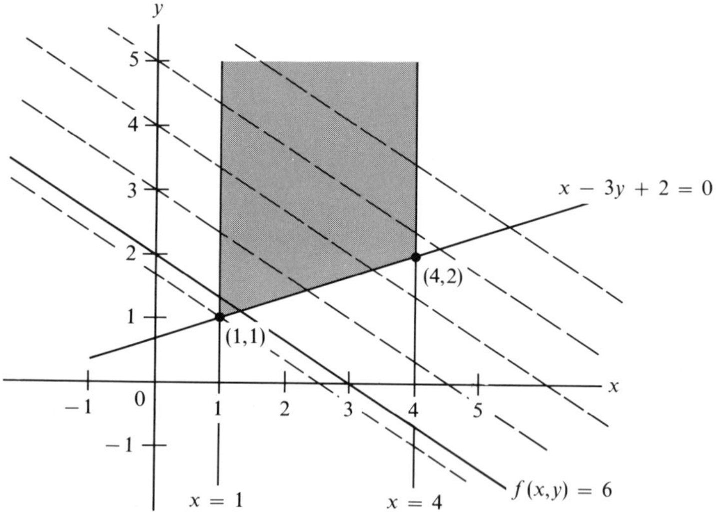

Figure 6.34

The graph of the line $f(x, y) = 2x + 3y$ for $f(x, y) = 6$ is drawn; several parallel lines are shown in Fig. 6.34. As indicated in the figure, the last point of the polygonal convex set that intersects the lines $f(x, y) = 2x + 3y$, where $f(x, y)$ is decreasing, is the point (1, 1). Hence, the minimum value of $f(x, y)$ is

$$f(1, 1) = 2(1) + 3(1) = 5$$

Also indicated, the lines $f(x, y) = 2x + 3y$ will continue to intersect the polygonal convex set for increasing values of $f(x, y)$. Hence, $f(x, y)$ will not take on a maximum value over the polygonal convex set.

As noted in Example 2, a function $f(x, y)$ defined over a polygonal convex set may take on a minimum value while no maximum value exists. Similarly, it is possible that a function may take on a maximum value while no minimum value exists; a function may take on neither a maximum nor a minimum value. Finally, note that a function may take on either a maximum or minimum value at infinitely many points of the polygonal convex set over which it is defined, as illustrated in the next example.

Example 3

Use the graphical method of solution to find the maximum value of the function $f(x, y) = x + 2y$ over the convex polygon defined by the following system of linear inequalities:

$$\begin{cases} x - y \geq 0 \\ x + 2y \leq 6 \\ x \leq 4 \\ y \geq 0 \end{cases}$$

The graph of the solution set of the system of linear inequalities is the convex polygon shaded in Fig. 6.35:

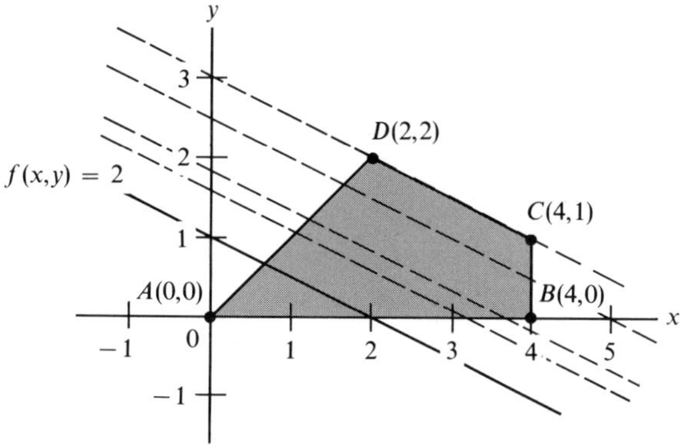

Figure 6.35

The graph of the line $f(x, y) = x + 2y$ for $f(x, y) = 2$ is drawn; several parallel lines are shown in Fig. 6.35. Considering increasing values of $f(x, y)$, the last member of the family of parallel lines that intersects the convex polygon does so along side CD of quadrilateral $ABCD$. Each point of line segment CD belongs to the feasible solution set. Thus, $f(x, y)$ takes on a maximum value for infinitely many points of the convex polygon. That is, the maximum value of $f(x, y)$ may be obtained by using the coordinates of points $C(4, 1)$ or $D(2, 2)$ or any point between them. For example, the maximum value of $f(x, y)$ is

$$f(4, 1) = (4) + 2(1) = 6$$

or

$$f(2, 2) = (2) + 2(2) = 6$$

Exercises

In Exercises 1 through 6 find the maximum and minimum values of each function over the polygonal convex set defined by the given system of inequalities.

1. $f(x, y) = 2x + y + 4;$

$$\begin{cases} x + y \leq 1 \\ x - y \geq 0 \\ y \geq 0 \end{cases}$$

2. $f(x, y) = x - y;$

$$\begin{cases} x \geq 0 \\ y \geq 0 \\ 2x + 3y \leq 6 \\ 2x + y \leq 4 \end{cases}$$

3. $f(x, y) = 2x + y;$

$$\begin{cases} x + 2y \leq 4 \\ x \geq 0 \\ y \geq 0 \end{cases}$$

4. $f(x, y) = 2x + 3y - 1;$

$$\begin{cases} x \geq 1 \\ y \geq 0 \\ x + 2y \leq 4 \\ 3x + y \leq 6 \end{cases}$$

5. $f(x, y) = x + y;$

$$\begin{cases} x + y - 1 \geq 0 \\ x - 2y + 1 \geq 0 \\ x - 1 \leq 0 \end{cases}$$

6. $f(x, y) = x - 3y + 2;$

$$\begin{cases} x + 3y - 6 \geq 0 \\ x + y - 4 \geq 0 \\ x - 6 \leq 0 \\ y - 4 \leq 0 \end{cases}$$

7. Consider each of the three functions $f(x, y) = 2x - y$, $g(x, y) = x + y$, and $h(x, y) = x - 2y + 3$ over the polygonal convex set defined by the following system of linear inequalities:

$$\begin{cases} 3x + 2y \leq 6 \\ x - y \geq 0 \\ x \leq 4 \end{cases}$$

Determine the maximum and minimum values of each function.

8. A company produces two grades of steel: low and high. The company can produce 500 units of steel per day. A unit of low-grade steel sells for \$20, and a unit of high-grade sells for \$25. A contract calls for delivery of at least 200 units of the low-grade steel and 100 units of the high-grade. Assuming that the total production for a day can be sold, how many units of each grade of steel should be produced for maximum profit?

9. A petroleum company produces daily 6,000 barrels of high-grade and 3,000 barrels of low-grade gasoline. The company sells two gasoline products: an expensive mixture of fuel containing three parts of high-grade gasoline for each part of low-grade gasoline and an intermediate-priced mixture of fuel containing equal parts of high- and low-grade gasoline. The expensive mixture

sells for \$150 per barrel; the other mixture sells for \$120 per barrel. (*a*) How much of each fuel should the company make daily to maximize sales revenue? (*b*) What is the maximum sales revenue?

10. A dietician must prepare a mixture of two foods A and B that provides at least 40 units of vitamins and 36 units of minerals. Each unit of food A contains 4 units of vitamins and 2 units of minerals; each unit of food B contains 2 units of vitamins and 5 units of minerals. If the costs of units of A and B are 10 and 8 cents, respectively, how many units of each food should be contained in the mixture to meet the nutritional requirements while minimizing the cost?

6.7

The Tabular Method of Solution

The graphical method of solution can be conveniently applied only to linear programming problems involving two variables. The tabular method of solution, which is presented in this section, is somewhat less restrictive. Although the tabular method of solution is applicable to problems involving two or more variables, we shall illustrate the application of the method to the two-variable case only. (If more than two variables are involved, the tabular method of solution requires a considerable amount of labor; a procedure called the *simplex method* and computing facilities are used.

Consider again Examples 1 and 3 of Sec. 6.6. In each case we determined the maximum and minimum values of a linear function over a convex polygon. In Example 1, the points of the convex polygon for which $f(x, y)$ had a maximum or minimum value were vertices of the polygon. In Example 3, the point of the convex polygon for which $f(x, y)$ had a minimum value was a vertex of the polygon; the points for which $f(x, y)$ had a maximum value were also vertices, as well as the infinitely many points of the line segment determined by these vertices. Each of these examples illustrates the following fact:

> ***The linear function*** $f(x, y) = ax + by + c$ ***defined over a convex polygon takes on its maximum (minimum) value at a vertex of the convex polygon.***

To prove this statement, we first need to discuss the values of a linear function defined over a line segment. Let $A(x_1, y_1)$ and $B(x_2, y_2)$ be the end points of a line segment, as shown in Fig. 6.36. Let $P(r, s)$ be any point between A and B. The equation of line AB is either of the form $x = n$ or $y = mx + k$ (see Exercise 16 of Sec. 6.2). Assume that the equation of line AB is of the form $y = mx + k$. Now, consider a linear function $f(x, y) = ax + by + c$ defined over the points on line

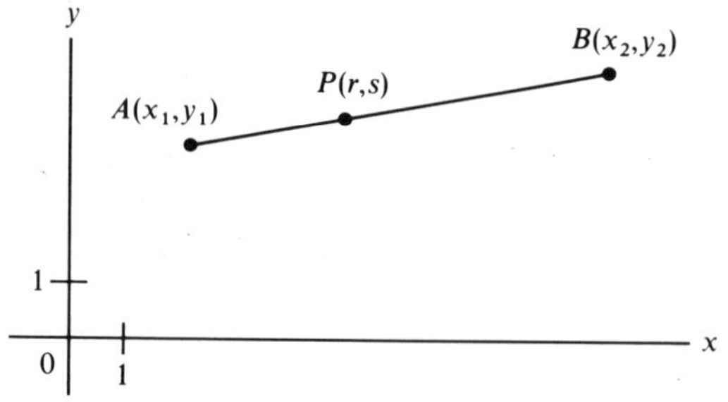

Figure 6.36

segment AB. We shall show that $f(r, s)$ lies between $f(x_1, y_1)$ and $f(x_2, y_2)$. Using the linear function,

$$\begin{aligned} f(x_1, y_1) &= ax_1 + by_1 + c \\ &= ax_1 + b(mx_1 + k) + c \\ &= (a + bm)x_1 + bk + c \end{aligned}$$

$$\begin{aligned} f(r, s) &= ar + bs + c \\ &= ar + b(mr + k) + c \\ &= (a + bm)r + bk + c \end{aligned}$$

$$\begin{aligned} f(x_2, y_2) &= ax_2 + by_2 + c \\ &= ax_2 + b(mx_2 + k) + c \\ &= (a + bm)x_2 + bk + c \end{aligned}$$

From these equations, we may obtain

$$\begin{aligned} f(r, s) - f(x_1, y_1) &= (a + bm)(r - x_1) \\ f(r, s) - f(x_2, y_2) &= (a + bm)(r - x_2) \end{aligned}$$

We can assume without any loss of generality that $x_1 < x_2$. Therefore, $x_1 < r < x_2$ since P lies between A and B. It follows that $r - x_1 > 0$ and $r - x_2 < 0$. If $a + bm > 0$, then $(a + bm)(r - x_1) > 0$ and $(a + bm)(r - x_2) < 0$. Thus,

$$\begin{aligned} f(r, s) - f(x_1, y_1) > 0 \quad &\text{or} \quad f(r, s) > f(x_1, y_1) \\ f(r, s) - f(x_2, y_2) < 0 \quad &\text{or} \quad f(r, s) < f(x_2, y_2) \end{aligned}$$

That is,

$$f(x_1, y_1) < f(r, s) < f(x_2, y_2)$$

In a similar manner, if $a + bm < 0$, then $(a + bm)(r - x_1) < 0$ and $(a + bm)(r - x_2) > 0$. Thus,

$$\begin{aligned} f(r, s) - f(x_1, y_1) < 0 \quad &\text{or} \quad f(r, s) < f(x_1, y_1) \\ f(r, s) - f(x_2, y_2) > 0 \quad &\text{or} \quad f(r, s) > f(x_2, y_2) \end{aligned}$$

In this case,

$$f(x_1, y_1) > f(r, s) > f(x_2, y_2)$$

In either case, the value of $f(x, y)$ at P lies between the values of $f(x, y)$ at A and B. Furthermore, if the equation of line AB is of the form $x = n$, the relationship among $f(x_1, y_1)$, $f(r, s)$, and $f(x_2, y_2)$ is the same. The proof is left as an exercise for the reader (Exercise 12).

In general, the line segment determined by any two points of a convex polygon lies entirely in that region. Now, consider the line segment AB, where A and B are points on the boundary of a convex polygon, as shown in Fig. 6.37; A is a

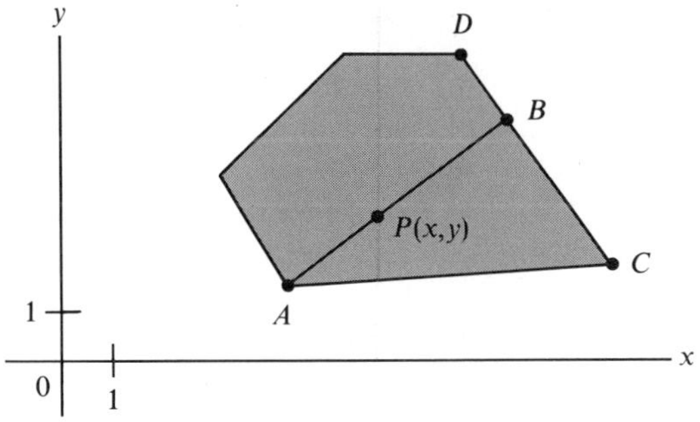

Figure 6.37

vertex. Let $P(x, y)$ be any interior point of line segment AB. If $f(x, y)$ is a given linear function defined over the convex polygon, then the value of $f(x, y)$ at P lies between the values of $f(x, y)$ at A and B. Thus, $f(x, y)$ takes on its maximum and minimum values at some points on the boundary of the convex polygon. Furthermore, since B lies between C and D, the value of $f(x, y)$ at B lies between the values of $f(x, y)$ at C and D. Therefore, as we have already stated, the linear function $f(x, y)$ defined over a convex polygon takes on its maximum and minimum values at vertices of the convex polygon.

It should be noted again that if a linear function takes on its maximum (minimum) value at two vertices, then it takes on that maximum (minimum) value for all points on the side of the polygon determined by those vertices.

Example 1

Find the maximum and minimum values of the function

$$f(x, y) = 4x + 7y + 1$$

over the convex polygon shown in Fig. 6.38.

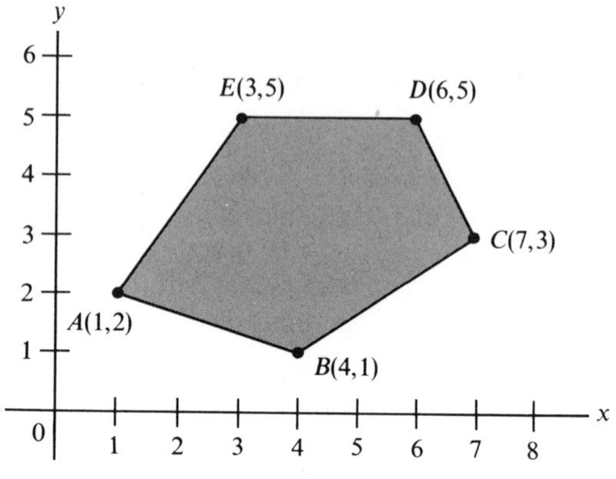

Figure 6.38

According to the theorem we have just proved, the linear function $f(x, y)$ must take on its maximum and minimum values at the vertices of the convex polygon $ABCDE$. Thus, we need only determine the values of $f(x, y)$ at A, B, C, D, and E. As indicated by the following table, the maximum value of $f(x, y)$ is 60 at point D; the minimum value of $f(x, y)$ is 19 at point A:

Vertex	A	B	C	D	E
x	1	4	7	6	3
y	2	1	3	5	5
$4x + 7y + 1$	19	24	50	60	48

The results of the theorem proved in this section suggest a tabular method of solution for a linear programming problem. The tabular method of solution may be described as follows:

To find the maximum (minimum) value of a linear function $f(x, y) = ax + by + c$ ***over a convex polygon defined by a system of linear inequalities,***

1. ***Find the coordinates of the possible vertices of the convex polygon by considering pairs of the inequalities of the system, treating them as equalities, and solving simultaneously.***

2. ***Test each possible vertex to determine whether or not the coordinates satisfy each inequality of the system; only those that satisfy each inequality represent vertices of the convex polygon.***
3. ***Evaluate the linear function $f(x, y)$ at each vertex.***
4. ***Choose the maximum (minimum) value of $f(x, y)$ obtained in part (3).***

Example 2

Use the tabular method of solution to find the maximum and minimum values of the function

$$f(x, y) = 5x + 3y$$

over the convex polygon defined by the following system of linear inequalities:

$$\begin{cases} 2x + y - 12 \leq 0 \\ x - 2 \geq 0 \\ x - y \geq 0 \\ y \geq 0 \end{cases}$$

To find the vertices, we consider pairs of the inequalities of the system, treat them as equalities, and solve simultaneously. The possible vertices are

$A(2, 8)$, the solution of $\begin{cases} 2x + y - 12 = 0 \\ x - 2 = 0 \end{cases}$

$B(4, 4)$, the solution of $\begin{cases} 2x + y - 12 = 0 \\ x - y = 0 \end{cases}$

$C(6, 0)$, the solution of $\begin{cases} 2x + y - 12 = 0 \\ y = 0 \end{cases}$

$D(2, 2)$, the solution of $\begin{cases} x - 2 = 0 \\ x - y = 0 \end{cases}$

$E(2, 0)$, the solution of $\begin{cases} x - 2 = 0 \\ y = 0 \end{cases}$

$F(0, 0)$, the solution of $\begin{cases} x - y = 0 \\ y = 0 \end{cases}$

Now, $A(2, 8)$ is not a vertex of the convex polygon because the coordinates of the point do not satisfy the inequality $x - y \geq 0$. The point $F(0, 0)$ is not a vertex of the convex polygon because the coordinates of the point do not satisfy the inequality $x - 2 \geq 0$. The vertices of the convex polygon are B, C, D, and E; the polygon is a quadrilateral.

The value of $f(x, y) = 5x + 3y$ at each vertex is given by the following table:

Vertex	B	C	D	E
x	4	6	2	2
y	4	0	2	0
$5x + 3y$	32	30	16	10

The maximum value of $f(x, y)$ is 32; the minimum value of $f(x, y)$ is 10.

A graph of the convex polygon is shown in Fig. 6.39, along with several representative members of the family of parallel lines $f(x, y) = 5x + 3y$.

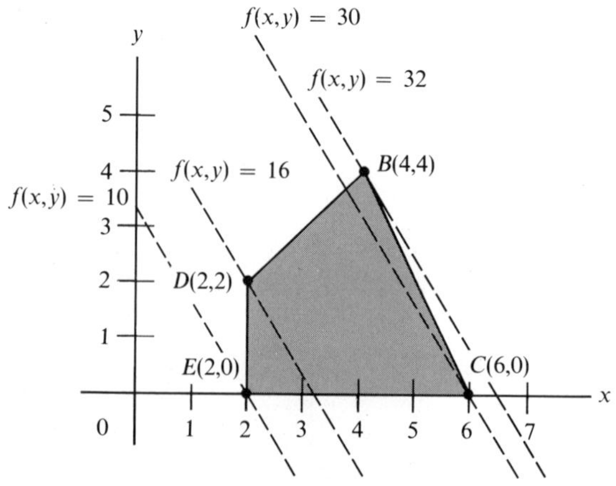

Figure 6.39

Exercises

In Exercises 1 through 4 find the maximum and minimum values of the given function over the convex polygon shaded in the figure.

1. $f(x,y) = 2x - 5y$

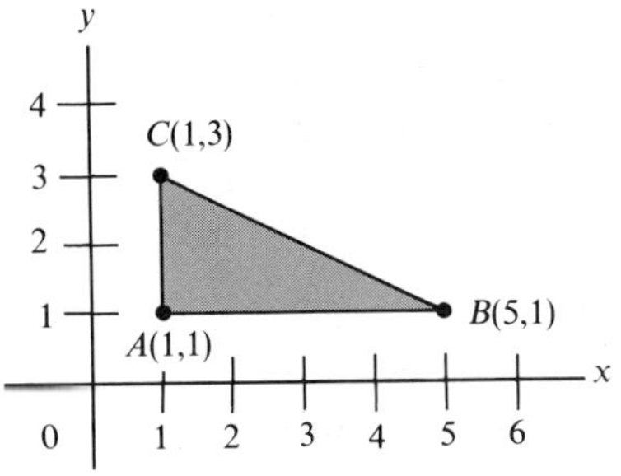

2. $f(x,y) = x - 4y + 20$

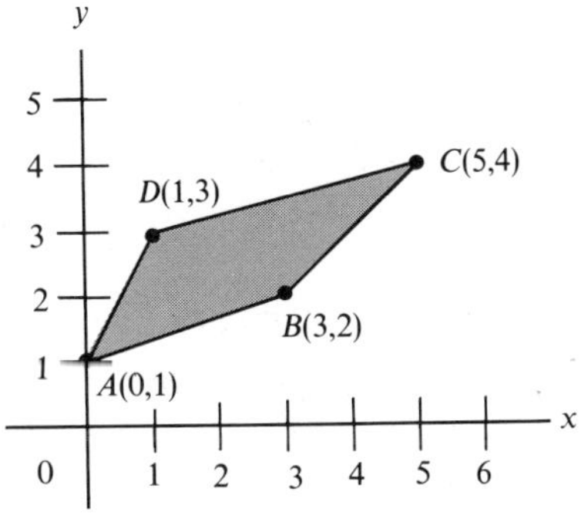

3. $f(x,y) = 3x + 3y - 7$

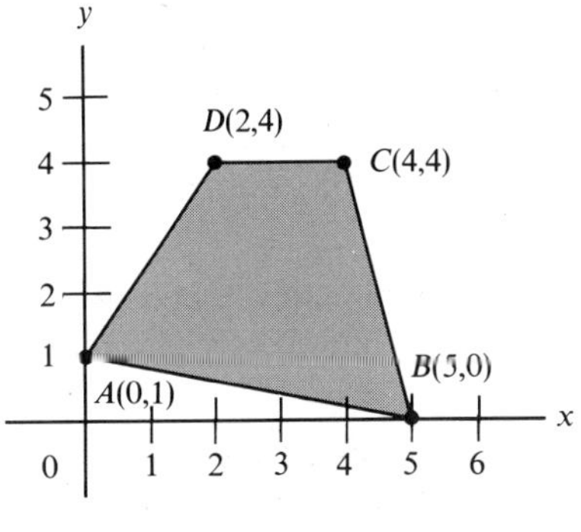

4. $f(x,y) = 5y - 2$

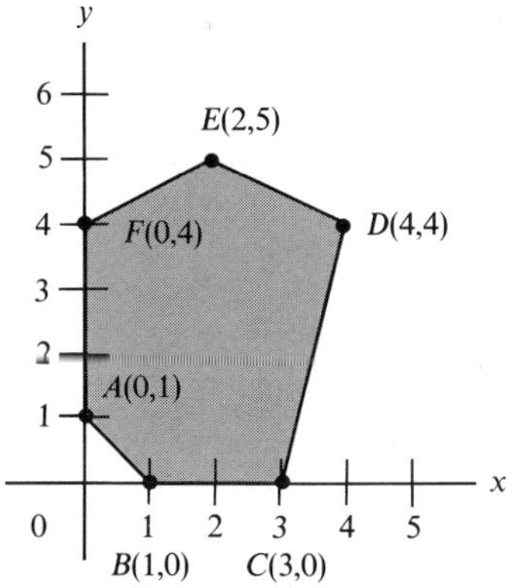

In Exercises 5 through 10 use the tabular method of solution to find the maximum and minimum values of each function over the convex polygon defined by the given system of linear inequalities.

5. $f(x, y) = x + y + 1;$

$$\begin{cases} x + 2y \leq 4 \\ 2x + \ y \leq 4 \\ x \geq 0 \\ y \geq 0 \end{cases}$$

6. $f(x, y) = x + 2y;$

$$\begin{cases} x + 2y \geq 4 \\ 2x + \ y \geq 4 \\ x \geq 0 \\ y \geq 0 \end{cases}$$

7. $f(x, y) = 3x - 6y + 4;$

$$\begin{cases} x + 2y + 2 \geq 0 \\ x + \ \ y - 4 \leq 0 \\ x - \ \ y + 2 \geq 0 \\ \qquad x - 3 \leq 0 \end{cases}$$

8. $f(x, y) = 5x + 2y + 8;$

$$\begin{cases} x \geq 3 \\ x \leq 10 \\ y \geq 0 \\ y \leq 15 \\ x + y \leq 20 \end{cases}$$

9. $f(x, y) = x + 3y;$

$$\begin{cases} 2x + \ \ y \geq 4 \\ x + \ \ y \leq 12 \\ 4x + 2y \leq 30 \\ x \geq 0 \\ y \geq 0 \end{cases}$$

10. $f(x, y) = 2x + y;$

$$\begin{cases} x + 3y - 12 \leq 0 \\ x + \ \ y - \ \ 4 \geq 0 \\ 2x + \ \ y - 10 \leq 0 \\ x \geq 0 \\ y \geq 0 \end{cases}$$

11. Which statement of inequality in Exercise 10 is unnecessary?

12. Let $A(n, y_1)$ and $B(n, y_2)$ be the end points of a line segment. Show that for any linear function $f(x, y) = ax + by + c$ defined over the line segment AB, the value of $f(x, y)$ at any interior point P lies between the values of $f(x, y)$ at A and B.

13. A contractor has 120 units of construction blocks and 200 units of wood available. He builds two types of houses: ranch and split-level. A ranch house requires 3 units of construction blocks and 2 units of wood; a split-level house requires 2 units of construction blocks and 8 units of wood. If the profit on a ranch house is \$3,000 and the profit on a split-level house is \$4,000, how many of each type should the contractor build to obtain the maximum profit?

14. In Exercise 13, how many of each type of house should the contractor build if the profit on a ranch house is \$5,000 and the profit on a split-level house is \$2,000?

15. A manufacturer uses machines A, B, and C to produce tricycles and bicycles. Each tricycle requires 2 hours on machine A, 1 hour on machine B, and 1 hour on machine C. Each bicycle requires 1 hour on machine A, 2 hours on machine B, and 3 hours on machine C. Machines A, B, and C are available 16, 11, and 15 hours/day, respectively. If a tricycle sells for \$30 and a bicycle for \$50, how many of each cycle should the manufacturer make to obtain the maximum amount of money?

16. In Exercise 15, how many tricycles and bicycles should the manufacturer produce if a tricycle sells for \$20 and a bicycle for \$50?

17. A farmer is considering the use of two different fertilizers on his crops. Each bag of Growfast fertilizer contains 2 units of chemical A, 12 units of chemical B, and 4 units of chemical C; each bag of Steadygrow fertilizer contains 6 units of chemical A, 2 units of chemical B, and 2 units of chemical C. The farmer's crops require 18, 24, and 16 units of chemicals A, B, and C, respectively. A bag of Growfast costs \$3.40; Steadygrow costs \$1.60. How

many bags of each type of fertilizer should the farmer buy for the most economical operation?

18. An architect estimates that the annual cost of heating a building is 75 cents/square foot of window as compared to 25 cents/square foot of wall. A zoning ordinance requires that the area of the windows in a certain type of building must be at least one-sixth as much as the area of the walls. What is the largest outside surface area (walls and windows) a building can have if the annual heating cost cannot exceed $720?
19. An athlete has decided to supplement his daily diet with two types of food, A and B. Each unit of food A contains 25 grams of protein and 5 grams of carbohydrates and costs 25 cents; each unit of food B contains 20 grams of protein and 20 grams of carbohydrates and costs 30 cents. What is the least expensive mixture of the two foods he must take in order to obtain at least 800 grams of protein and 400 grams of carbohydrates?
20. A company has 2,000 pounds of Brazilian coffee, 1,000 pounds of Columbian coffee, and 400 pounds of Mexican coffee. The company sells two types of coffee: regular and fine-ground. The regular type contains six parts Brazilian coffee and one part each of Columbian and Mexican coffee. The fine-ground type contains eight parts Brazilian coffee, five parts Columbian coffee, and three parts Mexican coffee. How many pounds of each type should the coffee company sell in order to maximize profits if it makes a profit of 40 cents for each pound of regular coffee and 45 cents for each pound of the fine-ground coffee?

6.8

Matrix Games

In many applied situations, the mathematical expectation of each of several events may be used as a basis for decision in choosing among alternatives. For example, in such cases one attempts to maximize expected profits, gains, or sales or minimize expected losses or costs. Sometimes it is possible to view a business, economic, sociological, or political situation that involves decision making based upon mathematical expectation as a "game" among two or more rational adversaries. In this section we present the basic ideas, terminology, and notations for doing so.

Every matrix A defines a game in the following way:

1. ***There are two players, $\mathscr{R}$ and $\mathscr{C}$.***
2. ***The game consists of $\mathscr{R}$ choosing a row i of A and $\mathscr{C}$ choosing a column j of A. The choices, called* plays, *are made independently and simultaneously.***

3. ***At the end of each game, player $\mathscr{R}$ receives from $\mathscr{C}$ an amount, called a* payoff, *equal to the element a_{ij} of matrix A. A negative value for a_{ij} denotes a payment from $\mathscr{R}$ to $\mathscr{C}$.***

As an example of a matrix game, consider the matrix

$$\begin{array}{cc} & \text{Player } \mathscr{C} \\ \text{Player } \mathscr{R} & \begin{pmatrix} 5 & 2 & 3 \\ -1 & 0 & 6 \end{pmatrix} \end{array}$$

Player $\mathscr{R}$ has two possible plays: rows one or two. If $\mathscr{R}$ wishes to maximize his gain, he might consider playing row two to gain an amount equal to 6. However, $\mathscr{C}$ has three possible plays: columns one, two, or three. Assuming that $\mathscr{C}$ is a rational player, he never has any need to play column three. He can always do better by playing the second column, regardless of $\mathscr{R}$'s play. Indeed, column three need not be considered in any further analysis of the game. Column two is said to *dominate* column three. Thus, player $\mathscr{R}$ should never expect a payoff equal to 6. Because player $\mathscr{C}$ would like to maximize his own gain (or minimize his loss), he might consider playing column one to obtain a payoff equal to 1 from $\mathscr{R}$. However, $\mathscr{R}$ is also a rational player; anticipating $\mathscr{C}$'s desires, he would certainly not play row two. In fact, $\mathscr{R}$ can always do better by playing row one. Hence, it would seem reasonable for $\mathscr{R}$ to always play row one. Knowing this, player $\mathscr{C}$ must realize that regardless of his play, $\mathscr{R}$ will gain some amount. Therefore, since he desires to lose the minimum amount (2, not 5), it is apparent that $\mathscr{C}$'s best play is column two. In this example, $\mathscr{R}$ must decide to play row one because he can do no worse than gain an amount equal to 2; $\mathscr{C}$ must decide to play column two because he can do no better than lose an amount equal to 2. This game is "favorable" to player $\mathscr{R}$.

The decisions by $\mathscr{R}$ and $\mathscr{C}$ to play rows and columns of a matrix game A in some manner are called *strategies*. If a player decides to consistently choose a particular row or column when a game is played again and again, his decision represents a *pure strategy*. Players $\mathscr{R}$ and $\mathscr{C}$ follow pure strategies in the example above. It is possible for a player to play the various rows or columns with a given probability distribution; such a strategy is called a *mixed strategy* and will be discussed in Sec. 6.9.

The objective of game theory is to determine *optimal strategies*—strategies that are most profitable to both players. The payoff corresponding to these optimal strategies is called the *value* of the game. In the example above, the value of the game is 2.

Matrix games may be classified according to the number of plays available to each player. An $m \times n$ game is one in which $\mathscr{R}$ has m plays available to him and $\mathscr{C}$ has n plays available to him. The example above represents a 2×3 game. Games are also classified according to the number of players and the winnings and losses of the players after each play is completed. The example above represents a *two-*

person zero-sum game because only two players are involved and the sum of the winnings and losses of the players after each play is zero; that is, the amount won by one player equals the amount lost by the second player. The theory of games includes *n*-person and nonzero-sum games. An example of a nonzero-sum game would be a gambling establishment in which each player pays the owner of the establishment a percentage of his wager for the right to play. (It is assumed that the game played involves some skill in decision making and is not simply a game of chance.) War can also be considered a nonzero-sum game because even the "winner" loses something. The study of nonzero-sum games involves a considerable amount of difficulty and is beyond the scope of our brief introduction to game theory. We shall restrict our attention primarily to two-person zero-sum games. Every two-person zero-sum game is a matrix game, that is, can be represented by a matrix.

The matrix game

$$\begin{array}{cc} & \text{Payer } \mathscr{C} \\ \text{Player } \mathscr{R} & \begin{pmatrix} 5 & 2 & 3 \\ -1 & 0 & 6 \end{pmatrix} \end{array}$$

is an example of a *strictly determined game*. The distinguishing characteristic of a strictly determined game is that there exists an element of the matrix which is simultaneously the least value in its row and the greatest value in its column. For example, 2 and -1 are the least values of rows one and two, respectively; however, -1 is not the greatest value in column one, but 2 is the greatest value in column two. The element 2 is called the saddle point of the game. A *saddle point* simultaneously represents the maximum of the minimum gains for player $\mathscr{R}$ in each row and the minimum of the maximum losses for player $\mathscr{C}$ in each column. When a game has a saddle point, the optimal strategies for both players are pure strategies. Player $\mathscr{R}$'s strategy is to select the row in which the saddle point is a minimum; player $\mathscr{C}$'s strategy is to select the column in which the saddle point is a maximum. If a game is strictly determined, the saddle point is the value of the game.

Example

Find which of the following games are strictly determined. For the strictly determined game(s) find the value of the game(s) and the optimal strategies for players $\mathscr{R}$ and $\mathscr{C}$:

(*a*) $\begin{pmatrix} 2 & -5 & 1 \\ 1 & 8 & 6 \end{pmatrix}$ (*b*) $\begin{pmatrix} 0 & 1 & -1 & 2 \\ 1 & 2 & 1 & 4 \\ 2 & -1 & 0 & 3 \end{pmatrix}$

(*a*) The element of least value in each row is circled; the element of greatest value of each column is enclosed in a square:

$$\begin{pmatrix} \boxed{2} & \enclose{circle}{-5} & 1 \\ \enclose{circle}{1} & \boxed{8} & \boxed{6} \end{pmatrix}$$

Since no element is simultaneously a row minimum and a column maximum, the game has no saddle point and is *nonstrictly determined.*

(*b*) The element of least value in each row is circled; the element of greatest value of each column is enclosed in a square:

$$\begin{pmatrix} 0 & 1 & \enclose{circle}{-1} & 2 \\ \enclose{circle}{1} & \boxed{2} & \enclose{circle}{\boxed{1}} & \boxed{4} \\ \boxed{2} & \enclose{circle}{-1} & 0 & 3 \end{pmatrix}$$

Since the element a_{23} is simultaneously a row minimum and a column maximum, the game has a saddle point and is a strictly determined game. The saddle point is 1. Player $\mathscr{R}$'s optimal strategy is to play row two; player $\mathscr{C}$'s optimal strategy is to play column three. Since the value of the game is 1 (the saddle point), the game is favorable to $\mathscr{R}$.

Exercises

In Exercises 1 through 8 state whether or not the matrix game is a strictly determined game. For each strictly determined game, find the value of the game and the optimal strategies for $\mathscr{R}$ and $\mathscr{C}$.

1. $\begin{pmatrix} 2 & 0 \\ -2 & 6 \end{pmatrix}$

2. $\begin{pmatrix} 2 & -3 \\ 0 & -2 \end{pmatrix}$

3. $\begin{pmatrix} 5 & -5 \\ 4 & 1 \end{pmatrix}$

4. $\begin{pmatrix} 3 & 2 \\ 1 & 2 \end{pmatrix}$

5. $\begin{pmatrix} 5 & -2 & -5 \\ 4 & 3 & 2 \end{pmatrix}$

6. $\begin{pmatrix} 3 & 6 & 4 \\ 1 & 0 & 8 \end{pmatrix}$

7. $\begin{pmatrix} 3 & -2 \\ 5 & 1 \\ -2 & 4 \end{pmatrix}$

8. $\begin{pmatrix} -1 & 0 & -3 & 1 \\ 3 & 2 & -2 & 4 \\ 1 & 2 & 0 & 3 \end{pmatrix}$

9. Show that the matrix game $\begin{pmatrix} a & a \\ b & c \end{pmatrix}$ is a strictly determined game for all values of a, b, and c.

10. Show that if either one row or one column dominates the other row or column, respectively, of a 2×2 matrix game, then the game has a saddle point.

6.9

Nonstrictly Determined Games

When a game is played again and again and has a saddle point, $\mathscr{R}$ and $\mathscr{C}$ should always choose the row and column containing the saddle point as their optimal pure strategies, respectively. Now, consider the matrix game represented by

$$\text{Player } \mathscr{R} \overset{\text{Player } \mathscr{C}}{\begin{pmatrix} 1 & \mathbf{-1} \\ \mathbf{-3} & 5 \end{pmatrix}}$$

Neither of the minimum entries shaded in each row is a maximum for that column, so no saddle point exists and the game is nonstrictly determined. In such cases, game theory gives no optimal strategies when the game is played only once. There exists an optimal strategy called a mixed strategy for games without saddle points that are played many times. *A mixed strategy* consists of a player choosing the various rows or columns with a given probability distribution. For example, a strategy for player $\mathscr{R}$ may be represented by the row matrix

$$(p \quad 1-p)$$

where p gives the probability that $\mathscr{R}$ will play row one and $1-p$ gives the probability that $\mathscr{R}$ will play row two. A strategy for player $\mathscr{C}$ may be represented by the column matrix

$$\begin{pmatrix} q \\ 1-q \end{pmatrix}$$

where q gives the probability that $\mathscr{C}$ will play column one and $1-q$ gives the probability that $\mathscr{C}$ will play column two. The mathematical expectation of the game to $\mathscr{R}$ is given by the matrix product

$$(p \quad 1-p)\begin{pmatrix} 1 & -1 \\ -3 & 5 \end{pmatrix}\begin{pmatrix} q \\ 1-q \end{pmatrix}$$

That is, $\mathscr{R}$ receives the payoff 1 with probability pq; he receives the payoff -1 with probability $p(1-q)$; he receives the payoff -3 with probability $(1-p)q$; and he receives the payoff 5 with probability $(1-p)(1-q)$. If the strategies $(p \quad 1-p)$ and $\begin{pmatrix} q \\ 1-q \end{pmatrix}$ are optimal mixed strategies, the mathematical expectation for $\mathscr{R}$ is the value of the game. Regardless of the strategy followed by player $\mathscr{C}$, if $(p \quad 1-p)$ represents an optimal strategy for $\mathscr{R}$, then

$$(p \quad 1-p)\begin{pmatrix} 1 & -1 \\ -3 & 5 \end{pmatrix} \geq (v \quad v)^{*}$$

* The matrices $((a_{ij}))$ and $((b_{ij}))$ of order m by n satisfy the inequality $((a_{ij})) \geq ((b_{ij}))$ provided that $a_{ij} \geq b_{ij}$ for all pairs of values i, j.

where v represents the value of the game. This matrix inequality is equivalent to the following system of linear inequalities:

$$\begin{cases} 4p - 3 \geq v \\ -6p + 5 \geq v \end{cases}$$

The graph of this system of linear inequalities is the shaded region in Fig. 6.40:

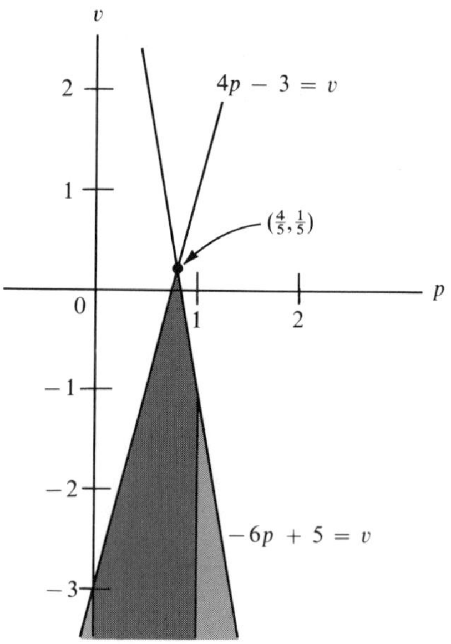

Figure 6.40

Since $0 \leq p \leq 1$, only those points within the heavily shaded region of Fig. 6.40 are of interest. Furthermore, since $\mathscr{R}$ chooses an optimal strategy so that v is a maximum, the point $(\frac{4}{5}, \frac{1}{5})$, that is, the intersection of $4p - 3 = v$ and $-6p + 5 = v$ tells us that $p = \frac{4}{5}$ and the value v of the game is $\frac{1}{5}$. Thus, $\mathscr{R}$'s optimal strategy is, $(\frac{4}{5} \;\; \frac{1}{5})$; that is, $\mathscr{R}$ should choose to play row one in some random fashion $\frac{4}{5}$ of the time and row two $\frac{1}{5}$ of the time. The game is favorable to $\mathscr{R}$ and he can do no worse than gain $\frac{1}{5}$ per play, regardless of $\mathscr{C}$'s strategy.

Player $\mathscr{C}$'s optimal strategy can be obtained by considering the matrix inequality

$$\begin{pmatrix} 1 & -1 \\ -3 & 5 \end{pmatrix} \begin{pmatrix} q \\ 1 - q \end{pmatrix} \leq \begin{pmatrix} v \\ v \end{pmatrix}$$

remembering that $\mathscr{C}$ chooses an optimal strategy so that v is a minimum. It can be shown that $q = \frac{3}{5}$. Thus, $\mathscr{C}$'s optimal strategy is to play column one $\frac{3}{5}$ of the time and column two $\frac{2}{5}$ of the time.

Example 1

Determine an optimal strategy for $\mathscr{R}$ and $\mathscr{C}$ for any nonstrictly determined 2×2 matrix game.

Let

$$\begin{array}{cc} & \text{Player } \mathscr{C} \\ \text{Player } \mathscr{R} & \begin{pmatrix} a & b \\ c & d \end{pmatrix} \end{array}$$

by any nonstrictly determined 2×2 matrix game with value v. If

$$(p \quad 1-p)$$

represents an optimal strategy for $\mathscr{R}$, then

$$(p \quad 1-p)\begin{pmatrix} a & b \\ c & d \end{pmatrix} \geq (v \quad v)$$

This matrix inequality is equivalent to the following system of linear inequalities:

$$\begin{cases} (a-c)p + c \geq v \\ (b-d)p + d \geq v \end{cases}$$

Remember that $\mathscr{R}$ chooses an optimal strategy so that v is a maximum. The maximum value of v that satisfies both inequalities of the system is the solution of the system of equations

$$\begin{cases} (a-c)p + c = v \\ (b-d)p + d = v \end{cases}$$

Thus,

$$p = \frac{d-c}{a+d-b-c} \qquad \text{and} \qquad v = \frac{ad-bc}{a+d-b-c}$$

Since $1-p = (a-b)/(a+d-b-c)$, player $\mathscr{R}$'s optimal strategy is

$$\left(\frac{d-c}{a+d-b-c} \quad \frac{a-b}{a+d-b-c}\right)$$

If

$$\begin{pmatrix} q \\ 1-q \end{pmatrix}$$

represents an optimal strategy for $\mathscr{C}$, then

$$\begin{pmatrix} a & b \\ c & d \end{pmatrix} \begin{pmatrix} q \\ 1-q \end{pmatrix} \leq \begin{pmatrix} v \\ v \end{pmatrix}$$

This matrix inequality is equivalent to the following system of linear inequalities:

$$\begin{cases} (a-b)q + b \leq v \\ (c-d)q + d \leq v \end{cases}$$

Player $\mathscr{C}$ chooses an optimal strategy to minimize v. The minimum value of v that satisfies both inequalities of the system is the solution of the system of equations

$$\begin{cases} (a-b)q + b = v \\ (c-d)q + d = v \end{cases}$$

Thus,

$$q = \frac{d-b}{a+d-b-c}$$

Since $1 - q = (a-c)/(a+d-b-c)$, player $\mathscr{C}$'s optimal strategy is

$$\begin{pmatrix} \dfrac{d-b}{a+d-b-c} \\ \\ \dfrac{a-c}{a+d-b-c} \end{pmatrix}$$

Example 2

Use the results of Example 1 to determine an optimal strategy for players $\mathscr{R}$ and $\mathscr{C}$ for the nonstrictly determined 2×2 matrix game

$$\begin{pmatrix} -2 & 6 \\ 3 & -9 \end{pmatrix}$$

Determine the value of the game.

Let $a = -2$, $b = 6$, $c = 3$, and $d = -9$. The optimal mixed strategy for $\mathscr{R}$ is $(p \quad 1-p)$, where

$$p = \frac{d-c}{a+d-b-c} = \frac{(-9)-(3)}{(-2)+(-9)-(6)-(3)} = \frac{3}{5}$$

$$1 - p = \frac{a-b}{a+d-b-c} = \frac{(-2)-(6)}{(-2)+(-9)-(6)-(3)} = \frac{2}{5}$$

Player $\mathscr{R}$ should choose in some random fashion to play row one $\frac{3}{5}$ of the time and row two $\frac{2}{5}$ of the time. The optimal mixed strategy for $\mathscr{C}$ is $\begin{pmatrix} q \\ 1-q \end{pmatrix}$, where

$$q = \frac{d-b}{a+d-b-c} = \frac{(-9)-(6)}{(-2)+(-9)-(6)-(3)} = \frac{3}{4}$$

$$1-q = \frac{a-c}{a+d-b-c} = \frac{(-2)-(3)}{(-2)+(-9)-(6)-(3)} = \frac{1}{4}$$

Player $\mathscr{C}$ should choose in some random fashion to play column one $\frac{3}{4}$ of the time and column two $\frac{1}{4}$ of the time.

The value v of the game is given by the equation

$$v = \frac{ad-bc}{a+d-b-c} = \frac{(-2)(-9)-(6)(3)}{(-2)+(-9)-(6)-(3)} = 0$$

Hence, the game is a fair game; that is, the game is not favorable to $\mathscr{R}$ or $\mathscr{C}$.

Example 3

Each of two players $\mathscr{R}$ and $\mathscr{C}$ simultaneously show one or two fingers. If the total number of fingers shown is even, $\mathscr{R}$ wins that sum from $\mathscr{C}$; if the total is odd, $\mathscr{C}$ wins that sum from $\mathscr{R}$.

(*a*) Construct the payoff matrix for this game.

(*b*) Determine optimal mixed strategies for $\mathscr{R}$ and $\mathscr{C}$.

(*c*) Show that the game is favorable to player $\mathscr{C}$.

(*a*) The payoff matrix is

		Player $\mathscr{C}$ one finger	Player $\mathscr{C}$ two fingers
Player $\mathscr{R}$	one finger	2	−3
	two fingers	−3	4

(*b*) Since the matrix game has no saddle point, each player's optimal strategy is a mixed strategy. By means of the formulas in Example 1, with $a = 2$, $b = -3$, $c = -3$, and $d = 4$,

$$p = \frac{d-c}{a+d-b-c} = \frac{(4)-(-3)}{(2)+(4)-(-3)-(-3)} = \frac{7}{12} \qquad 1-p = \frac{5}{12}$$

$$q = \frac{d-b}{a+d-b-c} = \frac{(4)-(-3)}{(2)+(4)-(-3)-(-3)} = \frac{7}{12} \qquad 1-q = \frac{5}{12}$$

Hence, $\mathscr{R}$'s optimal strategy is $(\frac{7}{12} \quad \frac{5}{12})$; $\mathscr{C}$'s optimal strategy is $\begin{pmatrix} \frac{7}{12} \\ \frac{5}{12} \end{pmatrix}$. That is, in some random fashion, $\mathscr{R}$ should play row one $\frac{7}{12}$ of the time and row two $\frac{5}{12}$ of the time; $\mathscr{C}$ should play column one $\frac{7}{12}$ of the time and column two $\frac{5}{12}$ of the time.

(*c*) The value v of the game is given by the formula in Example 1:

$$v = \frac{ad - bc}{a + d - b - c} = \frac{(2)(4) - (-3)(-3)}{(2) + (4) - (-3) - (-3)} = -\frac{1}{12}$$

Since the value of the game is a negative number, the matrix game is favorable to player $\mathscr{C}$.

Recall that in the example of Sec. 6.8, any column in which each element is greater than the corresponding element of another column can be eliminated from the analysis of the game because $\mathscr{C}$ can always do better by playing the other column. Similarly, any row in which each element is less than the corresponding element of another row can also be eliminated from the analysis of the game because $\mathscr{R}$ can always do better by playing the other row. This information can be used to reduce $m \times n$ matrix games to games of lower order. For example, consider the matrix game

$$\begin{pmatrix} 5 & -2 & 0 & 7 \\ -2 & 1 & 1 & 0 \\ 1 & 3 & 4 & 2 \end{pmatrix}$$

Row three is said to dominate row two because each element of row three is greater than the corresponding element of row two. Player $\mathscr{R}$ will therefore disregard row two and consider only the other rows. The matrix game is thereby reduced to

$$\begin{pmatrix} 5 & -2 & 0 & 7 \\ 1 & 3 & 4 & 2 \end{pmatrix}$$

Column two dominates column three because each element of column two is less than the corresponding element of column three. Player $\mathscr{C}$ will therefore disregard column three and consider only the other columns. The matrix game is thereby further reduced to

$$\begin{pmatrix} 5 & -2 & 7 \\ 1 & 3 & 2 \end{pmatrix}$$

Finally, column one dominates column three and the matrix game can be reduced to the nonstrictly determined 2×2 game

$$\begin{pmatrix} 5 & -2 \\ 1 & 3 \end{pmatrix}$$

Since the optimal strategies for $\mathscr{R}$ and $\mathscr{C}$ in this game can be shown to be $(\frac{2}{9} \;\; \frac{7}{9})$ and $\begin{pmatrix} \frac{5}{9} \\ \frac{4}{9} \end{pmatrix}$, respectively, the optimal strategies for $\mathscr{R}$ and $\mathscr{C}$ in the original matrix game

$$\begin{pmatrix} 5 & -2 & 0 & 7 \\ -2 & 1 & 1 & 0 \\ 1 & 3 & 4 & 2 \end{pmatrix}$$

are $(\frac{2}{9} \;\; 0 \;\; \frac{7}{9})$ and $\begin{pmatrix} \frac{5}{9} \\ \frac{4}{9} \\ 0 \\ 0 \end{pmatrix}$, respectively.

Thus far we have been primarily concerned with 2×2 matrix games. The next example illustrates a method by which every $2 \times m$ or $m \times 2$ matrix game can be studied. The method is based upon the fact, which we shall accept without proof, that every $2 \times m$ or $m \times 2$ game is equivalent to a reduced 2×2 matrix game.

Example 4

Find optimal strategies for players $\mathscr{R}$ and $\mathscr{C}$ for the 2×3 matrix game

$$\begin{pmatrix} 4 & -1 & 3 \\ 0 & 2 & 1 \end{pmatrix}$$

There are no dominant rows or columns and no saddle point. Hence the matrix cannot be immediately reduced, nor can optimal pure strategies be obtained. However, we assume that the given game is equivalent to one of the three matrix games

$$\begin{pmatrix} 4 & -1 \\ 0 & 2 \end{pmatrix} \qquad \begin{pmatrix} 4 & 3 \\ 0 & 1 \end{pmatrix} \qquad \begin{pmatrix} -1 & 3 \\ 2 & 1 \end{pmatrix}$$

with values

$$v_1 = \tfrac{8}{7} \qquad v_2 = 2 \qquad v_3 = \tfrac{7}{5}$$

respectively. If $\mathscr{C}$ must restrict himself to a mixed strategy involving only two columns, he will choose those columns for which the game has the least value, $v_1 = \frac{8}{7}$. For the matrix game

$$\begin{pmatrix} 4 & -1 \\ 0 & 2 \end{pmatrix}$$

the optimal strategies for $\mathscr{R}$ and $\mathscr{C}$ are

$$(\tfrac{2}{7} \;\; \tfrac{5}{7}) \qquad \text{and} \qquad \begin{pmatrix} \frac{3}{7} \\ \frac{4}{7} \end{pmatrix}$$

respectively. Therefore, the optimal strategies for $\mathscr{R}$ and $\mathscr{C}$ for the original 2×3 matrix game are

$$\begin{pmatrix} \frac{2}{7} & \frac{5}{7} \end{pmatrix} \quad \text{and} \quad \begin{pmatrix} \frac{3}{7} \\ \frac{4}{7} \\ 0 \end{pmatrix}$$

respectively.

To show that these strategies are indeed the optimal strategies for $\mathscr{R}$ and $\mathscr{C}$, note that

$$\begin{pmatrix} \frac{2}{7} & \frac{5}{7} \end{pmatrix}\begin{pmatrix} 4 & -1 & 3 \\ 0 & 2 & 1 \end{pmatrix} = \begin{pmatrix} \frac{8}{7} & \frac{8}{7} & \frac{11}{7} \end{pmatrix} \geq \begin{pmatrix} \frac{8}{7} & \frac{8}{7} & \frac{8}{7} \end{pmatrix}$$

and

$$\begin{pmatrix} 4 & -1 & 3 \\ 0 & 2 & 1 \end{pmatrix}\begin{pmatrix} \frac{3}{7} \\ \frac{4}{7} \\ 0 \end{pmatrix} = \begin{pmatrix} \frac{8}{7} \\ \frac{8}{7} \end{pmatrix} \leq \begin{pmatrix} \frac{8}{7} \\ \frac{8}{7} \end{pmatrix}$$

Player $\mathscr{R}$ can do no worse than gain $\frac{8}{7}$ and $\mathscr{C}$ can do no better than lose $\frac{8}{7}$. The value of the game is $\frac{8}{7}$.

Exercises

In Exercises 1 and 2, matrix A defines a two-person zero-sum game; R and C are the strategies for players $\mathscr{R}$ and $\mathscr{C}$, respectively. Calculate the mathematical expectation of $\mathscr{R}$ in each game.

1. $A = \begin{pmatrix} 5 & -2 \\ 0 & 1 \end{pmatrix} \qquad R = \begin{pmatrix} \frac{1}{8} & \frac{7}{8} \end{pmatrix} \qquad C = \begin{pmatrix} \frac{3}{8} \\ \frac{5}{8} \end{pmatrix}$

2. $A = \begin{pmatrix} 5 & 1 & -2 \\ -2 & 0 & 3 \end{pmatrix} \qquad R = \begin{pmatrix} \frac{1}{2} & \frac{1}{2} \end{pmatrix} \qquad C = \begin{pmatrix} \frac{2}{5} \\ \frac{2}{5} \\ \frac{1}{5} \end{pmatrix}$

In Exercises 3 through 12 find the value of each matrix game and determine the optimal strategies for $\mathscr{R}$ and $\mathscr{C}$.

3. $\begin{pmatrix} 1 & -2 \\ -2 & 4 \end{pmatrix}$

4. $\begin{pmatrix} 1 & -5 \\ -5 & 10 \end{pmatrix}$

5. $\begin{pmatrix} 3 & 0 \\ -1 & 4 \end{pmatrix}$

6. $\begin{pmatrix} 1 & -b \\ -b & c \end{pmatrix}$

7. $\begin{pmatrix} 0 & 1 & -2 \\ -1 & 2 & 1 \end{pmatrix}$

8. $\begin{pmatrix} 1 & 0 & 2 & 1 \\ -1 & 2 & -2 & 3 \end{pmatrix}$

9. $\begin{pmatrix} -3 & 3 & 1 \\ 3 & 4 & -4 \\ -2 & -1 & 2 \end{pmatrix}$

10. $\begin{pmatrix} 4 & 1 & 4 \\ -3 & -1 & 3 \\ 0 & 5 & 6 \end{pmatrix}$

11. $\begin{pmatrix} 3 & -1 \\ -2 & 1 \\ 2 & 0 \end{pmatrix}$

12. $\begin{pmatrix} 2 & 2 & 1 \\ -1 & 2 & 0 \\ -3 & -2 & 3 \end{pmatrix}$

13. Consider the matrix game $\begin{pmatrix} 5 & 2 \\ 3 & a \end{pmatrix}$. Determine optimal strategies for $\mathscr{R}$ and $\mathscr{C}$ and the value of the game if (*a*) $a < 2$; (*b*) $2 < a < 3$; (*c*) $a > 3$.

14. One of the most elementary matrix games is called Matching Pennies. In this game, $\mathscr{R}$ and $\mathscr{C}$ each put down a penny, either heads side up or tails side up. If both coins show heads or both show tails, $\mathscr{R}$ wins $\mathscr{C}$'s penny. If the coins do not match, $\mathscr{C}$ wins $\mathscr{R}$'s penny. Determine an optimal strategy for each player. What is the value of the game?

15. Consider a modified version of the game of Matching Pennies. If both coins show heads, $\mathscr{R}$ wins one penny; if both coins show tails, $\mathscr{R}$ wins three pennies; if the coins do not match, $\mathscr{C}$ wins two pennies. Construct the payoff matrix and determine an optimal strategy for each player. Is the game a fair game?

16. Each of two players $\mathscr{R}$ and $\mathscr{C}$ simultaneously show a penny, nickel, or dime. If the total amount of money shown is even, $\mathscr{R}$ wins $\mathscr{C}$'s coin; if the total is odd, $\mathscr{R}$ loses his coin to $\mathscr{C}$. (*a*) Construct the payoff matrix for this game. (*b*) Determine optimal mixed strategies for $\mathscr{R}$ and $\mathscr{C}$. (*c*) Determine the value of the game.

17. An airplane can search for an enemy submarine by visual means or by radar. The visual means is restrictive in terms of area; the use of radar extends the area of search but also warns the enemy through the use of radar detectors of the airplane's approach. The submarine captain has a choice between running deep or lying close to the surface. If the submarine runs deep, the submarine captain can decrease the probability of being discovered by visual means. However, a submarine running deep increases the probability of being discovered by radar because the submarine radar detector is not operable and evasive action is prevented. The "game" between the airplane and the submarine can be described by the following matrix:

		Submarine: Run deep with no radar detector	Submarine: Use radar detector near the surface
Airplane	Visual search	0.3	0.6
	Radar search	0.8	0.2

Each entry gives the probability of the airplane destroying the submarine under the given "plays" of the airplane pilot and the submarine captain. Find the optimal strategies for both the airplane pilot and the submarine captain.

18. An army must defend a city against an attacking enemy army. To win, the enemy must gain control of one of the two passes through the mountains surrounding the city. Both the defenders and the attackers have three divisions of men. Each army can deploy its forces in the following way: three divisions to one of the two passes or two divisions to one pass and one division to the other pass. A pass is gained or held according to whether or not the attacking army has the larger force. The "game" between the attacking army and the defending army can be described by the following matrix:

		Defending army	
		3 − 0	2 − 1
Attacking army	3 − 0	$\frac{1}{2}$	1
	2 − 1	1	$\frac{1}{2}$

Each entry gives the probability of the attacking army successfully capturing the city. The notations 3 − 0 and 2 − 1 indicate the deployment of forces. Determine the optimal strategy for each army. Is victory more likely to go to the attacking or defending army?

Keyed Summary

In this chapter we studied linear inequalities and systems of linear inequalities. Two methods of solution to the problem of finding the maximum and minimum values of a linear function over a convex polygon were presented. An introduction to matrix games, their values and strategies was included. The concepts and methods presented are part of the bodies of knowledge called *linear programming* and *game theory*.

6.1 Order Relations of Numbers

A number line is a representation of numbers by points on a line. A real number associated with a point is called the coordinate of the point; the point is called the graph of the number.

The set of real numbers is an ordered set of numbers. If $b - a$ is a positive number, then a is less than b and we write $a < b$; if $b - a$ is zero, then a is equal to b and we write $a = b$; if $b - a$ is a negative number, then a is greater than b and we write $a > b$.

The compound statement $a < b$ or $a = b$ is written $a \leq b$; the compound statement $a > b$ or $a = b$ is written $a \geq b$.

A half line consists of the points to the left or to the right of a particular point on a line. A ray is the union of a point on a line and one of the two half lines determined by that point.

A conditional statement is an equality or inequality involving at least one variable. The set of values of the variable(s) for which a conditional statement is true is called the solution set of the conditional statement. The graph of a conditional statement is the set of points whose coordinates are in the solution set.

Any conditional statement of inequality that can be expressed in one of the forms $ax + b > 0$, $ax + b < 0$, $ax + b \geq 0$, or $ax + b \leq 0$, where a and b are real numbers and $a \neq 0$, is called a linear inequality in one variable. The graphs of the solution sets of $ax + b > 0$ and $ax + b < 0$ are half lines; the graphs of $ax + b \geq 0$ and $ax + b \leq 0$ are rays.

The order relations $<$ and $>$ have the following properties, which are stated for the "less than" relation only:

Transitive property: If $a < b$ and $b < c$, then $a < c$.

Addition property: If $a < b$, then $a + c < b + c$.

Multiplication property: If $a < b$ and $c > 0$, then $ac < bc$; if $a < b$ and $c < 0$, then $ac > bc$.

6.2 *Graphs of Linear Equations*

The graph of the Cartesian set $R \times R$ consists of all points of the plane and is called the Cartesian plane. The graph of each element (a, b) of $R \times R$ can be obtained in terms of the coordinate axes: a horizontal number line called the x axis and a vertical number line called the y axis. The numbers a and b are called the x coordinate and the y coordinate of the point, respectively.

Any statement of equality that can be expressed in the form

$$ax + by + c = 0$$

where a, b, and c are real numbers and a and b are not both zero, is called a linear equation in the two variables x and y. Its graph is a straight line.

A set of lines with a common property is called a family of lines.

6.3 *Linear Inequalities in Two Variables*

Any conditional statement of inequality that can be expressed in one of the forms $ax + by + c > 0$, $ax + by + c < 0$, $ax + by + c \geq 0$, or $ax + by + c \leq 0$, where a, b, and c are real numbers and a and b are not both zero, is called a linear inequality in the two variables x and y.

Every straight line $ax + by + c = 0$ in a plane separates the plane into three disjoint sets of points:

1. The set of points either "above" or "to the left of" the line
2. The set of points on the line
3. The set of points either "below" or "to the right of" the line

The set of points described in (1) and (3) are called half planes; the line is the boundary of each half plane.

If $b > 0$, the graph of the linear inequality $ax + by + c > 0$ is the half plane above the line $ax + by + c = 0$; the graph of the linear inequality $ax + by + c < 0$ is the half plane below the line.

6.4 Systems of Linear Inequalities

A compound statement consisting of the conjunction of two or more linear inequalities is called a system of linear inequalities.

The intersection of n half planes with boundaries is called a polygonal convex set. A polygonal convex set with finite area consists of a polygon and its interior points; such a set is called a convex polygon.

6.5 Convex Sets and Applications

Linear programming is the study of problems in which several conditions have to be satisfied simultaneously. Each condition may be expressed by means of a linear inequality. The intersection of the solution sets of the given linear inequalities represents the feasible solutions. Subject to these conditions, the maximum or minimum value of a linear function must be found; that is, an optimal solution among the feasible solutions is desired.

6.6 Linear Functions Over Convex Polygons

The linear programming problem may be described more precisely as the problem of finding the maximum or minimum value of the linear function

$$f(x, y) = ax + by + c \qquad (a,\ b,\ \text{and } c \text{ are real numbers})$$

over the convex polygon defined by the system of linear inequalities

$$\begin{cases} a_{11}x + a_{12}y + a_{13} \leq 0 \\ a_{21}x + a_{22}y + a_{23} \leq 0 \\ a_{31}x + a_{32}y + a_{33} \leq 0 \\ \vdots \\ a_{n1}x + a_{n2}y + a_{n3} \leq 0 \end{cases}$$

Each a_{ij} for $1 \leq i \leq n$ and $1 \leq j \leq 3$ represents a real number.

A graphical method of solution for a linear programming problem may be described as follows. To find the maximum (minimum) value of a linear function $f(x, y) = ax + by + c$ over a convex polygon,

1. Sketch the convex polygon that represents the set of feasible solutions.
2. Graph the line $f(x, y) = ax + by + c$ for some convenient value of $f(x, y)$.
3. Use a straight edge to sketch lines parallel to the line in part (2), moving the straight edge in the direction of increasing (decreasing) values of $f(x, y)$.
4. Locate the last point of the convex polygon that intersects one of the lines in part (3).
5. The coordinates of the point determined in part (4) represent the values of x and y, respectively, for which $f(x, y)$ is a maximum (minimum).

6.7 *The Tabular Method of Solution*

The linear function $f(x, y) = ax + by + c$ defined over a convex polygon takes on its maximum (minimum) value at a vertex of the convex polygon.

A tabular method of solution for a linear programming problem may be described as follows. To find the maximum (minimum) value of a linear function $f(x, y) = ax + by + c$ over a convex polygon defined by a system of linear inequalities,

1. Find the coordinates of the possible vertices of the convex polygon by considering pairs of the inequalities of the system, treating them as equalities, and solving simultaneously.
2. Test each possible vertex to determine whether or not the coordinates satisfy each inequality of the system; only those that satisfy each inequality represent vertices of the convex polygon.
3. Evaluate the linear function $f(x, y)$ at each vertex.
4. Choose the maximum (minimum) value of $f(x, y)$ obtained in part (3).

6.8 *Matrix Games*

Every matrix A defines a game in the following way:

1. There are two players, $\mathscr{R}$ and $\mathscr{C}$.
2. The game consists of $\mathscr{R}$ choosing a row i of A and $\mathscr{C}$ choosing a column j of A. The choices, called plays, are made independently and simultaneously.
3. At the end of each game, player $\mathscr{R}$ receives from $\mathscr{C}$ an amount, called a payoff, equal to the element a_{ij} of matrix A. A negative value for a_{ij} denotes a payment from $\mathscr{R}$ to $\mathscr{C}$.

The decisions by $\mathscr{R}$ and $\mathscr{C}$ to play rows and columns of a matrix game A in some manner are called strategies. The objective of game theory is to determine optimal strategies—strategies that are most profitable to both players. The payoff corresponding to these optimal strategies is called the value of the game.

A saddle point of a matrix game simultaneously represents a row minimum and a column maximum. When a matrix game has a saddle point, the game is strictly determined and the optimal strategies for both players are pure strategies.

6.9 Nonstrictly Determined Games

If a matrix game has no saddle point, the game is nonstrictly determined and the optimal strategies for both players are mixed strategies. A mixed strategy consists of a player choosing the various rows or columns with a given probability distribution.

Given a nonstrictly determined 2×2 matrix game

$$\begin{pmatrix} a & b \\ c & d \end{pmatrix}$$

$\mathscr{R}$'s optimal mixed strategy is

$$\begin{pmatrix} \dfrac{d-c}{a+d-b-c} & \dfrac{a-b}{a+d-b-c} \end{pmatrix}$$

$\mathscr{C}$'s optimal mixed strategy is

$$\begin{pmatrix} \dfrac{d-b}{a+d-b-c} \\[2ex] \dfrac{a-c}{a+d-b-c} \end{pmatrix}$$

The value v of the game is given by the equation

$$v = \frac{ad-bc}{a+d-b-c}$$

Building Your Mathematical Vocabulary

Addition property (6.1)
Boundary (6.3)
Cartesian plane (6.2)
Conditional statement (6.1)
Convex polygon (6.4)
Coordinate (6.1)
Coordinate axes (6.2)
Family of lines (6.2)
Feasible solutions (6.5)
Graph (6.1)

Graph of a conditional statement (6.1)
Half line (6.1)
Half planes (6.3)
Line segment (6.1)
Linear equation (6.2)
Linear inequality (6.1, 6.3)
Linear programming (6.5)
Matrix games (6.8)
Mixed strategy (6.8, 6.9)
Multiplication property (6.1)
Nonstrictly determined game (6.8, 6.9)
Number line (6.1)
Optimal solution (6.5)
Optimal strategies (6.8)
Ordered set (6.1)
Payoff (6.8)
Plays (6.8)
Polygonal convex set (6.4)
Pure strategy (6.8)
Ray (6.1)
Saddle point (6.8)
Slope (6.2)
Solution set (6.1)
Strategies (6.8)
Strictly determined game (6.8)
System of linear inequalities (6.4)
Transitive property (6.1)
Two-person games (6.8)
Value of a game (6.8)
x axis (6.2)
x coordinate (6.2)
y axis (6.2)
y coordinate (6.2)
Zero-sum games (6.8)

Answers

Chapter 1

1.1

1. True.
3. False.
5. False.
7. True.
9. True.
11. False.
13. False.
15. True.
17. True.
19. True.
21. p, q, r.
23. $3 \not> 8$.
25. A rectangle does not have four sides.

1.2

1. True.
3. False.
5. True.
7. True.
9. True.
11. False.
13. Homer is a gentleman and a scholar.
15. Steve and Paul are good athletes.
17. He owns a yellow Dodge.
19. Four is an even number greater than three.
21. False.

1.3

1. True.
3. True.
5. False.
7. False.
9. False.
11. False.

13. True.
15. Lee, John, and Murray went to the party.
17. Lee and either John or Murray went to the party.
19. If p is true and q is true, then "p or q" is false; if p is true and q is false, then "p or q" is true; if p is false and q is true, then "p or q" is true; if p is false and q is false, then "p or q" is false.

1.4

1. Lake Michigan is not one of the Great Lakes of the United States. False.
3. At least one triangle does not have a right angle. True.
5. China does not have a larger population than Israel. False.
7. Some birds do not have feathers.
9. Something is correct.
11. Every movie is worth seeing.
13. At least one tree is not green.
15. At least one happy man is not rich.
17. False.
19. If p is true and q is true, then "(not p) and (not q)" is false; if p is true and q is false, then "(not p) and (not q)" is false; if p is false and q is true, then "(not p) and (not q)" is false; if p is false and q is false, then "(not p) and (not q)" is true.

1.5

1.

p	not (not p)
T	T
F	F

3.

p	q	not (p and q)
T	T	F
T	F	T
F	T	T
F	F	T

5.

p	q	(not p) or (not q)
T	T	F
T	F	T
F	T	T
F	F	T

7.

p	q	r	p or (q or r)
T	T	T	T
T	T	F	T
T	F	T	T
T	F	F	T
F	T	T	T
F	T	F	T
F	F	T	T
F	F	F	F

9.

p	q	r	(p and q) or r
T	T	T	T
T	T	F	T
T	F	T	T
T	F	F	F
F	T	T	T
F	T	F	F
F	F	T	T
F	F	F	F

11.

p	q	p or q, but not both
T	T	F
T	F	T
F	T	T
F	F	F

13. Paul is the freshman; Don majors in engineering.

1.6

11. p and (q or r).

13. (p and q) or [(not p) and r].

15. If $x > 5$ is true, then $x < 1$ is false; if $x < 1$ is true, then $x > 5$ is false. If $x = 2$, then $x(x - 2) = 0$ and $x^2 = 4$ are both true.

1.7

1. Our professor is either not a gentleman or not a scholar.
3. He is Mark's father or she is Henry's mother.
5. Students do not listen carefully, but they have an interest in the course.
7. Going to college is either interesting or a worthwhile experience.
11. Betty does not type fast.
13. Betty types fast but does not take dictation well.
15. Betty neither types fast nor takes dictation well.
17. Betty types fast or takes dictation well.

1.8

1. False.
3. True.
5. True.
7. False.
9. True.
11. True.
13. If all men were good, then there would be no wars.
15. If the alternate interior angles are equal, then the lines are parallel.
17. If a professional football player possesses very special skills, then he should be well paid.
19. If the sum of the angle measures of a geometric figure is 180 degrees, then the geometric figure is a triangle.

21. True.

23. True.

25. True.

29. Money is evil and wise men are not poor.

31. The team wins, but not every player performed well.

33. Every student studies well, but some failing grades will be given.

35.

p	q	r	If p, then (q and r)
T	T	T	T
T	T	F	F
T	F	T	F
T	F	F	F
F	T	T	T
F	T	F	T
F	F	T	T
F	F	F	T

1.9

1. If the baseball game is played Saturday, then it will not rain.

3. If one passes the course, then he attended class.

5. If an engine is a jet engine, then it is noisy.

7. If one is promoted to associate professor, then he has 5 years experience as an assistant professor.

9. (*a*) Tomorrow is Tuesday; (*b*) today is Monday.

11. (*a*) I watch television; (*b*) I am bored.

13. (*a*) He is wrong; (*b*) Adams will resign.

15. (*a*) We should arrive by morning; (*b*) all goes well.

17. The temperature is below freezing tonight, but we cannot go ice skating tomorrow.

19. The blocking is not good, but our fullback can score.

21. At least one girl at our school is not cute.

23. If p, then q.

1.10

1. The given statement is true. (*a*) If a person lives in Texas, then he lives in Dallas. False. (*b*) If a person does not live in Dallas, then he does not live in Texas. False. (*c*) If a person does not live in Texas, then he does not live in Dallas. True.

3. The given statement is false. (*a*) If a number is odd, then it is a prime. False. (*b*) If a number is not prime, then it is not odd. False. (*c*) If a number is not odd, then it is not prime. False.

5. The given statement is true. (*a*) If a number is greater than three, then it is greater than five. False. (*b*) If a number is not greater than five, then it is not greater than three. False. (*c*) If a number is not greater than three, then it is not greater than five. True.
7. Logically equivalent.
9. Logically equivalent.
11. Not logically equivalent.
13. (not q) or p.
15. q or (not p).
19. No.

1.11

1. Today is not Saturday.
3. Today is Saturday or I will not go to the movies.
5. It is not true that either I will go to the movies or I will go to the baseball game.
7. It is not raining or I will not go to the movies.
9. Either today is Saturday and I will go to the baseball game or it is raining.

11.

p	q	$p \to (\sim q)$
T	T	F
T	F	T
F	T	T
F	F	T

13.

p	q	$p \vee (\sim q)$
T	T	T
T	F	T
F	T	F
F	F	T

15.

p	q	$\sim(p \to q)$
T	T	F
T	F	T
F	T	F
F	F	F

17.

p	q	$[(p \to q) \wedge p] \to q$
T	T	T
T	F	T
F	T	T
F	F	T

19.

p	q	$[(p \vee q) \wedge (\sim q)] \to p$
T	T	T
T	F	T
F	T	T
F	F	T

21.

p	q	$[(p \vee q) \wedge p] \to (\sim q)$
T	T	F
T	F	T
F	T	T
F	F	T

23. $(p \vee q) \wedge r$.
25. $(p \wedge q) \wedge r$ or $p \wedge (q \wedge r)$.
27. $p \vee (q \wedge r)$.
29. $(p \vee q) \vee r$ or $p \vee (q \vee r)$.
31. $(p \vee q) \to r$.

1.12

9. Jerry is not over 21 years of age. Modus tollens.
11. You enjoy this course. Rule of detachment (modus ponens).
13. The committee is not well balanced. Modus tollens.
15. It is not a prime number. Disjunctive syllogism.
17. True.

1.13

3. Invalid. **5.** Invalid.
7. Valid.
9. Invalid; fallacy of denying the antecedent.
11. Valid; modus tollens.
13. Invalid; invalid disjunctive syllogism.
15. Invalid; fallacy of affirming the consequent.
17. Valid; disjunctive syllogism.
19. Valid; law of the syllogism.

1.14

1. Valid. **3.** Valid.
5. Valid. **7.** Invalid.
9. Invalid. **11.** Valid.
17. The geometric figure either does not have three sides or it does not have four sides.

Chapter 2

2.1

1. False. **3.** True.
5. {Canada, Mexico}.
7. $\{b, c, d, f, g, h, j, k, l, m, n, p, q, r, s, t, v, w, x, y, z\}$.
9. {0, 1, 2, 3, 4, 5}.
11. $\{x \mid x$ is an infield position in baseball$\}$.
13. $\{x \mid x$ is an odd whole number less than 14$\}$.
15. $\{x \mid x$ is an even prime number$\}$.

2.2

1. {2, 4, 6, 8, ...}.
3. {3, 6, 9, 12, ...}.
5. Finite set.
7. Finite set.
9. Finite set.
11. Infinite set.
13. Finite set.
15. Null set.
17. Null set.
19. Not the null set.
21. Not the null set.
23. Not the null set.

2.3

1. False.
3. True.
5. False.
7. True.
9. True.
11. Undetermined.
13. True.
15. True.
17. True.
19. False.
21. True.
23. True.
25. (*a*) $B \subset A$; (*b*) $C \subset A$; (*c*) $B \subset C$; (*d*) $D \subset A$; (*e*) $B = D$; (*f*) $D \subset C$.

2.4

1. (*a*) $\{a, b\}$, $\{a, c\}$, $\{a, d\}$, $\{a, e\}$, $\{b, c\}$, $\{b, d\}$, $\{b, e\}$, $\{c, d\}$, $\{c, e\}$, $\{d, e\}$;
(*b*) $\{a\}$, $\{b\}$, $\{c\}$, $\{d\}$, $\{e\}$;
(*c*) $\{a, b, c\}$, $\{a, b, d\}$, $\{a, b, e\}$, $\{a, c, d\}$, $\{a, c, e\}$, $\{a, d, e\}$, $\{b, c, d\}$, $\{b, c, e\}$, $\{b, d, e\}$, $\{c, d, e\}$,
(*d*) { };
(*e*) $\{a, b, c, d\}$, $\{a, b, c, e\}$, $\{a, b, d, e\}$, $\{a, c, d, e\}$, $\{b, c, d, e\}$;
(*f*) $\{a, b, c, d, e\}$.
3. (*a*) 3; (*b*) 7; (*c*) 15; (*d*) $2^n - 1$.
5. { }, {1}, {2}, {1, 2}, {4}, {1, 4}, {2, 4}, {1, 2, 4}, {8}, {1, 8}, {2, 8}, {1, 2, 8}, {4, 8}, {1, 4, 8}, {2, 4, 8}, {1, 2, 4, 8}.

2.5

1. $\{b, d\}$.
3. $\{a\}$.
5. $\{c, e\}$.
7. $\{c, e\}$.
9. {0}.
11. {1, 3, 5, 7, ...}.
13. {0, 2, 4, 6, ...}.
15. {1, 2, 3, 4, ...}.
17. (*a*) $\varnothing$; (*b*) U.
19. {1, 3, 4}.

2.6

1.

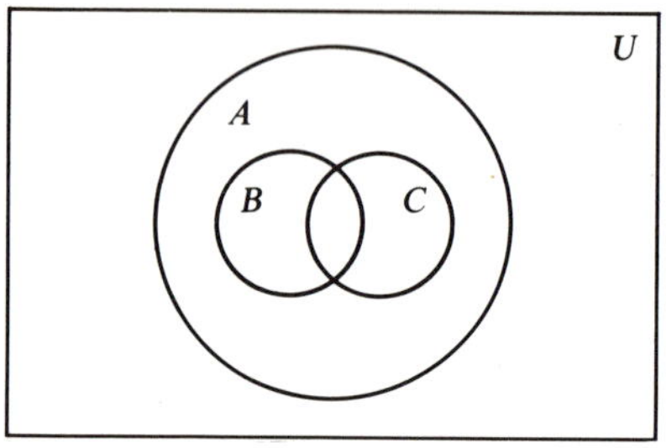

3.

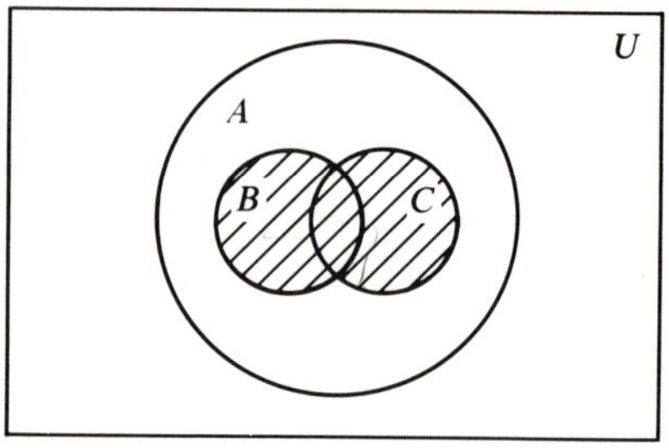

5.

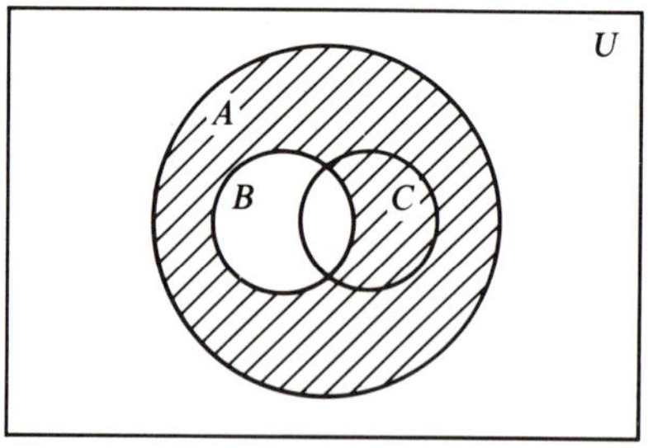

7.

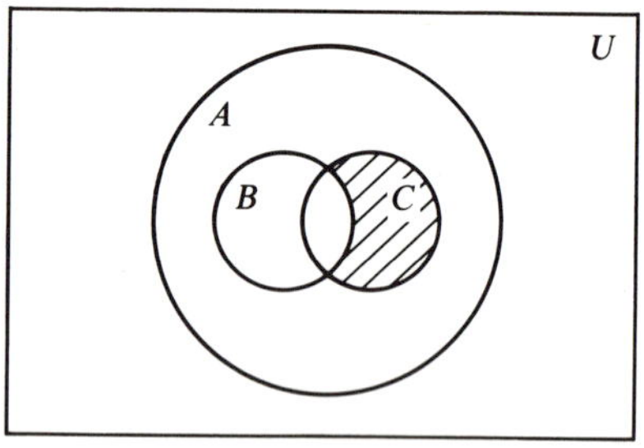

9.

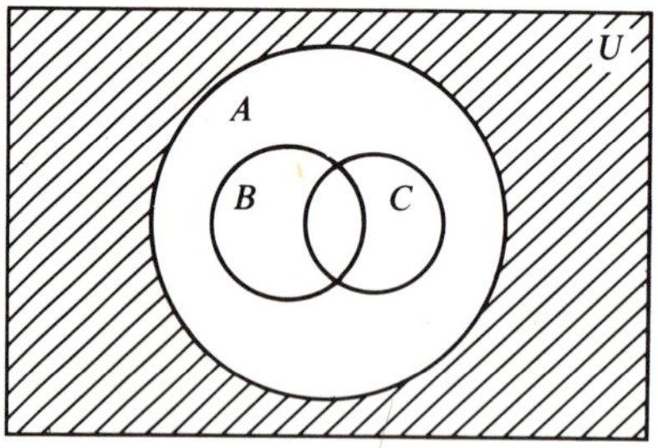

11.

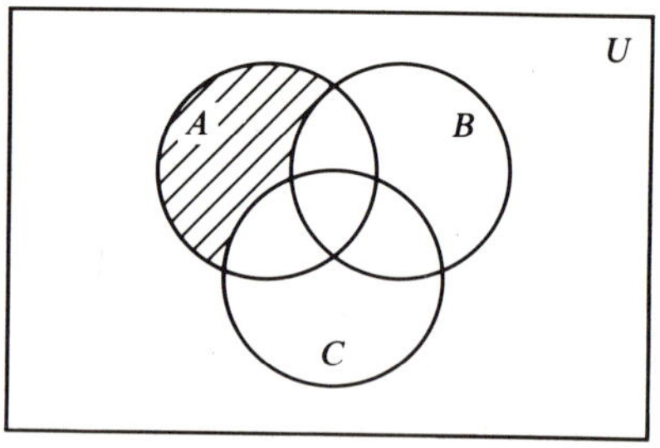

13.

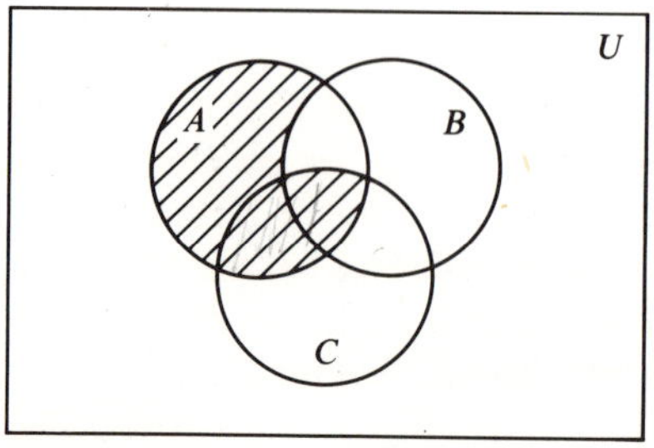

15.

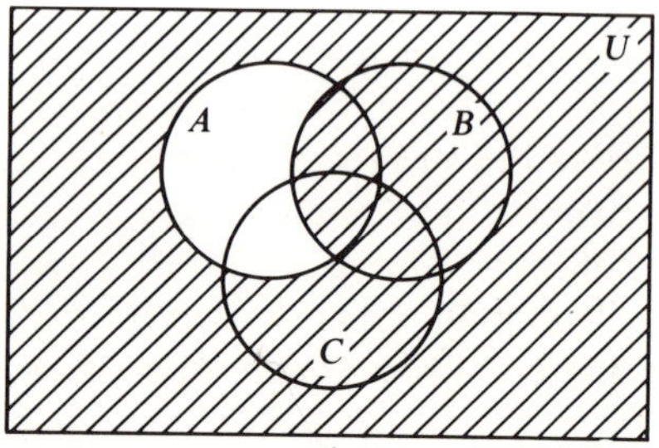

17.

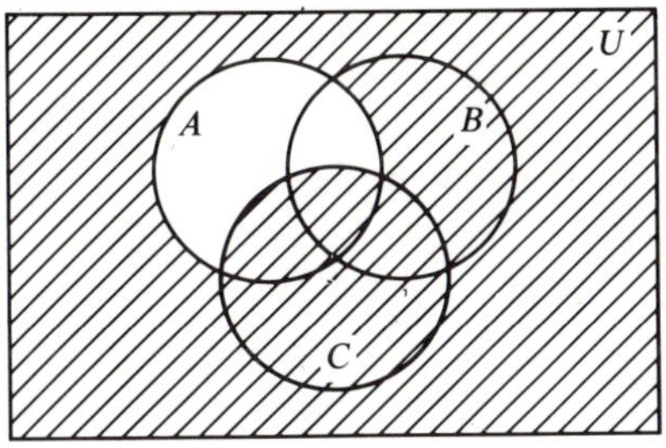

19. Cleveland vs. Detroit; Chicago vs. Oakland; Boston vs. Baltimore; New York vs. Minnesota.

2.7

1. {1, 4, 7}.
3. {4, 7, 10}.
5. {4, 7}.
7. {2, 5}.
9. {8, 9}.
13. {1, 2, 3, 6}; the set of common factors of 24 and 30.
15. (*a*) U; (*b*) $\varnothing$.

17.

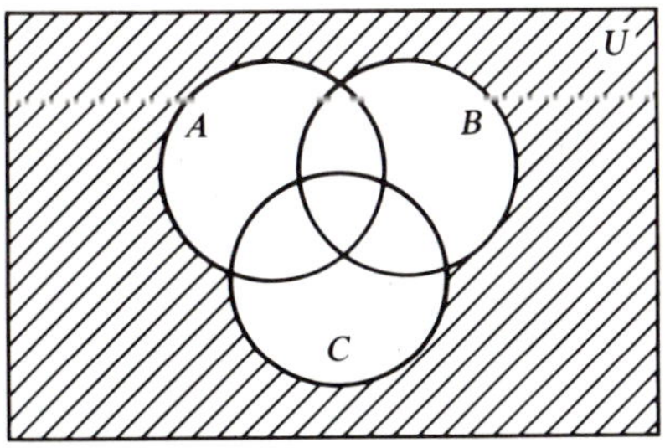

19.

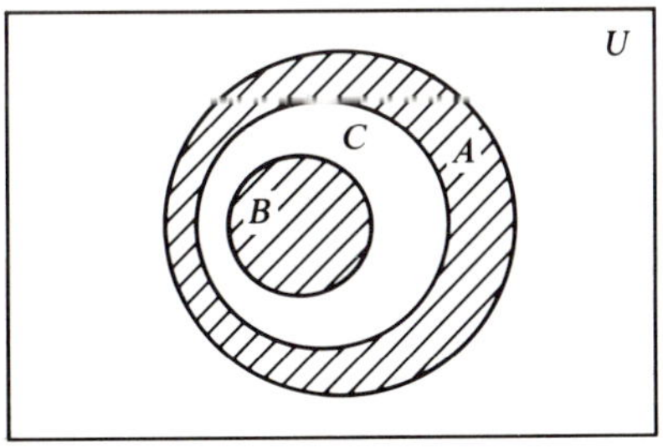

21. $\bar{A} \cap \bar{B}$.
23. $B - (A \cap C)$.
27. (*a*) The empty set; (*b*) the set of members of the House of Representatives; (*c*) the set of senators.
29. (*a*) The set of United States citizens who live in Chicago; (*b*) the set of all men; (*c*) the set of men who live in Chicago; (*d*) the set of United States citizens; (*e*) the set of people who are not United States citizens and live in Chicago; (*f*) the set of all female United States citizens; (*g*) the set of male United States citizens who do not live in Chicago; (*h*) the set of United States citizens except those male citizens who do not live in Chicago.

2.8

1. $\{a, b, c, d, f, g\}$.
3. $\{a, b, d, f, g\}$.
5. $\{e, h\}$.
7. $\{a, b, d, e, f, g, h\}$.
9. $\{a, b, c, d, f, g\}$.
17. (*a*) A; (*b*) A; (*c*) A; (*d*) A.

19.

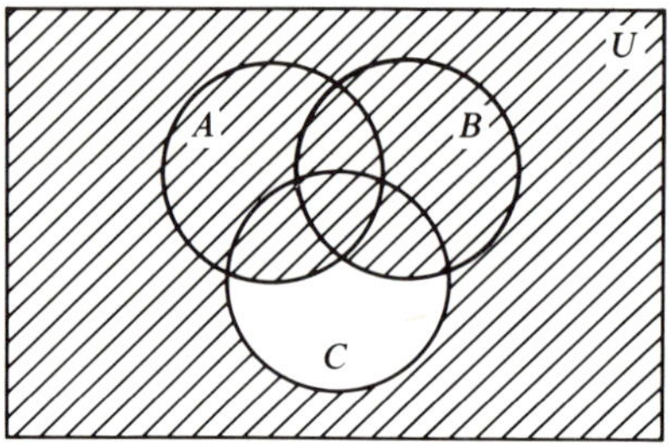

21.

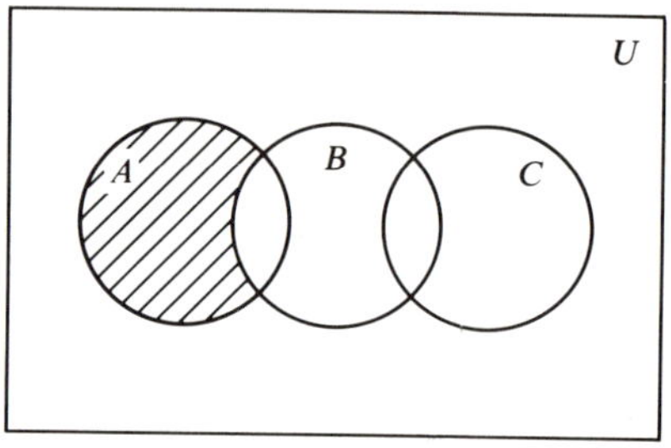

23.

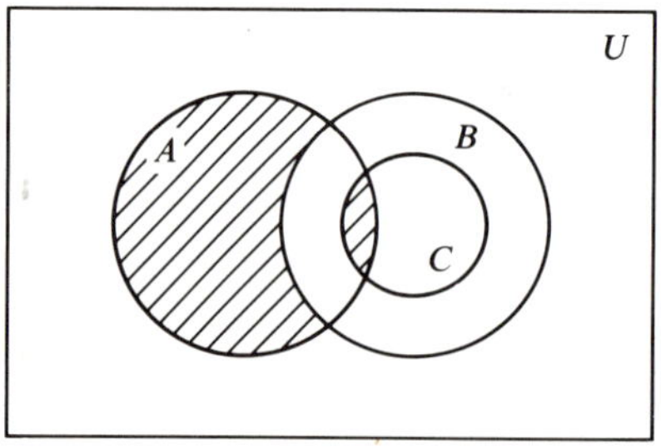

25. $A \cup \bar{B}$.
27. $A \cup (B - \bar{C})$.

2.9

1. (*a*) 46; (*b*) 27
3. 3,030.
7. $A \cap B = \varnothing$.

2.10

1. $\{(1, 3), (1, 4), (2, 3), (2, 4)\}$.
3. $\{(m, m), (m, n), (m, r)\}$.
5. {(apple, blue), (apple, green), (apple, yellow), (pear, blue), (pear, green), (pear, yellow), (lemon, blue), (lemon, green), (lemon, yellow), (orange, blue), (orange, green), (orange, yellow)}.
7. $\{(3, 1), (4, 1), (3, 2), (4, 2)\}$; $A \times B \neq B \times A$.
9. $\{(a, a), (a, b), (a, c), (b, a), (b, b), (b, c), (c, a), (c, b), (c, c)\}$.

11. (*a*) {(1, 1), (2, 2), (3, 3), (4, 4)}; (*b*) {(1, 2), (1, 3), (1, 4), (2, 1), (2, 3), (2, 4), (3, 1), (3, 2), (3, 4), (4, 1), (4, 2), (4, 3)}; (*c*) {(3, 1), (3, 2), (3, 3), (3, 4)}; (*d*) {(2, 1), (3, 1), (4, 1), (3, 2), (4, 2), (4, 3)}.

13. {(r, r, r), (r, r, s), (r, s, r), (r, s, s), (s, r, r), (s, r, s), (s, s, r), (s, s, s)}.

2.11

1. (*a*) Is shorter than; (*b*) is the child of; (*c*) is 1 less than; (*d*) is parallel to; (*e*) is congruent to; (*f*) is younger than.

3. $\mathcal{R}$: is equal to the square of

A		B
1	$\longrightarrow$	1
4	$\longrightarrow$	2
9	$\longrightarrow$	3
16	$\longrightarrow$	4

5. Is a city of.

7. Is a team in the sport of.

9. Domain: {1, 3, 4, 7}; range: {5, 2, 6}.

11. Domain: {0}; range; {1, 2, 3, 4, 5}.

13. Domain: {1, 2, 3, 4, ...}; range: {1, 4, 9, 16, ...}.

15. {(1, 1), (3, 3)}; domain: {1, 3}; range: {1, 3}.

17. {(1, 2), (1, 4), (1, 6), (2, 1), (2, 3), (2, 5), (3, 2), (3, 4), (3, 6), (4, 1), (4, 3), (4, 5), (5, 2), (5, 4), (5, 6), (6, 1), (6, 3), (6, 5)}; domain: {1, 2, 3, 4, 5, 6}; range: {1, 2, 3, 4, 5, 6}.

19. {(1, 1), (2, 8), (3, 27), (4, 64)}; domain: {1, 2, 3, 4}; range: {1, 8, 27, 64}.

21. {(3, 2), (1, 5), (2, 7)}.

23. $\mathcal{R}$ = {(1, 3), (2, 5), (3, 7), (4, 9)};
$\mathcal{R}'$ = {(3, 1), (5, 2), (7, 3), (9, 4)}.

2.12

1. A function.

3. Not a function. **5.** Not a function.

7. (*a*) 1; (*b*) 25; (*c*) $5a^2 + 2a + 1$; (*d*) $5a^2 + 22a + 25$; (*e*) $5a^2 + 2a + 26$; (*f*) 3,176.

9. No.

11. (*a*) b; (*b*) $a + b$; (*c*) $2a + b$; (*d*) $3a + b$; (*e*) a; (*f*) a; (*g*) a.

13. (*a*) 7; (*b*) 7; (*c*) $a^2 + ab + b^2$; (*d*) $a^2 + ab + b^2$.

Chapter 3

3.1

1. $1 \leftrightarrow a$
$2 \leftrightarrow b$
$3 \leftrightarrow c$
$4 \leftrightarrow d.$

3. $a \leftrightarrow y$
$b \leftrightarrow z$
$c \leftrightarrow x.$

5. $1 \leftrightarrow 6$
$2 \leftrightarrow 2$
$3 \leftrightarrow 8$
$4 \leftrightarrow 4.$

3.2

1. 1.

3. 1.

5. $(0 + 0) \cdot [1 + (0 \cdot 1)] = 0.$

7. $[(0 \cdot 1) \cdot 0)] + (1 \cdot 0) = 0.$

9. $A \cup A = A$. The statement and its dual are both true.

11. $(A \cap B) \cap C = A \cap (B \cap C)$. The statement and its dual are both true.

13. $p \vee p \Leftrightarrow p$; a valid statement.

15. $(p \wedge q) \wedge r \Leftrightarrow p \wedge (q \wedge r)$; a valid statement.

17. $\sim(p \vee q) \Leftrightarrow (\sim p) \wedge (\sim q)$; a valid statement.

3.3

1. $B(1) = 1.$

3. $B(1, 1) = 1.$

5. $B(0, 1, 1) = 1.$

11. $x + x \cdot y = x.$

3.4

5. (*a*) x; (*b*) x.

7. x.

9. $y \cdot x$.

11. x.

13. x.

3.5

1. (*a*) No; (*b*) yes.

3. (*a*) No; (*b*) no.

5. (*a*) Yes; (*b*) yes.

7.

3.6

1. $x \cdot x + y$.

3. $x \cdot y + x \cdot z$.

5. $x + (y + z)$.

7.

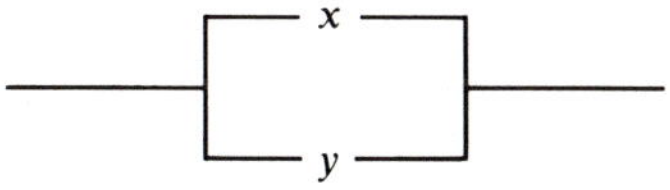

9.

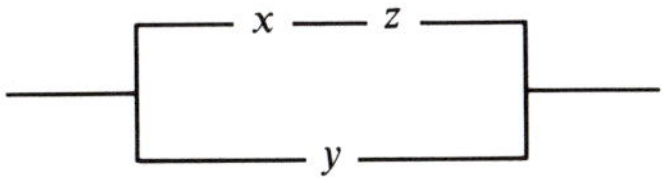

11.

13.

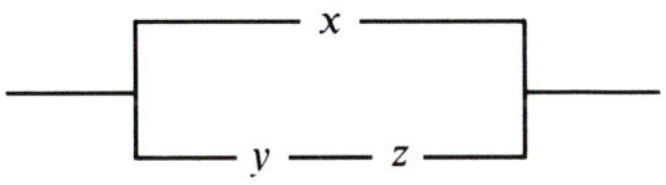

15.

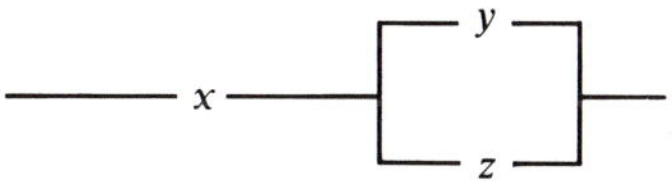

3.7

1. (*a*) 0; (*b*) 1; (*c*) 0; (*d*) 0.

3. (*a*) 0; (*b*) 1; (*c*) 0; (*d*) 1.

13. x.

15. x.

17. $z \cdot (x + y')$.

19. $x + y$.

21. $w' + x' + y' + z'$.

23. x'.

25. Exclusive "or".

3.8

1. $x \cdot (x' + y)$.

3. $x \cdot y' + z \cdot (x' \cdot y' + x \cdot y)$.

5.

7.

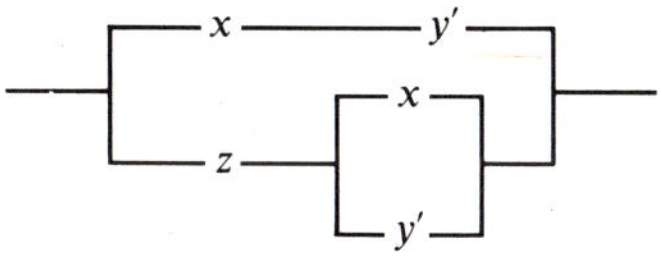

9.

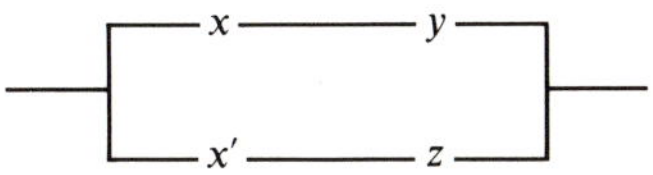

11.

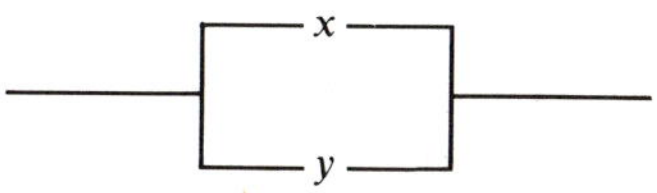

3.9

1. $x + y.$

3. $x \cdot (y + z) + x' \cdot y' \cdot z'.$

5.

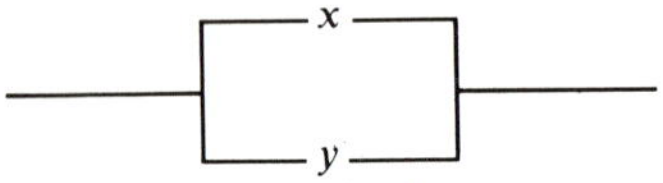

7.

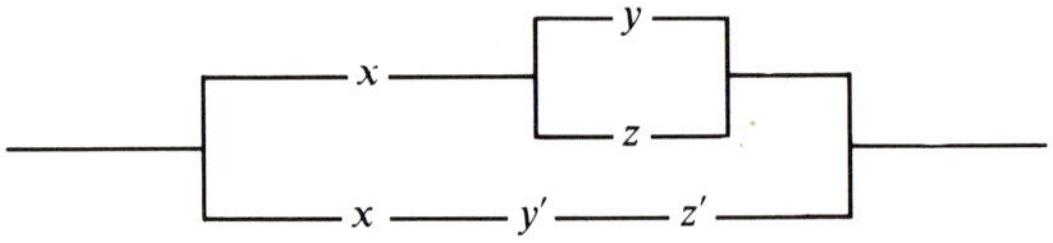

9.

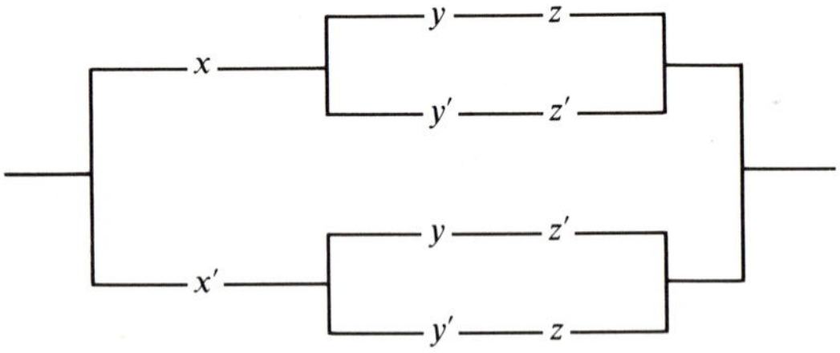

11.

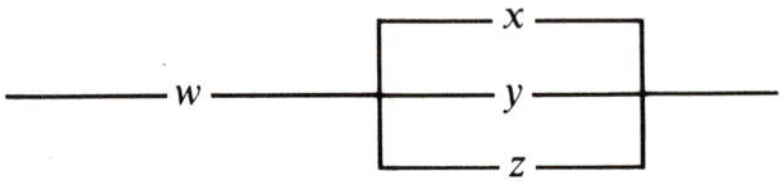

13. $x \cdot y + x \cdot y' + x' \cdot y.$

15. $x \cdot y \cdot z + x \cdot y' \cdot z.$

19.

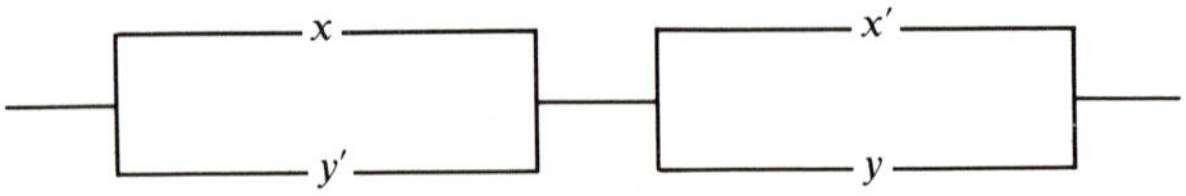

21.

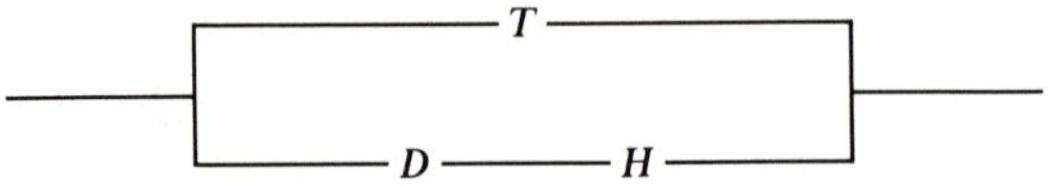

23.

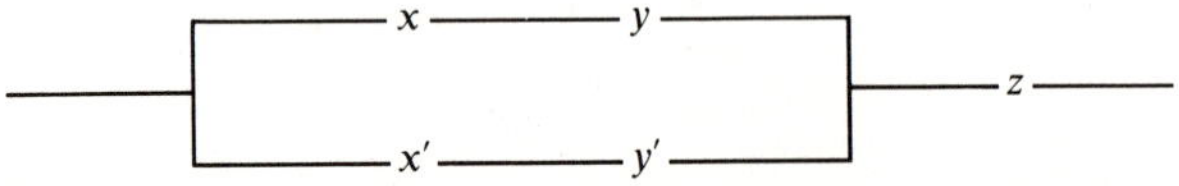

Chapter 4

4.1

1. 15.
3. 10.
5. 30.
7. 60.
9. 20.
11. 1,000 (including "000").
13. 243.
15. 280.
17. 2,600.

4.2

1. 48.
3. 25.
5. 64.
7. 340.
9. 1,405.

4.3

1. 5.
3. 990.
5. 120.
7. 5.
9. 56.
11. 42.
13. 840.
15. 362,880.
17. 6,840.
19. 2,880.
21. 151,200.
27. 840.
29. 2,520.

4.4

1. 5.
3. 28.
5. 1.
7. 1.
9. 56.
11. 1,140.
13. 21.
15. 66.
17. 21.
19. 60.
21. 93.

4.5

1. (*a*) $\frac{1}{6}$; (*b*) $\frac{1}{6}$; (*c*) 0; (*d*) $\frac{1}{3}$; (*e*) $\frac{5}{6}$; (*f*) 1.
3. (*a*) 25; (*b*) $\frac{1}{25}$; (*c*) $\frac{2}{25}$; (*d*) $\frac{1}{5}$.
5. $\frac{17}{35}$.
7. $\frac{1}{4}$.
9. (*a*) $\frac{1}{495}$; (*b*) 0; (*c*) $\frac{2}{55}$; (*d*) $\frac{14}{99}$.
11. $\frac{2}{35}$.

4.6

1. (*a*) {1, 2, 3, 4, 5}; (*b*) $\frac{1}{5}$; (*c*) $\frac{3}{5}$.
3. (*a*) {(1, 1), (1, 2), (1, 3), (1, 4), (1, 5), (2, 1), (2, 2), (2, 3), (2, 4), (2, 5), (3, 1), (3, 2), (3, 3), (3, 4), (3, 5), (4, 1), (4, 2), (4, 3), (4, 4), (4, 5), (5, 1), (5, 2), (5, 3), (5, 4), (5, 5)}.
   (*b*) $\frac{1}{25}$; (*c*) $\frac{1}{5}$.
5. $\frac{3}{8}$.
7. (*a*) $\frac{1}{6}$; (*b*) $\frac{1}{18}$; (*c*) $\frac{5}{36}$; (*d*) $\frac{1}{18}$.
9. $\frac{1}{6}$.
11. $\frac{1}{13}$.
13. $\frac{3}{13}$.
15. $\frac{1}{108{,}290}$.
17. $\frac{3}{8}$.
19. $\frac{1}{6}$.

4.7

1. (*a*) 1; (*b*) 0.
3. 0.6.
5. The event that at least one tail appears.
7. $\frac{7}{8}$.
9. $\frac{3}{4}$.
11. (*a*) $\frac{10}{11}$; (*b*) $\frac{19}{33}$.
13. $\frac{5}{6}$.

4.8

1. (*a*) 5:7; (*b*) 7:5; (*c*) 1:2; (*d*) 1:3.
3. (*a*) 7:1; (*b*) 1:3.
5. (*a*) 4:1; (*b*) 44:1; (*c*) 8:1; (*d*) 3.
7. 63,999:1.

4.9

1. Mutually exclusive.
3. Not mutually exclusive.
5. Mutually exclusive.
7. $\frac{5}{6}$.
11. $\frac{7}{36}$.
13. $\frac{1}{8}$.
15. $\frac{4}{13}$.
17. $\frac{7}{16}$.
19. (*a*) $\frac{55}{153}$; (*b*) $\frac{76}{153}$.
21. (*a*) $\frac{2}{5}$; (*b*) 1:4; (*c*) 2:3.

4.10

1. $\frac{6}{13}$.
3. $\frac{1}{3}$.
5. $\frac{4}{7}$.
7. $\frac{3}{4}$.
9. $\frac{2}{11}$.
11. (*a*) $\frac{12}{25}$; (*b*) $\frac{3}{10}$; (*c*) $\frac{18}{25}$; (*d*) $\frac{1}{6}$; (*e*) $\frac{1}{3}$; (*f*) $\frac{12}{25}$.

4.11

1. $\frac{1}{36}$.
3. $\frac{4}{663}$.
5. $\frac{25}{44}$.
7. $\frac{1}{26}$.
9. 0.92
11. $\frac{365}{729}$.
13. $\frac{8}{9}$.
15. 7.
17. $1 - \frac{364}{365} \cdot \frac{363}{365} \cdot \frac{362}{365} \cdots \frac{341}{365}$.

4.12

1. (*a*) \$0.10; (*b*) −\$0.90; (*c*) \$0.00; (*d*) −\$1.80.
3. 0 points.
5. \$200.
7. 7.
9. (*a*) 6; (*b*) 6.
11. \$0.089.
13. (*a*) 25 points; (*b*) 0 points.
15. (*a*) \$0.00; (*b*) \$0.025.
17. \$0.20.
19. (*a*) $\dfrac{1}{2-p}$; (*b*) $\dfrac{1-p}{2-p}$; (*c*) \$0.50; (*d*) $(1-p)^{n-1}$.

4.13

1. 0.16.
3. 0.7.
5. 0.72.
7. 46.
9. Approximately 0.33.

4.14

1. (*a*) {0, 1, 2, 3, 4, 5}; (*b*) {0, 1, 2, 3, . . . , *n*}.
3.

x	0	1	2	3	4	5
$f(x)$	$\frac{1}{32}$	$\frac{5}{32}$	$\frac{10}{32}$	$\frac{10}{32}$	$\frac{5}{32}$	$\frac{1}{32}$

5. A probability function.

7. A probability function.

9. A probability function.

11.

x	1	2	3	4	5	6
$f(x)$	$\frac{1}{21}$	$\frac{2}{21}$	$\frac{1}{7}$	$\frac{4}{21}$	$\frac{5}{21}$	$\frac{2}{7}$

13.

x	1	2	3
$f(x)$	$\frac{1}{25}$	$\frac{12}{25}$	$\frac{12}{25}$

15. 7.

17. (*a*) $f(x) = \frac{1}{6}$; (*b*) 3.5.

4.15

1. $\frac{21}{128}$.

3. $\frac{3}{16}$.

5. $\frac{64}{81}$.

9. $\frac{8}{27}$.

11. Three.

Chapter 5

5.1

1. (*a*) 2 by 3; (*b*) 2; (*c*) 3; (*d*) 3 by 2.

3. 1 by 24, 2 by 12, 3 by 8, 4 by 6, 6 by 4, 8 by 3, 12 by 2, and 24 by 1.

5. $\begin{pmatrix} 3 & 7 \\ -1 & 0 \end{pmatrix}$.

7. $\begin{pmatrix} 3 & 2 & 1 \\ 6 & 5 & 4 \end{pmatrix}$.

9. (*a*) The elements are in the third row; (*b*) the elements are in the second column; (*c*) the elements lie along a line starting in the upper left-hand corner and running diagonally down to the right; (*d*) the elements lie in a triangular region in the lower left-hand portion of the matrix, below the line in part (*c*).

11. (*a*) $x = 5$; (*b*) $x = 2$, $y = 8$; (*c*) no values of x exist; (*d*) no values of x and y exist.

13. $\begin{pmatrix} 0 & 0 & 0 \\ 0 & 0 & 0 \\ 0 & 0 & 0 \end{pmatrix}$.

5.2

1. $\begin{pmatrix} 5 & 8 \\ 6 & 7 \\ 8 & 2 \end{pmatrix}$.

3. (*a*) $m = 2$ and $n = 3$; (*b*) under no conditions.

5. $\begin{pmatrix} 2 & 1 & 4 \\ -2 & 0 & 3 \\ -4 & 2 & 7 \end{pmatrix}$.

7. (*a*) $\begin{pmatrix} 8 & 4 & 4 \\ 4 & 3 & -3 \end{pmatrix}$; (*b*) $\begin{pmatrix} 4 & 2 & -2 \\ 6 & -1 & -5 \end{pmatrix}$.

9. $\begin{pmatrix} a_{11} & a_{12} & a_{13} \\ a_{21} & a_{22} & a_{23} \end{pmatrix} = \begin{pmatrix} 3 & -5 & 2 \\ 0 & 7 & -3 \end{pmatrix}$.

11. $A = \begin{pmatrix} 3 & 2 \\ 1 & 1 \end{pmatrix}$ and $B = \begin{pmatrix} 5 & -1 \\ 4 & 0 \end{pmatrix}$.

13. (*a*) $\begin{pmatrix} 3{,}520 & 1{,}300 \\ 2{,}600 & 650 \\ 8{,}770 & 2{,}320 \end{pmatrix}$; (*b*) $\begin{pmatrix} 650 & 350 \\ 2{,}250 & 475 \\ 5{,}580 & 1{,}310 \end{pmatrix}$;

(*c*) $\begin{pmatrix} 4{,}170 & 1{,}650 \\ 4{,}850 & 1{,}125 \\ 14{,}350 & 3{,}630 \end{pmatrix}$; (*d*) $\begin{pmatrix} 2{,}870 & 950 \\ 350 & 175 \\ 3{,}190 & 1{,}010 \end{pmatrix}$.

5.3

1. $\begin{pmatrix} 4 & -14 & 0 \\ 6 & 10 & -8 \end{pmatrix}$,

3. $\begin{pmatrix} 7 & -6 & 4 \\ 13 & 12 & -6 \end{pmatrix}$,

5. $(r + s)A = rA + sA = \begin{pmatrix} 8 & 2 & 6 \\ 4 & -2 & 0 \end{pmatrix}$.

7. $r(sA) = (rs)A = \begin{pmatrix} 36 & 12 \\ 6 & 0 \\ 24 & 6 \end{pmatrix}$.

9. Yes, provided that $A \neq 0$.

11. (*a*) $(511 \quad 36 \quad 71 \quad 12)$; (*b*) $\left(\frac{511}{630} \quad \frac{36}{630} \quad \frac{71}{630} \quad \frac{12}{630}\right)$.

5.4

1. $\begin{pmatrix} 35 & 7 & 28 \\ 11 & -2 & 4 \end{pmatrix}$.

3. AB does not exist; $BA = \begin{pmatrix} 1 & 15 \\ 8 & 0 \\ -4 & 6 \end{pmatrix}$.

5.
$$AB = BA = \begin{pmatrix} 1 & 0 & 0 \\ 0 & 1 & 0 \\ 0 & 0 & 1 \end{pmatrix}.$$

7.
$$(A + B)(A + B) = \begin{pmatrix} 27 & 24 \\ 12 & 19 \end{pmatrix};$$

$$A^2 + 2AB + B^2 = \begin{pmatrix} 12 & 34 \\ 12 & 34 \end{pmatrix};$$

$(A + B)(A + B) \neq A^2 + 2AB + B^2$.

9.
$$(AB)C = \left[\begin{pmatrix} 2 & 0 & 1 \\ -1 & 3 & 4 \end{pmatrix}\begin{pmatrix} 1 & 2 & 0 \\ -1 & 3 & 4 \\ 0 & 2 & -1 \end{pmatrix}\right]\begin{pmatrix} 0 & 2 \\ 1 & 2 \\ 0 & 5 \end{pmatrix}$$
$$= \begin{pmatrix} 2 & 6 & -1 \\ -4 & 15 & 8 \end{pmatrix}\begin{pmatrix} 0 & 2 \\ 1 & 2 \\ 0 & 5 \end{pmatrix} = \begin{pmatrix} 6 & 11 \\ 15 & 62 \end{pmatrix};$$

$$A(BC) = \begin{pmatrix} 2 & 0 & 1 \\ -1 & 3 & 4 \end{pmatrix}\left[\begin{pmatrix} 1 & 2 & 0 \\ -1 & 3 & 4 \\ 0 & 2 & -1 \end{pmatrix}\begin{pmatrix} 0 & 2 \\ 1 & 2 \\ 0 & 5 \end{pmatrix}\right]$$
$$= \begin{pmatrix} 2 & 0 & 1 \\ -1 & 3 & 4 \end{pmatrix}\begin{pmatrix} 2 & 6 \\ 3 & 24 \\ 2 & -1 \end{pmatrix} = \begin{pmatrix} 6 & 11 \\ 15 & 62 \end{pmatrix};$$

$(AB)C = A(BC)$.

11. CA, DB.

13. The product of two nonsquare matrices will be a square matrix when they are conformable for multiplication and the number of rows of the first matrix is equal to the number of columns of the second matrix.

15. (*a*) Any matrix of the form $\begin{pmatrix} 0 & 0 \\ a & b \end{pmatrix}$, where a and b are real numbers;

(*b*) any matrix of the form $\begin{pmatrix} 3a + b & 2a \\ a & b \end{pmatrix}$, where a and b are real numbers.

17. $\begin{pmatrix} 1 & 21 \\ 0 & 8 \end{pmatrix}$.

19. $A(B + C)D = (AB + AC)D = ABD + ACD$.

21.

	A	*B*	*C*	*D*	*E*	*F*	*G*	*H*
A	*A*	*B*	*C*	*D*	*E*	*F*	*G*	*H*
B	*B*	*D*	*A*	*C*	*G*	*H*	*F*	*E*
C	*C*	*A*	*D*	*B*	*H*	*G*	*E*	*F*
D	*D*	*C*	*B*	*A*	*F*	*E*	*H*	*G*
E	*E*	*H*	*G*	*F*	*D*	*A*	*B*	*C*
F	*F*	*G*	*H*	*E*	*A*	*D*	*C*	*B*
G	*G*	*E*	*F*	*H*	*C*	*B*	*D*	*A*
H	*H*	*F*	*E*	*G*	*B*	*C*	*A*	*D*

23. $AD = \begin{pmatrix} 72 & 50 \\ 132 & 150 \\ 108 & 25 \end{pmatrix}$. The matrix represents the number of different typewriters produced by companies C_1 and C_2 each week.

5.5

1. a_{11}, a_{22}, and a_{33}.

3. (*a*) $\begin{pmatrix} 2 & 0 \\ 5 & -2 \end{pmatrix}$; (*b*) $\begin{pmatrix} 4 & 1 \\ 1 & 3 \end{pmatrix}$.

7. (*a*) $\begin{pmatrix} 4 & 2 \\ 7 & -1 \end{pmatrix}$; (*b*) $\begin{pmatrix} 3 & 6 \\ 3 & 0 \end{pmatrix}$.

9. False.

11. False.

13. True.

15. (*a*) 0; (*b*) 1; (*c*) 0; (*d*) 1.

17. $\begin{pmatrix} 2 & 1 & 1 \\ 4 & 6 & 4 \end{pmatrix}$.

19.
$$A^2 - 5A - 2I = \begin{pmatrix} 1 & 2 \\ 3 & 4 \end{pmatrix}\begin{pmatrix} 1 & 2 \\ 3 & 4 \end{pmatrix} - 5\begin{pmatrix} 1 & 2 \\ 3 & 4 \end{pmatrix} - 2\begin{pmatrix} 1 & 0 \\ 0 & 1 \end{pmatrix}$$
$$= \begin{pmatrix} 7 & 10 \\ 15 & 22 \end{pmatrix} - \begin{pmatrix} 5 & 10 \\ 15 & 20 \end{pmatrix} - \begin{pmatrix} 2 & 0 \\ 0 & 2 \end{pmatrix} = \begin{pmatrix} 0 & 0 \\ 0 & 0 \end{pmatrix}.$$

21.
$$(AB)^T = \left[\begin{pmatrix} 5 & 0 \\ 3 & 1 \end{pmatrix}\begin{pmatrix} 2 & 2 \\ 1 & 0 \end{pmatrix}\right]^T = \begin{pmatrix} 10 & 10 \\ 7 & 6 \end{pmatrix}^T = \begin{pmatrix} 10 & 7 \\ 10 & 6 \end{pmatrix};$$
$$B^T A^T = \begin{pmatrix} 2 & 2 \\ 1 & 0 \end{pmatrix}^T \begin{pmatrix} 5 & 0 \\ 3 & 1 \end{pmatrix}^T = \begin{pmatrix} 2 & 1 \\ 2 & 0 \end{pmatrix}\begin{pmatrix} 5 & 3 \\ 0 & 1 \end{pmatrix} = \begin{pmatrix} 10 & 7 \\ 10 & 6 \end{pmatrix};$$

hence, $(AB)^T = B^T A^T$ since $\begin{pmatrix} 10 & 7 \\ 10 & 6 \end{pmatrix} = \begin{pmatrix} 10 & 7 \\ 10 & 6 \end{pmatrix}$.

23. (*a*) Symmetric matrix;

(*b*)

	Orlando	Tampa	Tallahassee
Miami	229	264	476
Orlando	. . .	78	247
Tampa	. . .	. . .	253

25. (*a*)

	Donna	Ann	Edith	Betty	Teri	Cathy
Donna	0	1	1	0	0	0
Ann	0	0	0	0	0	1
Edith	0	1	0	1	1	0
Betty	0	1	0	0	1	1
Teri	1	0	0	0	0	0
Cathy	1	0	0	0	1	0

	Donna	Ann	Edith	Betty	Teri	Cathy
Donna	0	1	0	1	1	1
Ann	1	0	0	0	1	0
Edith	1	1	0	0	1	2
Betty	2	0	0	0	1	1
Teri	0	1	1	0	0	0
Cathy	1	1	1	0	0	0

(*b*) Edith, Betty, Donna, Cathy, Teri, Ann.

27. (*a*)

	Left front	Left rear	Right front	Right rear	Spare
Left front	0	1	0	0	0
Left rear	0	0	1	0	0
Right front	0	0	0	1	0
Right rear	0	0	0	0	1
Spare	1	0	0	0	0

(*b*)

	Left front	Left rear	Right front	Right rear	Spare
Left front	0	0	1	0	0
Left rear	0	0	0	1	0
Right front	0	0	0	0	1
Right rear	1	0	0	0	0
Spare	0	1	0	0	0

Left front tire to right front; left rear tire to right rear; right front tire to spare; right rear tire to left front; spare tire to left rear.

5.6

1. $\begin{pmatrix} 5 & 3 \\ 3 & 2 \end{pmatrix}\begin{pmatrix} 2 & -3 \\ -3 & 5 \end{pmatrix} = \begin{pmatrix} 2 & -3 \\ -3 & 5 \end{pmatrix}\begin{pmatrix} 5 & 3 \\ 3 & 2 \end{pmatrix} = \begin{pmatrix} 1 & 0 \\ 0 & 1 \end{pmatrix}.$

3. Does not exist.

5. $\begin{pmatrix} -1 & 2 \\ \frac{3}{2} & -\frac{5}{2} \end{pmatrix}.$

7. $\begin{pmatrix} \frac{1}{4} & 0 \\ 0 & 1 \end{pmatrix}.$

9. $\begin{pmatrix} \dfrac{a}{a^2+b^2} & \dfrac{-b}{a^2+b^2} \\ \dfrac{b}{a^2+b^2} & \dfrac{a}{a^2+b^2} \end{pmatrix}.$

11. $(A^{-1})^T = \begin{pmatrix} 3 & -5 \\ -1 & 2 \end{pmatrix}^T = \begin{pmatrix} 3 & -1 \\ -5 & 2 \end{pmatrix};$

$(A^T)^{-1} = \begin{pmatrix} 2 & 1 \\ 5 & 3 \end{pmatrix}^{-1} = \begin{pmatrix} 3 & -1 \\ -5 & 2 \end{pmatrix};$

$(A^{-1})^T = (A^T)^{-1}.$

13.
$$\begin{aligned} AB &= AC \\ A^{-1}(AB) &= A^{-1}(AC) \\ (A^{-1}A)B &= (A^{-1}A)C \\ IB &= IC \\ B &= C. \end{aligned}$$

15. -18.

17. $$\begin{pmatrix}1 & 6 & 6\\ 0 & 2 & 12\\ 0 & 0 & 3\end{pmatrix}\begin{pmatrix}1 & -3 & 10\\ 0 & \frac{1}{2} & -2\\ 0 & 0 & \frac{1}{3}\end{pmatrix} = \begin{pmatrix}1 & -3 & 10\\ 0 & \frac{1}{2} & -2\\ 0 & 0 & \frac{1}{3}\end{pmatrix}\begin{pmatrix}1 & 6 & 6\\ 0 & 2 & 12\\ 0 & 0 & 3\end{pmatrix} = \begin{pmatrix}1 & 0 & 0\\ 0 & 1 & 0\\ 0 & 0 & 1\end{pmatrix}.$$

19. You are right.

5.7

1. $x = 0$ and $y = 3$.

3. $x = 1$ and $y = -2$.

5. $x = 1$ and $y = 2$.

7. $x = 2$ and $y = -1$.

9. $$\begin{pmatrix}\frac{11}{3} & -3 & \frac{1}{3}\\ -\frac{7}{3} & 3 & -\frac{2}{3}\\ \frac{2}{3} & -1 & \frac{1}{3}\end{pmatrix}\begin{pmatrix}1 & 2 & 3\\ 1 & 3 & 5\\ 1 & 5 & 12\end{pmatrix}\begin{pmatrix}x\\ y\\ z\end{pmatrix} = \begin{pmatrix}\frac{11}{3} & -3 & \frac{1}{3}\\ -\frac{7}{3} & 3 & -\frac{2}{3}\\ \frac{2}{3} & -1 & \frac{1}{3}\end{pmatrix}\begin{pmatrix}1\\ 2\\ -2\end{pmatrix}$$
$$\begin{pmatrix}1 & 0 & 0\\ 0 & 1 & 0\\ 0 & 0 & 1\end{pmatrix}\begin{pmatrix}x\\ y\\ z\end{pmatrix} = \begin{pmatrix}-3\\ 5\\ -2\end{pmatrix};$$
$x = -3$, $y = 5$, and $z = -2$.

11. 30 wins and 12 losses.

13. 7 hours.

15. 5.5 percent on the stock investment; 4.5 percent on the savings and loan investment.

17. The price of each evergreen is \$6.50 and the price of each pound of grass seed is \$0.50.

5.8

1. $$\begin{cases} 3x + 5y - 3z = 2\\ 2x - y + z = 3\\ x - 3y + z = -1. \end{cases}$$

3. $x = 3$, $y = 2$, and $z = 4$.

5. $x = 1$, $y = 1$, and $z = 1$.

7. $x = -2$, $y = 0$, and $z = 2$.

9. $w = 5$, $x = 1$, $y = t - 6$, and $z = t$, where t is any real number.

11. (*a*) $t = 9$, $r \neq 12$; (*b*) $t \neq 9$; (*c*) $t = 9$, $r = 12$.

15. $\begin{pmatrix}0 & 1\\ 1 & 0\end{pmatrix}$, $\begin{pmatrix}k & 0\\ 0 & 1\end{pmatrix}$, $\begin{pmatrix}1 & 0\\ 0 & k\end{pmatrix}$, $\begin{pmatrix}1 & k\\ 0 & 1\end{pmatrix}$, and $\begin{pmatrix}1 & 0\\ k & 1\end{pmatrix}$.

17. 2 trucks of type 1, 3 trucks of type 2, and 1 truck of type 3; or 1 truck of type 1, 3 trucks of type 2, and 3 trucks of type 3; or 3 trucks of type 2 and 5 trucks of type 3.

19. 7.5 gallons of 30% nitric acid solution, 15 gallons of 50% nitric acid solution, and 2.5 gallons of 10% nitric acid solution.

5.9

1. 1.

3. 0.01.

5. $x = 2$.

7. $x = -3$.

13. 0.

15. 27.

19. $\begin{vmatrix} 5-x & 1 \\ 2 & 3-x \end{vmatrix} = 0$

$x^2 - 8x + 13 = 0;$

$$\begin{pmatrix} 5 & 1 \\ 2 & 3 \end{pmatrix}\begin{pmatrix} 5 & 1 \\ 2 & 3 \end{pmatrix} - 8\begin{pmatrix} 5 & 1 \\ 2 & 3 \end{pmatrix} + 13\begin{pmatrix} 1 & 0 \\ 0 & 1 \end{pmatrix} = \begin{pmatrix} 0 & 0 \\ 0 & 0 \end{pmatrix}$$

$$\begin{pmatrix} 27 & 8 \\ 16 & 11 \end{pmatrix} + \begin{pmatrix} -40 & -8 \\ -16 & -24 \end{pmatrix} + \begin{pmatrix} 13 & 0 \\ 0 & 13 \end{pmatrix} = \begin{pmatrix} 0 & 0 \\ 0 & 0 \end{pmatrix}$$

$$\begin{pmatrix} 0 & 0 \\ 0 & 0 \end{pmatrix} = \begin{pmatrix} 0 & 0 \\ 0 & 0 \end{pmatrix}.$$

5.10

1. True.

3. -4.

5. 4.

7. 12.

9. 0.

17. k^4m.

21. -30.

5.11

1. $\begin{pmatrix} 1 & -3 & 2 \\ -3 & 3 & -1 \\ 2 & -1 & 0 \end{pmatrix}.$

3. $\begin{pmatrix} \frac{10}{3} & -\frac{17}{3} & \frac{1}{6} \\ -\frac{5}{3} & \frac{10}{3} & -\frac{1}{3} \\ \frac{1}{3} & -\frac{2}{3} & \frac{1}{6} \end{pmatrix}.$

5. $\begin{pmatrix} \frac{2}{3} & \frac{1}{3} & \frac{2}{3} \\ -\frac{2}{3} & \frac{2}{3} & \frac{1}{3} \\ \frac{1}{3} & \frac{2}{3} & -\frac{2}{3} \end{pmatrix}.$

7. $\begin{pmatrix} -\frac{1}{2} & 1 & \frac{1}{2} \\ \frac{3}{2} & -\frac{3}{2} & \frac{1}{2} \\ \frac{1}{2} & -\frac{1}{2} & -\frac{1}{2} \end{pmatrix}.$

5.12

1. (*a*)

$$\text{Today}\begin{cases}\text{Orlando}\\ \text{Miami}\end{cases}\overset{\overbrace{\text{Orlando}\quad\text{Miami}}^{\text{Tomorrow}}}{\begin{pmatrix}0 & 1\\ \frac{1}{5} & \frac{4}{5}\end{pmatrix}}$$

(*b*) $\frac{1}{6}$ of his time in Orlando; $\frac{5}{6}$ of his time in Miami.

3. $\frac{8}{9}$ liquid and $\frac{1}{9}$ vapor.

5. $\frac{2}{3}$.

7. (*a*) $\begin{pmatrix}\frac{3}{4} & \frac{1}{5} & \frac{1}{20}\\ \frac{1}{2} & \frac{2}{5} & \frac{1}{10}\\ \frac{1}{4} & \frac{1}{2} & \frac{1}{4}\end{pmatrix}$; (*b*) $\frac{3}{8}$; (*c*) $\frac{1}{2}$; (*d*) $\frac{27}{40}$.

Chapter 6

6.1

1. (*a*) $x < y$; (*b*) $x \geq 4$.

3.

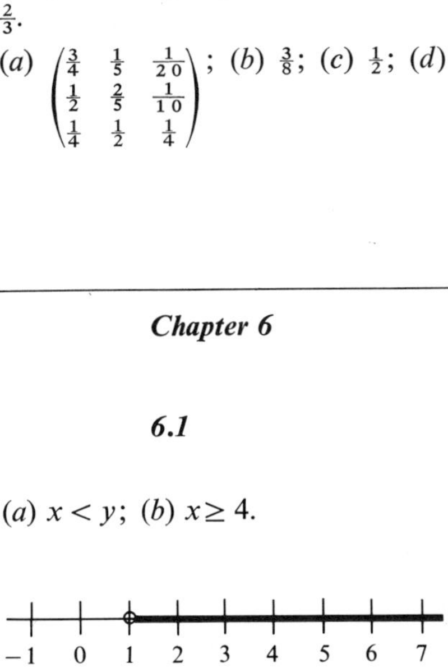

5.

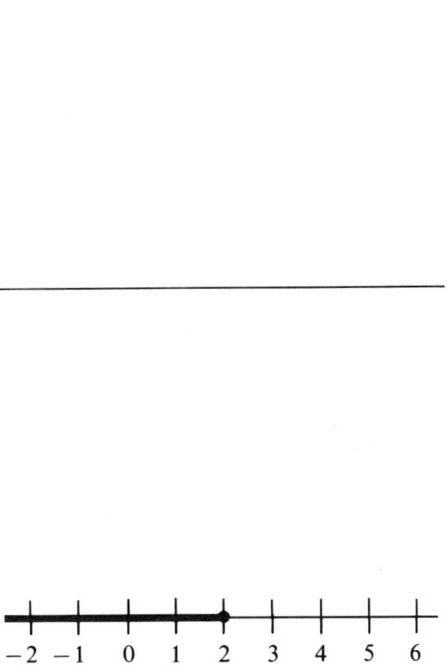

7.

9.

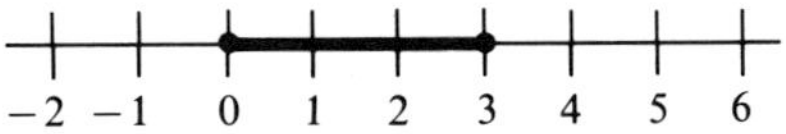

11.

13.

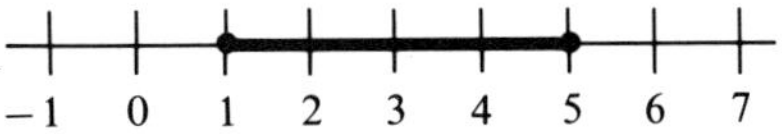

15.

17.

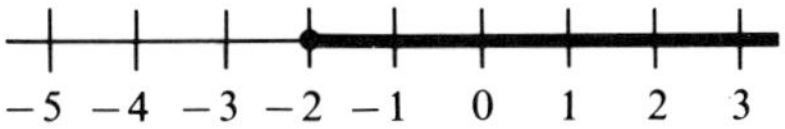

19.

21.

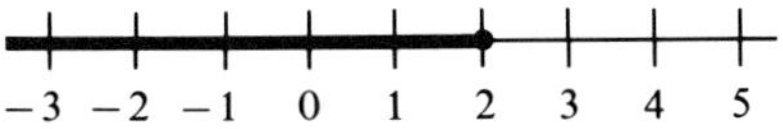

23. (*a*) $5 \leq x \leq 8$; (*b*) $-3 \leq x \leq 2$.

25. True.

27. True.

6.2

1. $A(4, 2)$, $B(-2, 1)$, $C(0, 3)$, $D(3, -3)$, $E(2, 0)$, $F(-4, -2)$.

3.

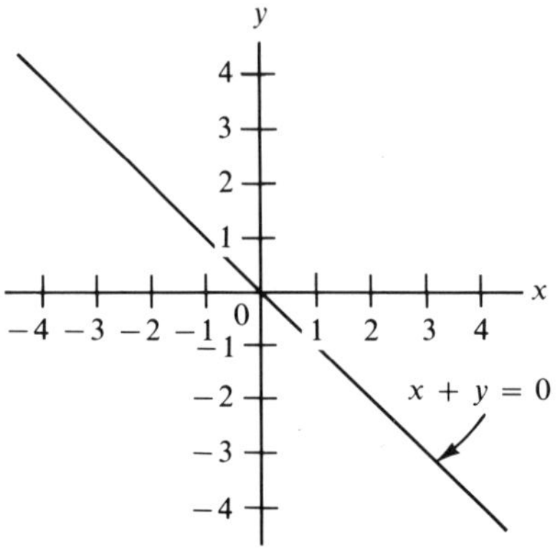

5.

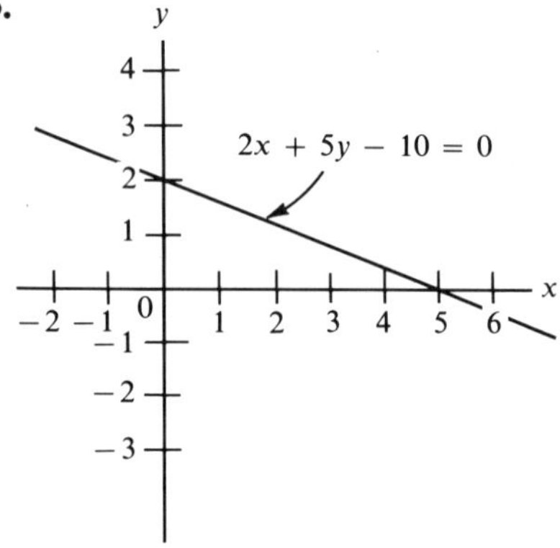

7.

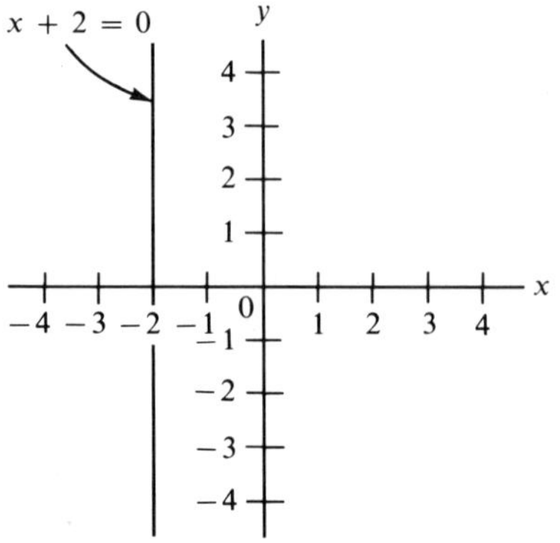

9.

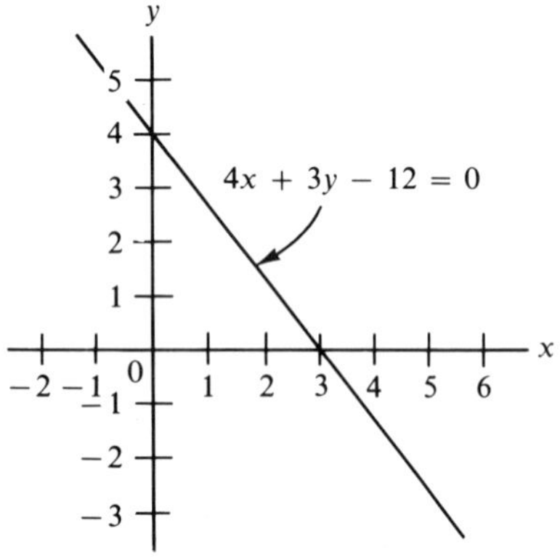

11. (*a*) $y = 0$; (*b*) $x = 0$.

13. (*a*) $x = 2$; (*b*) $x = 5$.

15. (*a*) -2; (*b*) 3.

17.

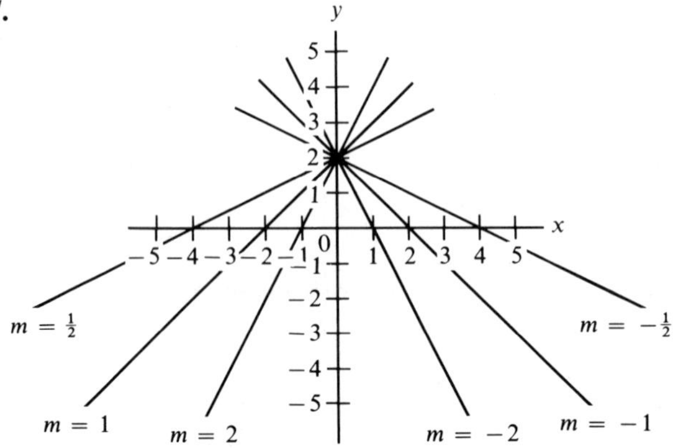

6.3

1.

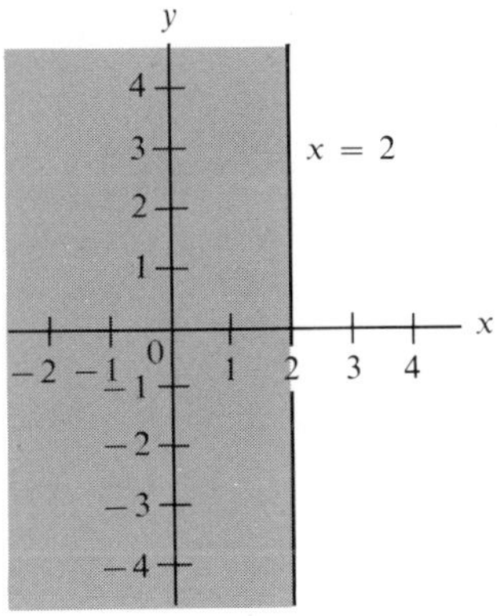

3.

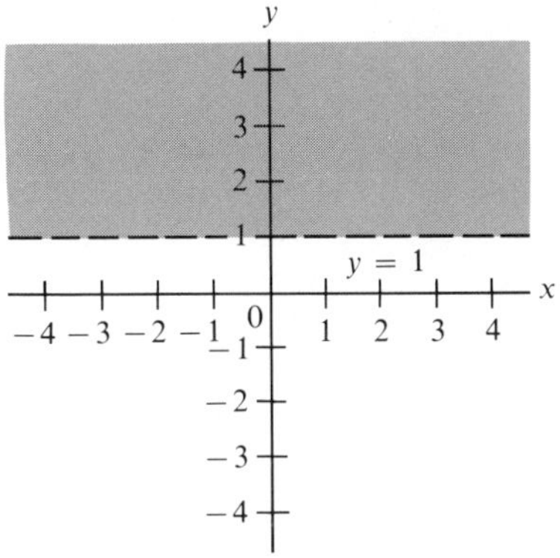

5.

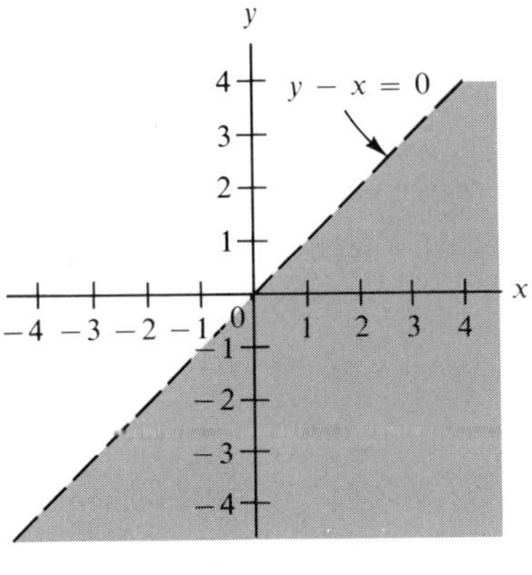

7.

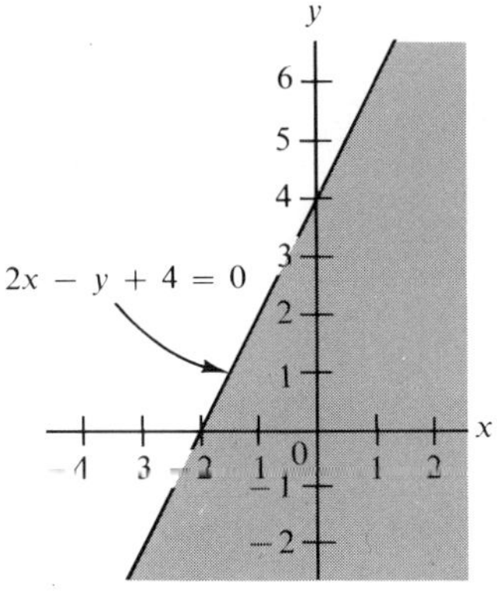

9.

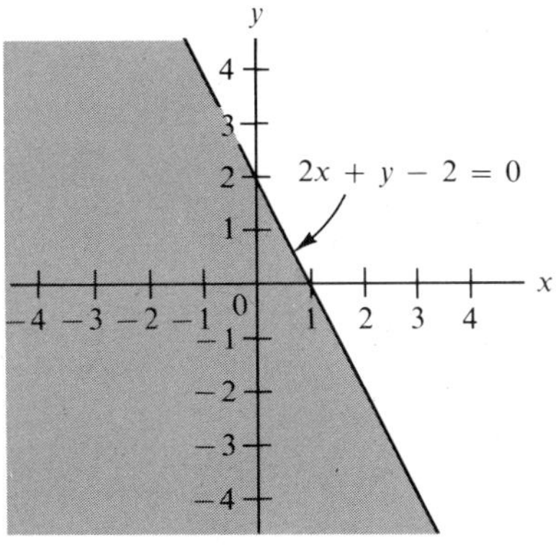

11.

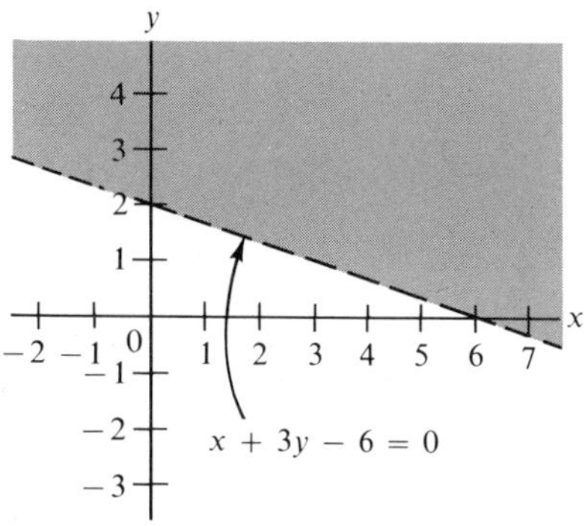

13.

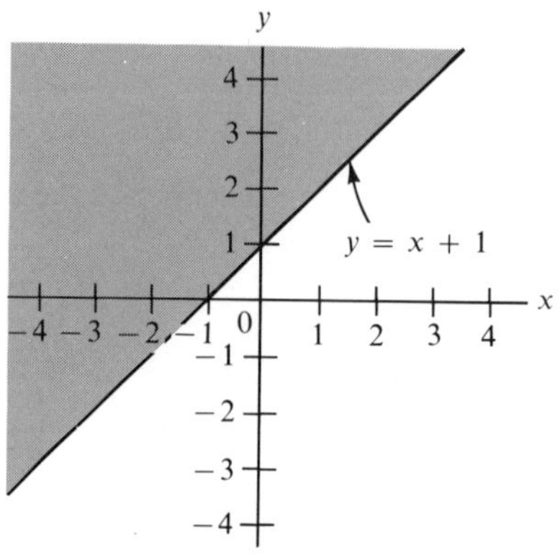

15.

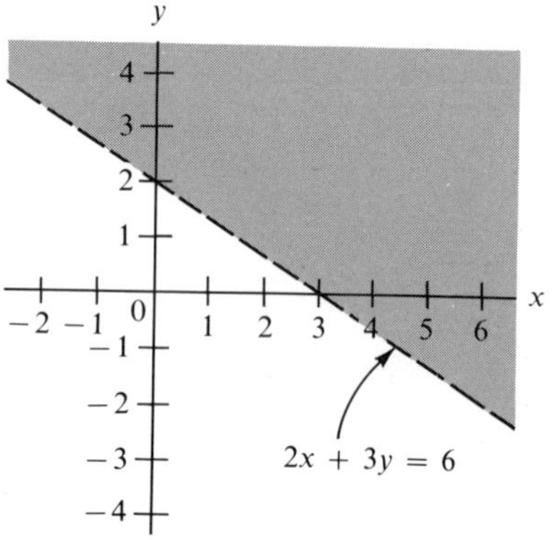

6.4

1.

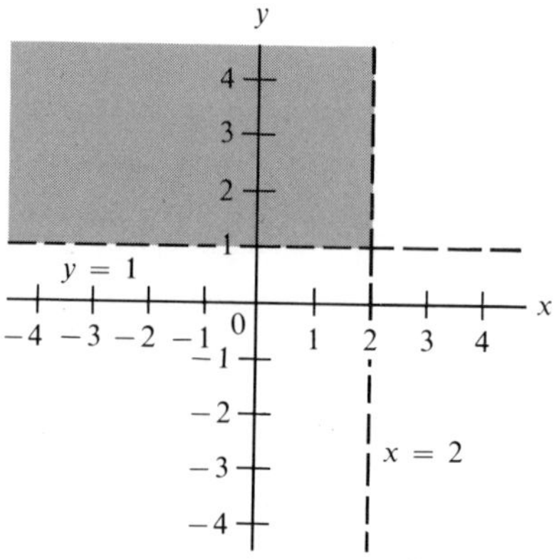

3.

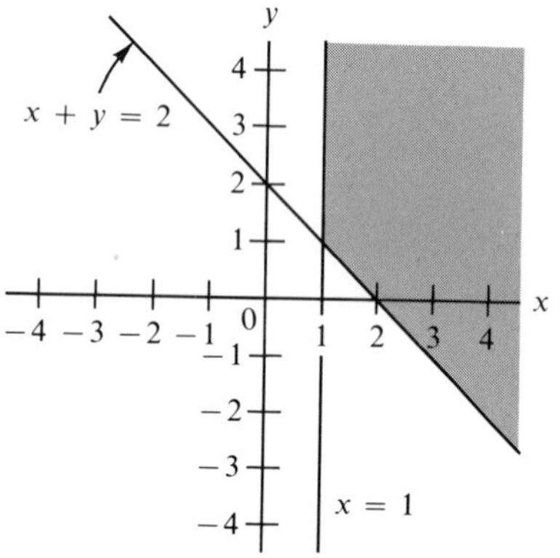

5.

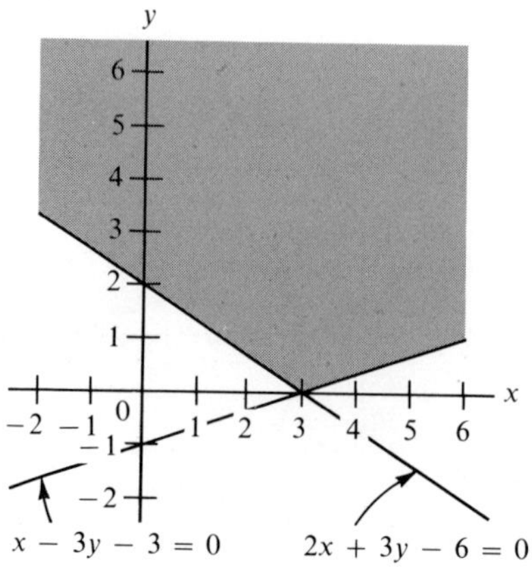

7.

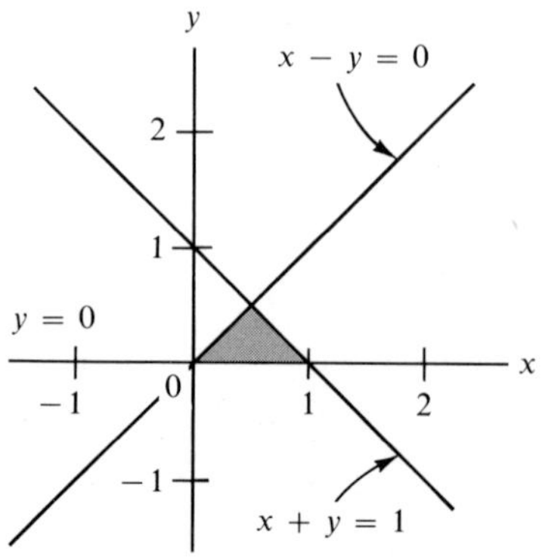

9.

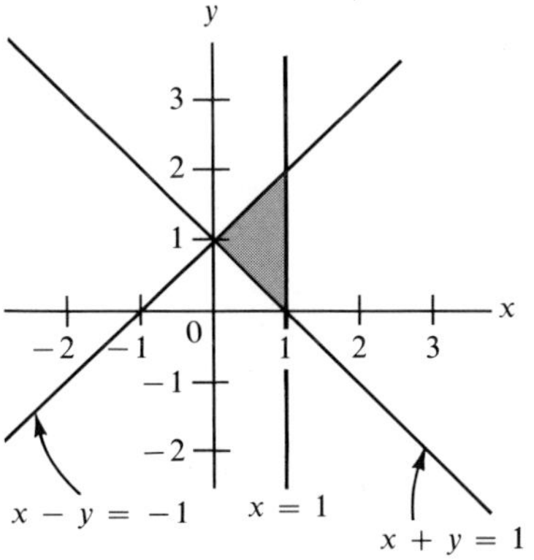

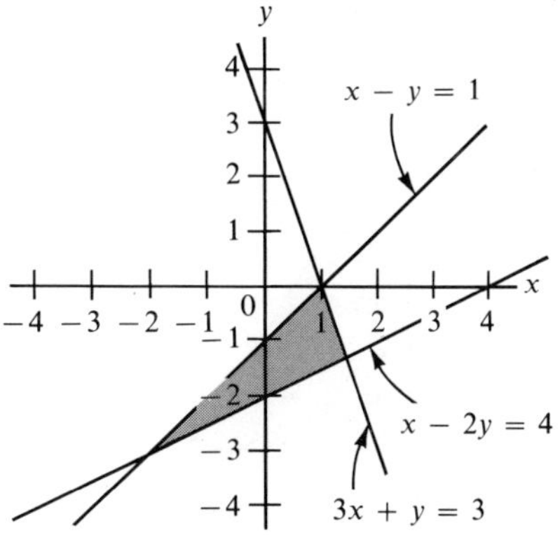

13.

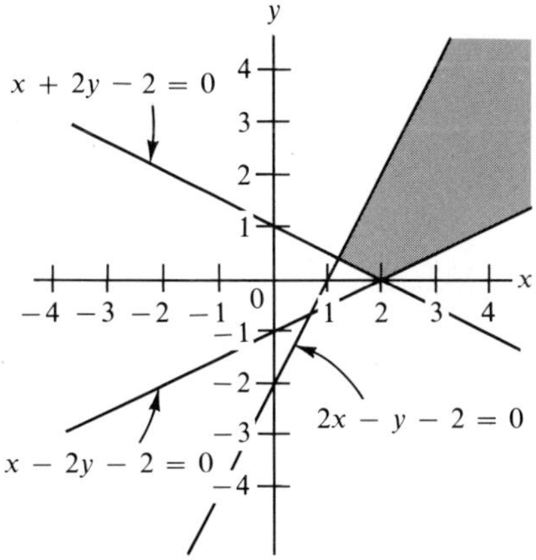

15.

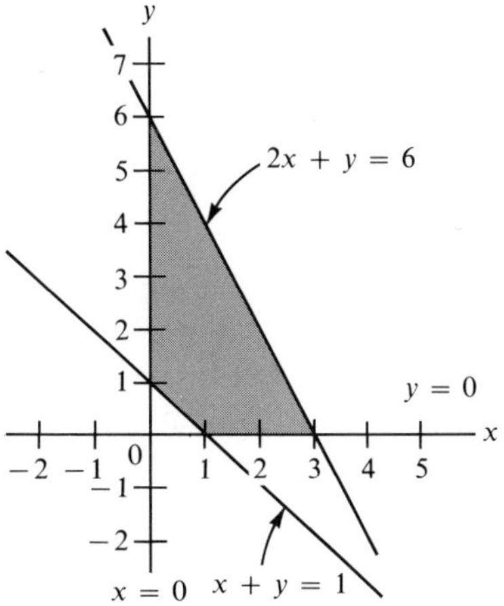

17. $\begin{cases} x \geq 0 \\ y \geq 0. \end{cases}$

19. $\begin{cases} y \geq 0 \\ x + y \leq 0. \end{cases}$

6.5

1.

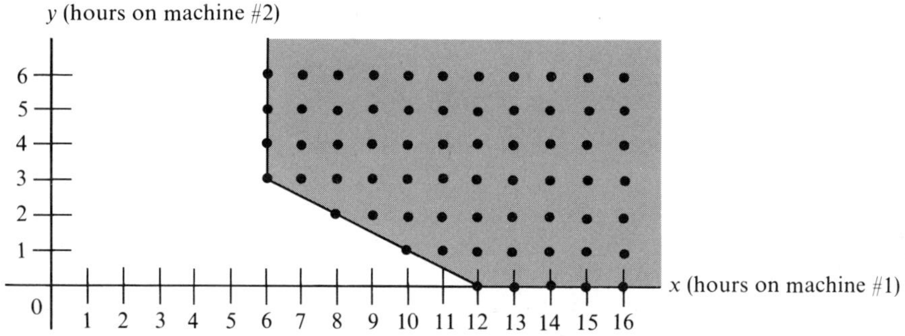

3.

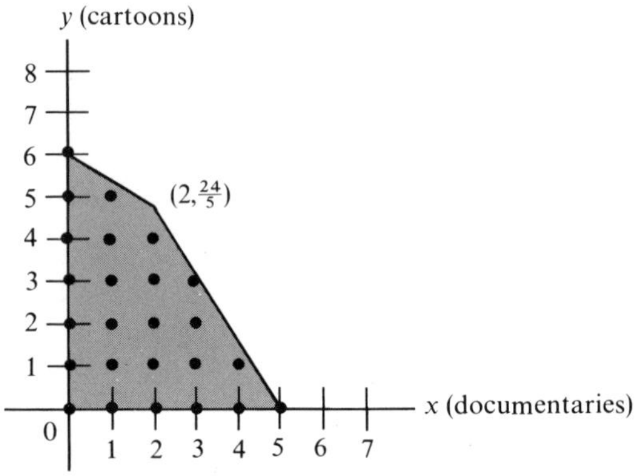

5. Six cartoons and no documentaries.
7. 10 bulldozers and 12 tractors.

6.6

1. Maximum, 6; minimum, 4.
3. Maximum, 8; minimum, 0.
5. Maximum, 2; minimum, 1.
7. $f(x, y)$ has neither a maximum nor a minimum value;
$g(x, y)$ has a maximum value of $\frac{12}{5}$, but no minimum value;
$h(x, y)$ has a minimum value of $\frac{9}{5}$, but no maximum value.
9. (*a*) 6,000 barrels of the expensive mixture of fuel and 3,000 barrels of the intermediate-priced mixture of fuel; (*b*) \$1,260,000.

6.7

1. Maximum, 5; minimum, -13.
3. Maximum, 17; minimum, -4.
5. Maximum, $\frac{11}{3}$; minimum, 1.
7. Maximum, 28; minimum, -11.
9. Maximum, 36; minimum, 2.
11. $x \geq 0$.
13. 28 ranch houses and 18 split-level houses.
15. Seven tricycles and two bicycles.
17. One bag of Growfast and six bags of Steadygrow.
19. 20 units of food A and 15 units of food B.

6.8

1. A nonstrictly determined game.
3. A strictly determined game with value 1. The optimal strategies for $\mathscr{R}$ and $\mathscr{C}$ are to always play row two and column two, respectively. The game is favorable to $\mathscr{R}$.

5. A strictly determined game with value 2. The optimal strategies for $\mathscr{R}$ and $\mathscr{C}$ are to always play row two and column three, respectively. The game is favorable to $\mathscr{R}$.
7. A nonstrictly determined game.

6.9

1. $\frac{5}{8}$.
3. Value of the game, 0. $\mathscr{R}$'s optimal strategy is $(\frac{2}{3} \ \ \frac{1}{3})$; $\mathscr{C}$'s optimal strategy is $\begin{pmatrix} \frac{2}{3} \\ \frac{1}{3} \end{pmatrix}$. The game is fair.
5. Value of the game, $\frac{3}{2}$. $\mathscr{R}$'s optimal strategy is $(\frac{5}{8} \ \ \frac{3}{8})$; $\mathscr{C}$'s optimal strategy is $(\frac{1}{2} \ \ \frac{1}{2})$. The game is favorable to $\mathscr{R}$.
7. Value of the game, $-\frac{1}{2}$. $\mathscr{R}$'s optimal strategy is $(\frac{1}{2} \ \ \frac{1}{2})$; $\mathscr{C}$'s optimal strategy is $\begin{pmatrix} \frac{3}{4} \\ 0 \\ \frac{1}{4} \end{pmatrix}$. The game is favorable to $\mathscr{C}$.
9. Value of the game, $-\frac{2}{11}$. $\mathscr{R}$'s optimal strategy is $(0 \ \ \frac{4}{11} \ \ \frac{7}{11})$; $\mathscr{C}$'s optimal strategy is $\begin{pmatrix} \frac{6}{11} \\ 0 \\ \frac{5}{11} \end{pmatrix}$. The game is favorable to $\mathscr{C}$.
11. Value of the game, $\frac{2}{5}$. $\mathscr{R}$'s optimal strategy is $(0 \ \ \frac{2}{5} \ \ \frac{3}{5})$; $\mathscr{C}$'s optimal strategy is $\begin{pmatrix} \frac{1}{5} \\ \frac{4}{5} \end{pmatrix}$. The game is favorable to $\mathscr{R}$.
13. (*a*) Value of the game, 2. $\mathscr{R}$'s optimal strategy is the pure strategy $(1 \ \ 0)$; $\mathscr{C}$'s optimal strategy is the pure strategy $\begin{pmatrix} 0 \\ 1 \end{pmatrix}$.

    (*b*) Value of the game, a. $\mathscr{R}$'s optimal strategy is the pure strategy $(0 \ \ 1)$; $\mathscr{C}$'s optimal strategy is the pure strategy $\begin{pmatrix} 0 \\ 1 \end{pmatrix}$.

    (*c*) Value of the game, $(5a - 6)/a$. $\mathscr{R}$'s optimal strategy is $\left(\frac{a-3}{a} \ \ \frac{3}{a}\right)$; $\mathscr{C}$'s optimal strategy is $\begin{pmatrix} \frac{a-2}{a} \\ \frac{2}{a} \end{pmatrix}$.
15. Player $\mathscr{C}$

$$\text{Player } \mathscr{R} \quad \begin{array}{c} \\ H \\ T \end{array} \begin{array}{c} \begin{array}{cc} H & T \end{array} \\ \begin{pmatrix} 1 & -2 \\ -2 & 3 \end{pmatrix} \end{array}.$$

$\mathscr{R}$'s optimal strategy is $(\frac{5}{8} \ \ \frac{3}{8})$; $\mathscr{C}$'s optimal strategy is $\begin{pmatrix} \frac{5}{8} \\ \frac{3}{8} \end{pmatrix}$. The value of the game is $-\frac{1}{8}$; hence, the game is favorable to $\mathscr{C}$.

17. The optimal strategy for the airplane pilot is to use a visual search $\frac{2}{3}$ of the time and a radar search $\frac{1}{3}$ of the time in some random fashion. The optimal strategy for the submarine captain is to run deep with no radar detector $\frac{4}{9}$ of the time and use the radar detector near the surface $\frac{5}{9}$ of the time in some random fashion.

Index

Ann Cameron

THE SEED

ILLUSTRATED BY

Beth Cannon

PANTHEON BOOKS

 Published in the United States by Pantheon Books, a division of Random House, Inc., and simultaneously in Canada by Random House of Canada Limited, Toronto.

Library of Congress Cataloging in Publication Data

Cameron, Ann, 1943- The seed. Summary: Describes the terror and delight of growth as a seed becomes a young tree. [1. Germination—Fiction. 2. Trees—Fiction] I. Cannon, Beth, illus. II. Title. PZ7.C1427Se [E] 74-20978 ISBN 0-394-83087-3 ISBN 0-394-93087-8 lib. bdg.
Manufactured in the United States of America. 0 9 8 7 6 5 4 3 2 1

Designed by Harriet Banner

For Paula

"I'm here!" said the Seed.

She was awake.
She was wrapped up tight
in her hard coat.
"I'm here," she said,
"and I'm all alone in this dark world."
She was quiet for a long time,
crouched in her coat.

Way above her,
in the world she couldn't see,
things were happening.
The sun shone on the earth.
The rain fell on the earth
and trickled down to the Seed.
"I feel something warm," she said.

It was the warm rainwater.
And the secret, whispering voice of the rain sang to the Seed.
"Seed, little Seed, send down your root. Seed, little Seed, send down your root."

The Seed sent down her long root.
She could not help it,
the voice of the rain was so beautiful.
The root grew so long
it passed straight through her coat.
"Oh dear!" said the Seed.
"I've broken my beautiful coat!"
But the rain didn't listen.
It went on singing its own song.

"Drink me, little Seed, drink me and drink me."
And the Seed did what the rain said to do
and drank the rain with her root.
And then she heard the deep voice of the earth
saying, "Taste my riches.
Taste them and taste them."
And the rainwater carried
the earth's rich minerals to the Seed,
and she tasted them.

Then the sun came out, and the sunlight fell on the ground, and the sun sang to the Seed in a golden voice. "Send your stem upward, send up your stem."

And the Seed began to do that.

Her stem grew.
And it grew more.
It grew straight through her coat.
Then the rain stopped,
and the sun set,
and nobody sang to her.
And all around her it was dark.
The dark stretched for
thousands of miles.
It was the giant darkness of the earth.

And the Seed could feel it.
She felt very small.
"I'm here!" said the Seed.
"I'm here all alone!
Isn't there anyone here at all?"
But the rain had gone away.
It could not answer her.
And the clouds could not answer her.
They were too far away.
And the sun could not hear her.

It was shining on the other side of the world.
The Seed had relatives, but they could not answer her.
They did not know where she was.
"There *is* no one here but me," she said. "I am alone."
And she listened to the silence.

But still all night she went on growing.
She could not help it.
And in the middle of the night
something big slid by her root.
"Who are you?" said the Seed.
"I'm meeeee," said the creature.
"I'm me, too," said the Seed.
"It's a pleasure to meet someone at last.
Would you mind telling me where we are?"
"We're in the earth," the slow voice answered,
and pushing its long head through the ground,
the creature moved away.

"I don't understand," said the Seed.
"Where is the earth?"
But the creature was gone.
Even though she didn't understand,
the Seed kept on growing.
Her stem grew longer.
And it spread out two leaves.
"I like it here, I like it in the earth,"
said the Seed.
Under the ground she changed every day.
"You're taller," said the creature
in its slow voice, every time it crawled by.

One night the Seed's stem reached higher than ever—it was almost at the surface of the ground. And that night, above the Seed, in the world she couldn't see, there was a storm. The sky cracked open with lightning. And the Seed heard the thunder crash right overhead.

The rain pounded on the ground, shouting in a giant's voice, "NOTHING CAN STOP ME, NOTHING CAN STOP ME, NOTHING CAN STOP ME."

The Seed held tight in the ground with her root, for fear she would be torn loose.

Then the rain's voice stopped, and the night was quiet.

The Seed was very wet, and frightened.

"I won't grow any more, I won't," she said.

But she couldn't help growing.

The next morning the creature came sliding through the wet ground.

"That was a hard night," it said.

"It was! It was!" said the Seed.

The creature touched the Seed with its long head.

"You're taller," it said. "Soon you'll see something you'll never forget."

"No!" said the Seed. "I don't want to see anything! I'm staying right here!"

"Nevertheless," said the creature, "something great will happen to you very soon."

"What?" asked the Seed. "What?" But the creature had crawled away.

The Seed wondered where she was going.
She knew that, just like the creature,
she was pushing herself through the ground.
"Why am I pushing this way?" she wondered.

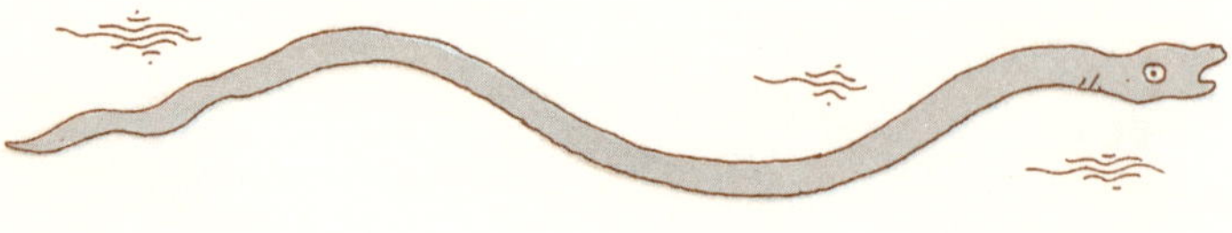

"Push," said the creature, coming by again.
"Just push."
She did. And suddenly she broke through
the earth's surface.

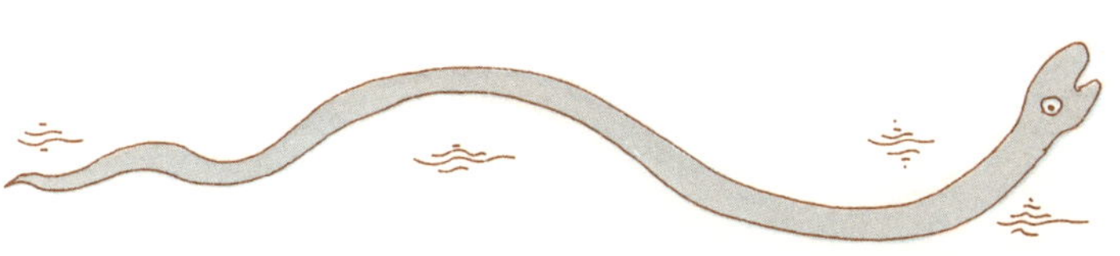

She felt her stem in the air. She felt a breeze.
She felt the sunlight. She felt grass touch her.
She heard the birds singing.
And from far away the great trees that were her relatives touched her with their shadows.
"She's here!" they said to each other. "She's here!"
"I'm not a seed any more," she said. "I'm something else!"
"You're going to be a tree," said a bird, above her.

"Someday birds will sit in your branches and they'll see up and down, see towns and cities and hills and valleys, and they'll sing to you."

"Really?" said the Seed. Her stem trembled in the breeze.

"Yes," said the bird. "Really."

And all the other birds sang it. "Really, really, really."

"Today and today and tomorrow," sang the wind.

"You're a tree."

"Here I am! I'm here,"
sang the Little Tree.
"Yes!" sang her relatives.
The song blew through their leaves.